Coastal Geomorphology

Second Edition

Coastal Geomorphology

An Introduction

Second Edition

Eric Bird
Principal Fellow in Geomorphology
University of Melbourne, Australia

John Wiley & Sons, Ltd

GEOSTUDIES

Other Wiley Editorial Offices

John Wiley & Sons Inc., 111 River Street, Hoboken, NJ 07030, USA

Jossey-Bass, 989 Market Street, San Francisco, CA 94103-1741, USA

Wiley-VCH Verlag GmbH, Boschstrasse 12, D-69469 Weinheim, Germany

John Wiley & Sons Australia Ltd, 33 Park Road, Milton, Queensland 4064, Australia

John Wiley & Sons (Asia) Pte Ltd, 2 Clementi Loop #02-01, Jin Xing Distripark, Singapore 129809

John Wiley & Sons Canada Ltd, 6045 Freemont Boulevard, Mississauga, Ontario, L5R 4J3, Canada

Wiley also publishes its books in a variety of electronic formats. Some content that appears in print may not be available in
electronic books.

Library of Congress Cataloguing-in-Publication Data

Bird, E. C. F. (Eric Charles Frederick), 1930-
   Coastal geomorphology : an introduction / Eric Bird. — Second edition
       p. cm.
   Includes bibliographical references and index.
   ISBN 978-0-470-51729-1 (cloth) — ISBN 978-0-470-51730-7 (pbk.)
   1. Coasts.   I. Title.
GB451.2.B55 2007
551.45′7–dc22   2007039323

British Library Cataloguing in Publication Data

A catalogue record for this book is available from the British Library

ISBN 978-0-470-51729-1 (HB)   ISBN 978-0-470-51730-7 (PB)

Typeset in 10.5/12.5pt Minion by Aptara Inc., New Delhi, India

This book is printed on acid-free paper responsibly manufactured from sustainable forestry
in which at least two trees are planted for each one used for paper production.

Contents

Preface to the Second Edition

This is the second edition of an introduction to the study of coastal geomorphology that provides a background for people interested in learning how coastal features (such as cliffs, beaches, spits or deltas) have developed, and how they are changing. It is intended for people coming newly to the subject, for students and for ecologists, engineers, planners and developers concerned with the coast.

Coastal geomorphology is a broad subject that has developed rapidly, and now generates about 400 publications each year. It has become difficult to produce an introductory textbook, for topics covered in chapters in previous textbooks have subsequently been dealt with at book length, as in the Wiley *Coastal Morphology and Research* series. A comprehensive treatise on coastal geomorphology would now require a massive volume that would certainly be too expensive for students. This book provides a concise introduction that draws attention to unsolved problems and matters on which there are differences of opinion, and gives references to more detailed research work. The coverage is necessarily selective, and somewhat personal, drawing upon my studies of coasts in various parts of the world over the past five decades.

The book discusses the shaping of coastal landforms and examines the changes that are taking place in response to coastal processes. It demonstrates the dynamic nature of coastal landforms and provides a background for ana-

lytical planning and management decisions in coastal areas subject to continuing change. One of the problems in producing an introductory textbook on coastal geomorphology is the need to be selective in quoting examples of coastal features and process relationships, bearing in mind that most readers come from Britain, Europe, North America or Australasia, and are likely to be more interested in local and accessible examples. Reference can be made to *The World's Coasts Online*, produced by Springer in 2003, for examples from various other coasts. Place names in England are identified by county, in the USA and Australia by state, and elsewhere by country.

The book begins with an introduction to concepts and terminology, and the factors that have affected coastal evolution and coastline changes (Chapter 1). This is followed by a discussion of waves, tides, currents and other nearshore processes (Chapter 2), and a study of the effects of land and sea level changes, notably the Holocene marine transgression, which has played a major part in shaping modern coastlines and can be regarded as a unifying theme in coastal geomorphology (Chapter 3). Cliffs are discussed in Chapter 4 and the shore platforms that border them in Chapter 5. Chapter 6 deals with the origin of beaches and the changes taking place on them, and Chapter 7 with the beach erosion problem. Spits, barriers and bars are discussed in Chapter 8 and the formation of coastal dunes in

Chapter 9. Intertidal wetlands, including mud-flats, salt marshes and mangroves, are dealt with in Chapter 10, followed by estuaries and lagoons, including other inlets (rias, fiords, fiards, calanques, sharms and sebkhas) in Chapter 11. Chapter 12 considers deltas produced by deposition at river mouths, and Chapter 13 deals with the various kinds of reef built by corals, algae and other organisms on the shore and in coastal waters. The final chapter reviews the response of coastlines to the predicted world-wide rise in sea level, resulting from global warming by the enhanced greenhouse effect, and documented by the Intergovernmental Panel on Climate Change

(2007). A list of references provides a guide to more detailed information, including many pre-2000 publications that remain relevant.

Supplementary material, including a classification of coastal landforms (which appeared as an appendix in the first edition), will appear on this book's companion website www.wileyeurope.com/college/bird, along with a glossary, a bibliography, case studies and many more illustrations. It will also be useful to refer to two recent reference works, the *Encyclopedia of Geomorphology* (Goudie, 2004) and the *Encyclopedia of Coastal Science* (Schwartz, 2005).

Eric Bird

Acknowledgements

I am greatful to Juliet Bird for advice during this revision of the text, and to Chandra Jayasuriya and Blaise Vinot for help with maps and diagrams. Lew Ward provided help with computer processing. I would also like to thank the staff of the Earth Sciences Library in the University of Melbourne for assistance in the course of exploring the coastal literature.

Eric Bird
Black Rock, March 2007

List of Figures

List of Tables and Panels

1

Introduction

1.1 Coastal geomorphology

More than half the world's population lives in coastal regions, and many people visit the coast frequently. Most come for seaside recreation, but some also wonder about the origins of coastal scenery. A walk along the shore or a coastal footpath prompts questions about how such features as cliffs, rocky outcrops, beaches and dunes formed, and how and why they are changing. A coastal journey is likely to encounter estuaries, lagoons and river deltas that have evolved over longer periods, and it soon becomes clear that sea level has not always been where it is now.

Coastal geomorphology deals with the shaping of coastal features (landforms), the processes at work on them and the changes taking place. Coastal geology is concerned with the rock formations and structures seen in cliff and shore outcrops, and the sediments that have been deposited in coastal regions. It provides the background for coastal geomorphology.

Apart from incidental comments by classical Greek and Roman observers and by Leonardo da Vinci, the first systematic attempts to explain coastal landforms were by 19th century scientists such as Charles Lyell and Charles Darwin, and the pioneer American geomorphologist William Morris Davis. While a great deal of work was done in the 20th century on various parts of the world's coastline, particularly in Europe and North America, it is only in the past few decades that coastal research has become widespread, and there is still plenty of opportunity for original contributions.

Coastal geomorphology has several themes, each of which will be discussed in this book.

(a) The shaping of landforms in relation to geology, processes, variations in climate and the relative levels of land and sea.

(b) Coastline changes measured over specified periods, with analyses of their causes.

(c) Nearshore processes and responses, particularly on beaches.

(d) Evidence of geological history, notably changes in land and sea level and climatic variations.

(e) The sources and patterns of movement of coastal sediment.

(f) The array of weathering processes in the coastal zone.

Coastal Geomorphology: An Introduction, Second Edition. Eric Bird.
© 2008 by John Wiley & Sons, Ltd

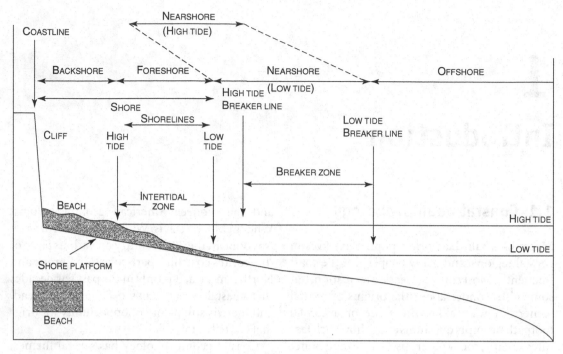

Figure 1.1   Coastal terminology

1.2 Terminology

The coast consists of a number of zones (Figure 1.1). The shore is the zone between the water's edge at low tide and the upper limit of effective wave action, usually extending to the cliff base. It includes the foreshore, exposed at low tide and submerged at high tide, and the backshore, extending landward from the normal high tide limit, but inundated by exceptionally high tides or by large waves during storms. The shoreline is strictly the water's edge, migrating to and fro as the tide rises and falls.

The nearshore zone, comprising the surf zone (with breaking waves) and the swash zone (covered as each wave runs up the foreshore), also migrates to and fro as the tides rise and fall. The breaker zone (where waves are disrupted) is bordered seaward by the offshore zone, extending to an arbitrary limit in deep water. The terms offshore, onshore and longshore are also used

to describe directions of flow of wind, water or sediment.

A beach is an accumulation of loose sediment, such as sand, gravel or boulders, sometimes confined to the backshore but often extending across the foreshore as well. Some beaches extend down to, and below, low tide level. Shingle is beach gravel, especially where the stones are well rounded.

The coast is a zone of varying width, including the shore and the nearshore zone, out at least to the line where waves break, and extending inland to the limit of penetration of marine influences: the crest of a cliff, the head of a tidal estuary, or the rising ground behind coastal lowlands, or dunes, lagoons and swamps. The coast is thus the zone where land, sea and air (the lithosphere, hydrosphere and atmosphere) meet and interact. It is subject to an array of processes, including tectonic movements (upward, downward or laterally) of the land margin, changes in

Panel 1.1 Coastline or shoreline?

The coastline is defined as the edge of the land at the limit of normal high spring tides; the subaerial land margin, often marked by the seaward boundary of terrestrial vegetation. On cliffed coasts it is taken as the cliff foot at high spring tide level.

The shoreline is the water's edge, moving to and fro as the tides rise and fall, so that there is a low-tide shoreline, a mid-tide shoreline and a high-tide shoreline. Shorelines thus move to and fro as the tide rises and falls, whereas coastlines are submerged only in exceptional circumstances (e.g. during storm surges).

If coastline and shoreline are regarded as synonyms this distinction is lost. There is a difficulty where the tide range is large, as in NW Australia, where tides exceed 10 m and the distance between the coastline (high spring tide shoreline) and the low spring tide shoreline is up to 8 km. However, the term shoreline is often used for the coastlines of lakes, estuaries and lagoons, where the tide range is generally small and the intertidal zone narrow or non-existent.

Many American authors have preferred the term shoreline to coastline, but there are notable exceptions: Shepard and Wanless (1971) entitled their book *Our Changing Coastlines*, and the leading American journal is called the *Journal of Coastal Research*. In the United States the term shoreline is defined legally as mean high water (MHW), as shown on nautical charts produced by the National Oceanic and Atmospheric Administration (NOAA). Shorelines at other levels are simply called lines, e.g. the mean lower low water line, which is a private property seaward boundary in some eastern states (Parker, 2001). It should be noted that the American shoreline, thus defined, is not the margin of normally dry land.

Details of work cited (Shepard and Wanless, 1971; Parker, 2001) are given in the References section (pp. 387–404).

sea level, the effects of tides, waves and currents in the sea and variations in temperature, pressure and wind action in the atmosphere. Some coasts have been shaped primarily by erosion, others by deposition. Erosion is the removal of rock material, and the term denudation is used where surface rock is removed to expose underlying rock formations and structures to further erosion.

The term coastline indicates the land margin at normal high spring tide (behind the backshore zone), and may be the base of a cliff or the seaward margin of dunes or dry land. In American literature the term shoreline (or seaboard) is often used as a synonym for coastline, while the coast is elaborated to the coastal zone. The preference here is to maintain a distinction between coastline and shoreline (Panel 1.1), acknowledging that the shoreline moves to and fro as tides rise and fall, so that one can define a low-tide shoreline, a mid-tide shoreline, and a high-tide shoreline.

1.3 Ancient coastlines

Coastlines have existed since oceans first formed on the surface of a cooling Earth, about 4 000 million years ago, but it is difficult to find early coastlines because most of the evidence has been removed by erosion or concealed by deposition. Table 1.1 shows the geological column (the sequence of geological periods). Deposits indicating coastlines that existed in Mesozoic and Tertiary times can be found in the stratigraphy of southern Britain. An example is seen on the Haldon Hills, east of Dartmoor in SW England, where there are pebbly sands with corals and mollusc shells that represent a beach deposited in the Cretaceous, about 110 million years ago. Other fragments of ancient coastlines have been preserved far inland. In the Czech Republic there is a quarry on Kank Hill, near Kutna Hora, about 70 km east of Prague, where it is possible to stand on the Upper Cretaceous shore. A beach resting on an irregular wave-worn surface of

Table 1.1 The geological column: the sequence of rock formations arranged by age (my – million years).

Quaternary Period:	Holocene (Recent)
	———————— 10 000 years
	Pleistocene Epoch
	———————— 2.3 my
Tertiary Period:	Pliocene Epoch
	———————— 5 my
	Miocene Epoch
	———————— 23 my
	Oligocene Epoch
	———————— 36 my
	Eocene Epoch
	———————— 53 my
	Palaeocene Epoch
	———————— 65 my
Mesozoic Era	Cretaceous Period
	———————— 144 my
	Jurassic Period
	———————— 213 my
	Triassic Period
	———————— 248 my
Palaeozoic Era:	Permian Period
	———————— 290 my
	Carboniferous Period
	———————— 360 my
	Devonian Period
	———————— 405 my
	Silurian Period
	———————— 436 my
	Ordovician Period
	———————— 510 my
	Cambrian Period
	———————— 560 my
Pre-Cambrian Era	

In North America the Carboniferous Period is divided into upper (Pennsylvanian) and lower (Mississippian) Periods. Geologists recognise Formations within each Period, based on rock type (lithology), e.g. the Old Red Sandstone Formation in the Devonian and the Chalk Formation in the Cretaceous, and when these are shown on maps and in sections they are useful for geomorphology. Alternatively, they divide each period into a number of stages, based on their fossil content, but these may not correspond to lithological units.

Pre-Cambrian rock marks the limits of a Cretaceous sea that reached here about 95 million years ago (Ager, 1980). There have been many such transgressions of the sea over the land during geological time, probably related to changes in the size and shape of ocean basins, particularly during the splitting of the ancient supercontinent of Pangaea into several drifting continents, a process that began early in the Mesozoic era.

Evidence of former coastlines becomes clearer in the most recent of the geological periods, the Quaternary, which comprises the Pleistocene (which began about 2.3 million years ago) and the succeeding Holocene (the last 10 000 years). The Quaternary period was one of major global climate and sea level fluctuations, and Quaternary coastlines can be found above and below present sea level (Chapter 3). There are Late Pleistocene beaches and shore platforms standing above present sea level on many coasts, notably in SW England and around Scotland, while submerged Pleistocene coastlines (cliffs, shore platforms and beaches) have been detected on the sea floor, notably off California and Japan. Coastal plains built forward by deposition, as in the SE United States, may include stranded remnants of coastlines of Pleistocene and Holocene age, containing evidence of past conditions that has generally been lost on receding cliffed coasts.

During cold climate phases of the Quaternary, when glaciers and ice sheets became extensive, global sea level was much lower than it is now, and when the climate of the Ice Age gave place to milder conditions there was a major world-wide sea level rise. Existing coastal landforms have been largely shaped within the past 6000 years, when the sea has stood at or close to its present level, with global climate much as it is now. Some coasts have older (relict) features, inherited from earlier environments when the sea stood higher or lower, or when the climate was warmer or colder, wetter or drier, or stormier or calmer than it is now.

1.4 Coastline morphology

Maps and charts show that few of the world's coastlines are straight: even those of simple outline are typically gently curved. An example of an almost straight coastline is the north coast of Madura in Indonesia, which may be related to a major fault line. The almost straight 800 km east coast of Madagascar could also be fault guided, but it includes depositional sandy barriers shaped by Indian Ocean swell, and has not been produced directly by faulting. Probably the best example of a fault coast is seen in California north of San Francisco, where the coastline runs along the San Andreas Fault NW to the Bolinas Lagoon, and then follows the fault along the inner (eastern) shore of the Point Reyes peninsula bordering Tomales Bay.

There are often simple relationships between coastal outlines and the geology and topography of coastal areas. Headlands and promontories generally occur where there are outcrops of resistant rock at, above or below sea level, or where higher ground comes to the coast, as on interfluves between incised valleys. Bays have been excavated where softer rock outcrops are bordered by more resistant formations, particularly where lowlands have formed. Where there have been relatively recent tectonic movements (upward or downward, tilting or folding) of the land it is likely that uplifted sectors protrude seaward and that subsided areas have become bays.

There are distinctive cliff and shore features related to certain geological formations, such as chalk or granite, where they outcrop on the coast. However, there is not always a good correlation between coastal landforms and the outcrops of rock formations shown on geological maps, particularly when geological formations have been defined by mineralogy or palaeontology, rather than by rock type (lithology). Thus the cliffs and rocky shores on the coast of Aberdeenshire bear little relationship to several mapped divisions of the Dalradian schist, classified on the basis of their mineralogy (Ritchie, 2006). Some coasts of similar geology, latitude and aspect are compared in Panel 1.2.

Panel 1.2 Comparisons of coasts of similar geologies, latitudes and aspects

It may be useful to compare features on similar geological formations in similar latitudes and with similar aspects: for example coastal landforms on glacial drift deposits in the Danish archipelago with those in New England and around Puget Sound. The features of the Normandy coast (such as the landslides at Les Vaches Noires and Longues-sur-mer) are similar to those of the south coast of England, with contrasts related to the higher wave energy on the northern side of the English Channel. Davies (1980) suggested that features on coasts of varying aspect should be compared, for example the east and west or north and south coasts of islands such as Tasmania, Ireland or Sri Lanka. Contrasts related to aspect can be studied on islands, such as the coastal blowouts and parabolic dunes on the east and west coasts of King Island in Bass Strait (Jennings, 1957). Interpretation of the major dune formations of Fraser Island and Cooloola in SE Queensland is aided by comparisons with similar dune systems in equivalent latitudes on the coasts of southern Brazil and Mozambique.

On a smaller scale, there are contrasts related to local variations in exposure: on the coast of Tahiti beaches occur only inshore of gaps in the bordering coral reefs. On the west coast of the Galloway Peninsula, in Scotland, an emerged shore platform is backed by bluffs that became bolder as exposure through the 'window' to the Atlantic (between Islay and Ulster) increases at Bellochantuy Bay.

Certain rock sequences and structures produce the same kind of landform association. Similar landforms accompany particular rock sequences, as on the south coast of England where landslides occur as the Chalk gives place laterally to Upper Greensand and Gault Clay, as at Beer in Devon, White Nothe in Dorset, Freshwater Bay and Culver Cliff on the Isle of Wight, Holywell in Sussex and Folkestone in Kent. Similar features are seen at Bempton on the east coast of England, and near Boulogne and at Sainte Adresse in northern France.

The shaping of many coastlines has been influenced by upward or downward movements of sea level (Chapter 3). Embayed coastlines with valley-mouth inlets, as on the Atlantic coasts of the United States and western Britain, are the outcome of relatively recent marine submergence (Chapter 11). Where sea level has fallen there are often emerged coastal plains and smooth coastlines where the sea floor was relatively featureless near the coast. There are exceptions where the sea floor had an irregular topography that has emerged, as in the archipelago of SW Finland.

Many coasts formed by deposition of sediments have simple, often gently curved beach-fringed outlines (e.g. much of the Gulf Coast of the United States), as have some cliffed coasts cut in fairly soft rock formations (e.g. Lyme Bay in Dorset, Figure 6.24). There are exceptions where deposition at a river mouth has formed protruding deltas. Other coasts with geological diversity are more intricate, with headlands and embayments (e.g. South China), branching inlets and ramifying peninsulas (e.g. Sulawesi in Indonesia and the Kimberley coast in northern Australia) or numerous islands (e.g. the Dalmatian coast).

Rounded bays have formed where the sea has penetrated into volcanic craters or calderas, as in the South Shetland Islands, a chain of volcanic islands parallel to the west coast of the Antarctic Peninsula. Here Deception Island is a partly collapsed volcanic cone with a rim rising to 580 m, overlooking a deep caldera penetrated by the sea to form a circular embayment. Similar rounded bays were formed by explosive eruptions at Santorini in the Aegean Sea and Krakatau, between Java and Sumatra, but these now contain younger volcanoes. Theoretically a rounded coastal embayment could be formed by marine submergence of a breached meteorite crater, but no example has been demonstrated. There are bays in breached and drowned sinkholes on the limestones of the NW coast of Gozo

in Malta. Where a resistant geological formation running along the coast is backed landward by a weaker outcrop, penetration of the outer rampart by marine erosion may be followed by the excavation of a rounded embayment, as at Lulworth Cove on the south coast of England (Figure 4.28).

Smoothly curved coastlines have formed where incoming refracted waves have shaped the outlines of depositional coasts, as on the Ninety Mile Beach and in Discovery Bay, SE Australia. They are well developed on coasts exposed to ocean swell, but can also form on the shores of large bays and coastal lagoons (Section 6.9). On some coasts the smooth curvature extends across cliffed sectors in soft rock formations as well as along the intervening beaches, as in Hawke Bay on the east coast of North Island, New Zealand, and Te Waewae Bay in South Island, New Zealand. These are both shaped by refracted southerly ocean swell originating from storm centres in the Southern Ocean. On the south coast of England the Seven Sisters in Sussex are cliffs that truncate several valleys but are smooth in outline, with no inlets (Figure 1.2). The chalk cliffs are bordered by a gently sloping shore platform that is exposed at low tide, and has been cut across strata that dip gently seaward. At high tide this platform is submerged, and waves wash against the base of the cliff. Marine erosion has cut into the southern slopes of the South Downs to form vertical receding cliffs, the lower part of the cliff showing fresh white chalk recently scoured by waves armed with the chalk and flint boulders and cobbles that are strewn over the shore platform. In addition to wave abrasion, several other processes have contributed to the shaping of these coastal landforms. They include solution by rain water and sea spray, bioerosion by the plants and animals that inhabit the shore and frost shattering in cold winters. The cliffs undulate across dry valleys that were cut by streams when the climate was much colder during Pleistocene times. The

Figure 1.2 The Chalk cliff at Seven Sisters, Sussex

chalk surface was then disintegrated by freezing and thawing, and runoff from melting snow excavated valleys in the weathered rubble. Remnants of this rubble, known as Coombe Rock, underlie the dry valleys, and can be seen in the cliff at Birling Gap (marked by the building in the distance in Figure 1.2). See Chapter 4.

Various attempts have been made to describe the coastal outlines shown on maps and air photographs numerically, but without much success. The mathematician Mandelbrot (1967) saw coastlines as analogues of fractal curves, which retain the same general pattern regardless of how much they are magnified. Similar coastline features occur on a variety of scales. Beach cusps, for example (Section 6.10.7), maintain their shape as their dimensions increase or decrease in relation to incident wave heights, but a particular beach cusp is not subdivided into smaller, nested beach cusps, and the beaches on which they occur are not as a rule cuspate on a larger scale. It is

true that coastal promontories and embayments occur on various scales from continental down to a particular headland and cove, but their pattern is not maintained hierarchically as the scale changes. The Mandelbrot observations have not led to any advance in coastal geomorphology.

1.5 Coastline length

Measurements of coastline length are necessary for describing the proportions of various types of coastline around the world or the lateral extent of erosion and accretion on beaches. Such measurements can be made can be made by counting straight intercepts of a selected length (e.g. 1 km) on maps of uniform scale (e.g. 1:250 000), or by using computers to integrate the grid squares within which coastline segments occur, taking each grid square as representing a specific coastline length. It is difficult to make

Table 1.2 Coastal dimensions

Inman and Nordstrom (1971) described first-order coasts as having length, width and height dimensions of about 1000 km × 100 km × 1 km, and second-order coasts about 100 km × 10 km × 1 km. They introduced ranges (1–100 km long and 10–1000 m wide) for third-order coasts, but did not develop the series further. Suggested categories:

First-order features – about 1000 km long, 100 km wide and 10 km high (e.g. continental coasts, related to global tectonics).

Second-order features are about 100 km long, 10 km wide and 1 km high (e.g. deltas, fiords).

Third-order features about 10 km long, 1 km wide and 100 m high (e.g. coastal barriers).

Fourth-order features are about 1 km long, 100 m wide and 10 m high (e.g. foredunes).

Fifth-order features are about 100 m long, 10 m wide and 1 m high (e.g. beach berms, shore platforms, sand bars).

Sixth-order features are about 10 m long, 1 m wide and 10 cm high (e.g. beach cusps).

Seventh-order features are about 1 m long, 10 cm wide and 1 cm high (e.g. current ripples).

In each case the dimension given should be regarded as being within a range of from 50% of to five times the figure given (e.g. sixth-order features 5–50 m long, 0.5–5 m wide and 5–50 cm high).

precise measurements, and different results are obtained with variations in the starting point for segment measurements or the location of grid squares.

The total length of the world's coastline is certainly considerably longer than the figure of 439 700 km given by Inman and Nordstrom (1971) and is probably close to a million kilometres, including the coasts of the very many small islands. Information on coastline lengths is available on the Internet, and can be obtained from Wikipedia (http://en.wikipedia.org/wiki/list or from the United States Central Intelligence Agency World Factbook (https://www.cia.gov/library/publications/the-world-factbook/index.html). The latter source lists coastlines with a total length of 847 942.30 km, but as several (e.g. Finland) exclude archipelagoes and coastal indentations and several other islands are omitted the global total is probably indeed close to a million kilometres.

Table 1.2 shows a classification of coastal dimensions. Variations in geomorphology around the world's coastline were illustrated by Bird and Schwartz (1985) and documented in *The World's Coasts: Online* (Bird, 2003).

1.6 Coastal evolution

The shaping of coastal landforms has been influenced by a range of morphogenic factors. These include geology, which determines the pattern of rock outcrops on the coast, on the sea floor and in the hinterland, and movements of the Earth's crust, which result in uplift, tilting, folding, faulting and subsidence of coastal rock formations. Climatic factors have influenced the wind and wave regimes that shape coastal features, and the weathering processes that decompose and disintegrate coastal rock outcrops vary from tropical to arctic and from humid to arid environments. Climate also conditions coastal vegetation and fauna, which have produced features ranging from salt marshes and mangrove swamps to shelly beaches, coral reefs and stabilised dunes, and also the organisms that attack rock surfaces (the processes of bioerosion, Section 5.1.4).

Coastal processes include the effects of rising and falling tides and associated tidal currents, and are influenced by oceanographic factors such as sea temperature and salinity, determined by climate and the patterns of ocean currents.

The various processes are discussed in Chapter 2. Mention has been made of ancient coastlines, produced by past changes in the relative levels of land and sea, and these changes have continued to influence the evolution of existing coasts. Within historical times coastal evolution has also been modified by the effects of various human activities on the coast and in the hinterland.

Evolution of coastal landforms can be considered in terms of morphogenic (morphodynamic) systems, within which various factors influence the processes acting upon the coast (Short, 1999). There is an input of energy (e.g. wind, tide, living organisms) and materials (e.g. water, rock, sediment) that interact to generate the coastal landforms, and there is feedback in the sense that the developing morphology modifies geomorphological processes, and thus becomes a factor influencing subsequent changes. These can be studied in terms of response to various coastal processes operating over specified periods: that is, as process-response systems. Attempts have been made to quantify the various inputs and to describe and analyse the interactions mathematically (Scheidegger, 1991), but the ideal of a complete quantitative understanding of a coastal system is more easily advocated than achieved. It is realistic to formulate and attempt to solve specific problems, and establish empirical relationships between process and change that can be put to practical use in coastal management.

1.7 Changing coastlines

While some coastlines have changed little over the past 6000 years, most have advanced or retreated, and some have shown alternations of advance and retreat. A coastline advances where the deposition of sediment exceeds the rate of erosion, or where there is emergence due to uplift of the land or a fall in sea level, and retreats as the result of erosion exceeding deposition, or where there is submergence due to land subsidence or a sea level rise (Figure 1.3). The high

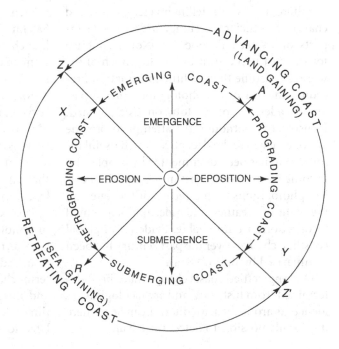

Figure 1.3 Analysis of coastline changes in terms of emergence and submergence, progradation and retrogradation, as proposed by Valentin (1952)

tide shoreline may advance or retreat independently of the low tide shoreline as the intertidal zone widens or narrows and the transverse gradient flattens or steepens.

Coastlines have changed at varying rates in response to coastal processes, with sudden changes during storms, earthquakes and volcanic eruptions (these attract media attention), and more gradual changes over quieter intervening periods (apt to pass unnoticed until someone produces historical photographs that can be used to demonstrate them). Coastline changes can be measured over various timescales, ranging from the past few thousand years down through recent centuries or decades to the annual or seasonal fluctuations and short term changes related to the various tidal cycles or caused by particular weather events. Some changes are cyclic over varying periods; others continue as erosion or deposition proceeds.

Measurement of coastline changes can be made by comparing historical maps and charts, providing these were based on accurate surveys, with the configuration shown on modern maps, air photographs or satellite imagery. Maps and charts of sufficient accuracy are available for parts of western Europe and North America for the past two centuries, but for much of the world's coastline there is little information preceding the era of air photography in the past few decades. A coastal tour on Google Earth is instructive, although the attempt to provide oblique views can be deceptive. Much useful information has been documented by people who become interested in coastline changes and collect photographs with a record of the date and the state of weather and tide: undocumented recollections are unreliable. Evidence of global coastline changes over the past century has been summarised by Bird (1985a).

On long-settled coasts changes have been determined from historical and archaeological evidence, as around the Mediterranean Sea, where it is locally possible to detect the advance or retreat of parts of the coastline over at least 2000 years (Kraft, Aschenbrenner and Rapp, 1988) (Figure 1.4). Changes since present sea level was established (within the past 6000 years) may be determined from evidence of the preceding land surface intersecting the sea floor (Section 4.9) or from stratigraphical and sedimentological analyses of coastal depositional formations, using radiometric and other forms of dating as well as palaeontological and archaeological evidence (Carter and Woodroffe, 1994).

Traditional methods of observing, mapping and measuring changes on the coast and the processes that cause them have recently been supplemented by new techniques, including various electronic measuring instruments and the application of modelling. Computers are used to process and extend field survey data, generating serial models of beach or coastal dune topography from which the pattern of gains and losses can be mapped and quantified (Section 6.14). Air photographs have been used for some time as an aid to the mapping and measurement of coastal changes, and colour photography has extended these studies to the nearshore sea floor. Satellite imagery has been used to trace coastline changes over the past three decades. Short term changes, which range from a few minutes to a few hours (as on beaches or dunes during a storm), require monitoring by repeated field surveys, the use of micro-erosion meters, serial photo-recording or photogrammetry. Ground surveys of coastal landforms can be made using a global positioning system (GPS) in traverses that can be translated into morphological maps by computer.

In recent years increasing use has been made of remote sensing techniques such as airborne laser terrain mapping (ALTM) and light detection and ranging (LIDAR) to measure short term changes on beaches, dunes, marshland and intertidal and nearshore areas. Reflection time is used to calculate altitudes that are related to a selected datum such as the high tide

Figure 1.4 Archaeological evidence of coastline change

shoreline. Vertical changes of as little as ±10 cm have been measured (Leatherman, Whitman and Zhang, 2005; Davidson-Arnott, 2005; Finkl, 2005).

Some coastline changes have resulted from human activities, such as reclamation (also known as land claim), the making of new ground by enclosing or filling nearshore areas, which in places has advanced the coastline several kilometres (French, 1997). The Netherlands has a long history of winning land from the sea by building dykes (sea walls) to enclose areas that were previously beneath the sea (at least at high tide) and draining these to form polder lands, thereby advancing the coastline seaward. New land has also been created on densely populated coasts in SE Asia, as in Tokyo Bay and Hong Kong, and Singapore has increased its land area by 10% in recent decades by landfill.

Coastlines have also been modified by the introduction of structures such as groynes and breakwaters, intended to stabilise features that were changing in ways considered unacceptable, notably where erosion threatened seaside towns, ports, or other developed coastal areas. The dredging of harbour entrances and the dumping of material on the coast and offshore have also modified coastal topography. In consequence, many coastlines have become largely or entirely artificial, and the extent of these is increasing rapidly. Appropriate coastal management may succeed in maintaining or enhancing the coastal environment, but there have been mistakes that could have been avoided if those concerned had understood the principles of coastal geomorphology.

1.8 Summary

Coastal geomorphology deals with the shaping of coastal landforms, the processes at work on them and the changes taking place. It uses a defined set of terms to describe coastal features, past and present. Former coastlines exist above (emerged) and below (submerged) present sea level. Coastal outlines are related to geology and processes of erosion and deposition. Coastline length can be measured by such methods as 1 km intercepts: the world's coastline is about a million kilometres long. Coastal evolution is treated in terms of geology, climate, organisms, changes in land and sea level and processes in coastal waters. Coastline changes resulting from erosion or deposition and changes in sea level relative to the land can be studied over various timescales, and are ongoing. Some are directly or indirectly due to human activities, notably land reclamation and the building of artificial structures such as sea walls and breakwaters. An understanding of coastal evolution is an essential basis for coastal management.

2

Coastal processes

2.1 Introduction

Processes at work in coastal waters include winds, waves, tides and currents, which together provide the energy that shapes and modifies a coastline by eroding, transporting and depositing sediment. Although waves, tides and currents interact, one process augmenting or diminishing the effects of another, it is convenient to discuss them separately. The various kinds of current are treated incidentally (indexed on Panel 2.1, p. 14)

2.2 Waves

Waves are undulations on a water surface produced by wind action. The turbulent flow of the wind blowing over water produces stress and pressure variations on the surface, initiating waves that grow as the result of the pressure contrast between their driven (upwind) and advancing (downwind) slopes. Waves consist of orbital movements of water that diminish rapidly from the surface downwards, until the motion is very slight where the water depth (d) equals half the wavelength (L) (Figure 2.1). The depth at which waves become imperceptible is termed

the wave base, and in theory erosion by waves could ultimately reduce the world's land areas to a planed-off surface at this level, providing they remained tectonically stable. Orbital motion in waves is not quite complete, so that water particles move forward as each wave passes, producing a slight drift of water in the direction of wave advance.

Wave height (H) is the vertical distance between successive crests and troughs, wave steepness the ratio between the height and the length (H/L) and wave velocity (C) the rate of movement of a wave crest. Wave height is proportional to wind velocity, and wave period (T, the time interval between the passage of successive wave crests) to the square root of wind velocity. Wave dimensions are also determined partly by fetch (the extent of open water across which the wind is blowing) and by the duration and strength of the wind. Large waves are generated by severe storms, and in mid-ocean the largest storm waves, generated by prolonged strong winds over distances of at least 500 km, can be more than 20 m high, travelling at more than 80 km/hr. Waves transmitted across the oceans from storm centres become long and regular, and are known as ocean swell. In coastal waters waves are diminished by friction with the shallowing sea floor, but locally generated

Coastal Geomorphology: An Introduction, Second Edition. Eric Bird.
© 2008 by John Wiley & Sons, Ltd

Panel 2.1   Currents

Currents are generated in various ways, and some currents are of multiple origin. Some are discussed in sections on waves and tides, but to avoid repetition the various kinds of current are listed here and indexed to the text.

1. *Rip currents* flow back into the sea through breaking waves at intervals along the shore. See Section 2.2.7

2. *Wave-generated currents* flow alongshore when waves arrive at an angle to the shoreline. See Section 2.2.8

3. *Tidal currents* are ebb and flow (flood) currents generated by falling and rising tides. See Section 2.3.1

4. *Ocean currents* are slow mass movements of water in response to variations in water temperature and salinity, atmospheric pressure and wind stress. See Section 2.6.1

5. *Wind-generated currents* flow in the direction of the wind. See Section 2.6.2

6. *Fluvial currents* are the discharge where a river flows into the sea. See Section 2.6.3

There are also *density currents*, which occur where water of higher specific gravity (colder or more saline) moves to displace water of lower specific gravity, but these have no direct effect on coasts.

---

storm waves can still be several metres high when they break on the shore. On the Atlantic coast of the United States, for example, occasional hurricanes generate waves up to 5 m high when they break. Such storm waves can cause erosion or deposition well above the level of the highest tides.

Simple equations indicate the relationships between wave parameters. In deep water, wave velocity (C_o) is the ratio (L_o/T) of wavelength (measured in metres) to wave period (measured in seconds). Wavelength (L_o) in deep water (where $d > L/2$) can be used to calculate wave velocity (C_o) from the following formula, in which g is the gravitational acceleration (about

980.62 cm/sec^2 at latitude 45°):

$$C_o{}^2 = \frac{g L_o}{2\pi}$$

from which, since $L_o = C_o T$,

$$C_o = \frac{g T}{2\pi} \text{ or } 1.56\,T \text{ in m/sec}$$

so that $L_o = 1.56\,T^2$, providing a means of calculating wavelength from measurements of wave period in deep water.

Measurements of nearshore waves can be made using a staff with graduated electric wires, a pressure transducer on the sea floor or sonic devices mounted on a pier or platform. The problems of monitoring waves were discussed by Morang, Larson and Gorman (1997), and more detailed accounts of the nearshore wave field are given by Hardisty (1994) in relation to beaches and Sunamura (1992) on rocky shores.

2.2.1   Ocean swell

During storms strong winds generate irregular patterns of waves, varying in height, length and direction, which radiate from the generating area. The longest waves move most rapidly, and are most durable, so that as waves move across the ocean they become sorted into swell of more regular (gradually diminishing) height and (gradually increasing) length, which eventually arrives to break on a distant shore. There are major wave-generating storm regions in the Southern Ocean and in the northern parts of the Atlantic and Pacific Oceans.

Ocean swell consists typically of long, low waves with periods of 12–16 seconds. As they move towards a coast the wave crests gain in height and steepness, and as they enter shallow water they break to produce the surf observed on the shores of the Pacific, Atlantic and Indian

Figure 2.1 Wave terminology, and the pattern of currents as wave crests and troughs move shoreward

Oceans. Storms in the Southern Ocean initiate the SW swell that travels thousands of kilometres to arrive on the western and southern shores of Australia, New Zealand, the Americas and Africa. The long swell that breaks on the coast of California has travelled about 12 000 km across the Pacific Ocean.

SW swell generated by gales south of Africa is transmitted across the Indian Ocean to the southern coasts of India and Sri Lanka and the western coast of Thailand. It reaches the southern coasts of Sumatra and Java, and other Indonesian islands as far east as Timor. SW swell originating south of Australia and New Zealand moves across the Pacific Ocean to coasts between Chile, California and Alaska. On the way it breaks on the shores of many Pacific islands. The stormy waters south of South America produce SW swell across the Atlantic Ocean to West Africa and Western Europe (Portugal to the Hebrides), and up to the south coast of Iceland. Occasionally a SW ocean swell with wave periods up to 20 seconds arrives on the south coast of Britain, breaking heavily on the Loe Bar in

Cornwall and Chesil Beach in Dorset, and this was probably generated in the vicinity of the Falkland Islands. Similar swell has been recorded on the Cornish coast about four days after hurricane disturbances off Florida.

As a SW swell moves across the oceans it fans out to produce a weaker S and SE swell. Southerly swell occasionally reaches Iceland and the Aleutian Islands, and SE swell arrives on the coasts of South America (Argentina to Recife in Brazil), SE Africa and southern Arabia, SE India, SE Australia (eastern Tasmania north to Fraser Island) and the east coast of New Zealand. SE swell is often augmented by the effects of the SE monsoon and trade winds in coastal waters.

Storms in northern latitudes generate similar ocean swell, especially in winter, when a NW swell from the north Pacific arrives on shores between British Columbia, California and Central America. In the north Atlantic a NW swell extends to the coasts of Western Europe (Ireland to Portugal) and West Africa (Morocco to Senegal). It is frequently masked in high latitudes by locally generated storm waves. The NW swell is stronger in the northern winter, but fades in the summer months, whereas the SW swell from the Southern Ocean is stronger in the northern summer (southern winter) and weaker in the northern winter. This leads to seasonal alternations, the winter NW swell alternating with the summer SW swell on the coasts of Portugal and California. These seasonal contrasts are well known in Half Moon Bay, California.

The NW swell in the North Pacific diverges to form a weaker NE (or northerly) swell on the north and east coasts of Japan, extending to China, Vietnam, the north and east coasts of the Philippines and northern New Guinea. In the North Atlantic there is similar modification of the NW swell to produce a NE swell from Cape Hatteras south to the eastern islands of the Caribbean and the NE coast of Brazil. Again, the storm-generated waves radiating across these oceans may be augmented by waves produced by NE winds by the time they reach these coasts.

2.2.2 Storm waves

Apart from ocean swell, nearshore wave regimes depend on climatic conditions in coastal waters. Storm waves, generated by strong wind action, arrive frequently on west-facing coasts in latitudes subject to frequent westerly gales, as on the Atlantic coasts of NW Europe and the Pacific coasts of Canada and the NW United States. There are stormy coasts in Patagonia (southernmost Chile) and on the western seaboard of South Island, New Zealand. Although storms in the Southern Ocean generate the large waves that spread out across the Atlantic, Indian and Pacific Oceans, only small waves reach the coasts of Antarctica, even on parts that are ice free in summer.

Monsoon winds generate waves on the coasts of India and SE Asia. In Peninsular Malaysia the SW monsoon (May to September) produces waves along the west coast, and the NE (winter) monsoon generates strong wave action on the east coast, extending to SE Thailand and the coast of Vietnam.

Large waves generated by occasional tropical cyclones (also known as hurricanes or typhoons) are accompanied by storm surges (Section 2.4) on the SE coast of the United States, in the Caribbean, Madagascar and Mozambique, India and the Bay of Bengal, from Thailand to Vietnam, in southern China and southern Japan, and in northern Australia. By contrast, the coastal waters of equatorial regions (such as NE Brazil and Indonesia) are relatively calm, except where they receive ocean swell of distant origin, as in the Gulf of Guinea, southern Indonesia and the Pacific coast of Central America. Ocean swell dominates open coasts outside the stormy zones, although there may occasionally be strong locally generated wave action.

2.2.3 Nearshore waves

Ocean swell generally arrives as regular waves, breaking at intervals of 12 to 16 seconds on the shore. When ocean swell arrives from different sources there are variations in wave height as interacting sequences of waves break upon the shore. The idea that every seventh wave is larger is legendary, but occasional higher waves occur as the result of the merging of two or more sets of waves. Sometimes there is a steady increase in the height of successive waves to a combined phase maximum, followed by a diminution as the waves move out of phase. Known as surf beat, this interaction can produce maximum waves breaking at intervals of several minutes, accompanied by pulsations of current flow alongshore and onshore–offshore. Wave set-up is the raising of sea level close to the shore as the result of waves driving water in. It is roughly proportional to incident wave height, but also depends on shore gradient and beach texture, shingle absorbing more wave energy than sand.

Waves generated locally by winds (particularly onshore winds) blowing over coastal waters are typically shorter (wave period < 10 seconds) and less regular than ocean swell of distant derivation, and in stormy periods they are much steeper. They may be superimposed on ocean swell arriving in coastal waters, an onshore wind accentuating the swell and adding shorter waves to it, a cross-wind producing shorter waves that move at an angle through the swell. Offshore winds flatten swell to produce relatively calm conditions in the nearshore zone.

Ocean swell is usually strong enough to produce waves that shape the coastline, and waves generated in coastal waters by onshore winds exceeding about 20 km/hr (Beaufort scale 4 and over) also break on the shore with sufficient energy to erode coastal rock formations and transport sediment alongshore.

Geographical variation in nearshore wave conditions is related to fetch and coastal configuration. Coastlines protected by promontories, reefs or offshore islands receive swell (if at all) in a much modified and weakened form, so that locally wind-generated waves predominate. This is the case around landlocked seas, such as the Mediterranean and the Baltic, the Arabian Gulf and the Gulf of California, and in embayments with constricted entrances such as Port Phillip Bay in Australia. Around the British Isles wave regimes are determined largely by winds in coastal waters, ocean swell reaching the Atlantic coasts and penetrating the English Channel, the Irish Sea and the northern North Sea only to a limited extent.

On many coasts a particular wave direction is clearly dominant. The south coast of England, for example, is dominated by waves produced by the prevailing SW winds, and much of the east coast of Australia has a prevalence of SE wave action. Other coasts show greater variation: the North Norfolk coast in England has waves arriving from the NW, north and NE according to local wind conditions, which change with the passage of depressions and anticyclones. In some years NE waves are dominant; in others NW waves prevail. The east coast of Port Phillip Bay, Australia, has seasonal variations in dominant waves, with westerly and SW waves prevalent in summer and NW waves commoner in the winter months (Figure 6.15). Correlation with meteorological patterns can be used to determine long term dominant wave incidence (and resulting sediment movement) in such conditions.

2.2.4 Wave refraction

Ocean swell has parallel wave crests in deep water, but as the waves move into shallower water they begin to be modified by the sea floor: the free orbital motion of water is impeded and the frictional effects of the sea floor retard the advancing waves. Sea floor topography thus influences the pattern of swell moving towards the coast, the angle between the swell and the submarine contours diminishing so that the wave crests become realigned until eventually they are parallel to the submarine contours.

This is known as wave refraction (Figure 2.2). Where the angle between the ocean swell and the submarine contours is initially large, the adjustment is often incomplete by the time the waves arrive at the shore, so that they break at an angle (usually < 10°). Where the angular difference is initially small the waves are refracted in such a way that they fit the outline of a beach or of cliffs cut in soft formations, breaking synchronously along the coastline (Figure 2.2(a)). Sharp irregularities of the sea floor have stronger effects, a submerged bank retarding the waves while a submarine trough allows them to run on (Figure 2.2(b)). In Australia a bank or reef that causes local raising and steepening of wave fronts (useful for surfing) is known as a bombora. Islands or reefs awash at low tide interrupt the waves and produce converging patterns of wave refraction (Figure 2.2(c)), while waves that have passed

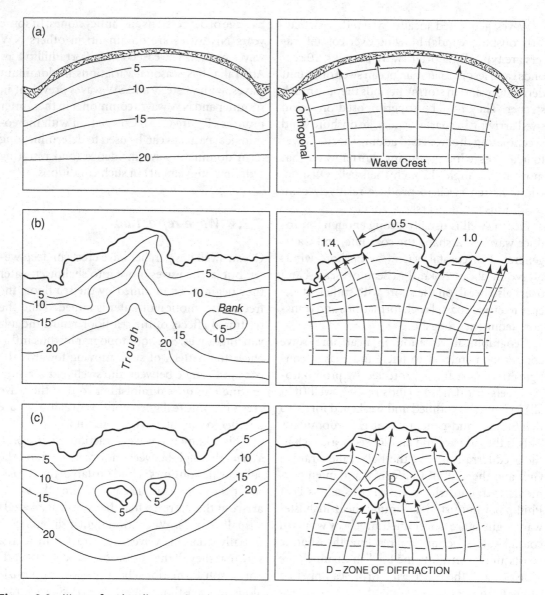

Figure 2.2 Wave refraction diagrams in relation to submarine topography: (a) refraction of waves moving into a bay; (b) refraction of waves over a trough (refraction coefficients indicated); (c) refraction round and diffraction between offshore islands

through narrow straits or entrances are modified by diffraction, spreading out in the wider water beyond. Waves are also retarded as they impinge on headlands, but run on into the intervening bays. In broad embayments they develop gently curved patterns, the waves in the middle of the bay moving on through deeper water while towards the sides, in shallower water, they are held back. In narrow embayments the refracted wave crests are more sharply curved.

WAVE CREST

Figure 2.3 The curved outline of the beach in Seven Mile Bay, Tasmania, has been shaped by gently refracted ocean swell, the wave crests breaking to form surf which sweeps sand on to the shore. This is a swash-dominated beach

These curved patterns of wave crests in coastal waters can be observed from headlands or seen on air photographs (Figure 2.3). Given knowledge of the direction of approach of waves in deep water, their length or period, and the detailed configuration (bathymetry) of the sea floor from nautical charts, it is possible to draw diagrams confirming the patterns of refraction that develop as waves enter nearshore waters and approach the coast. Such diagrams are now generated using a computer program. Wave refraction diagrams can also be used to predict the direction of longshore currents produced when waves arrive at an angle to the shore.

The shaping of curved coastlines by refracted swell is obvious on ocean coasts, but wave patterns generated by local winds on landlocked seas, lakes or coastal lagoons can also produce curved coastlines. On the shores of the Mediterranean Sea, for example, the curved outlines of Languedoc, the Venetian–Trieste coastline and the Bay of Benghazi have been shaped by waves generated locally. The same is true of curved coasts beside the Caspian Sea, the Sea of Azov, the Red Sea, the Arabian Gulf and much of the Gulf of Mexico. In Australia there are smoothly curved beaches in landlocked embayments, as on the east coast of Port Phillip Bay (Section 6.9) (Black and Rosenberg, 1992), or coastal lagoons such as Lake Wellington (Gippsland Lakes). Similar features are seen in the lagoons (termed limans) of the Ukrainian Black Sea coast. The curved outlines are determined by the resultant directions of waves that come from various directions, and are orthogonal to the maximum fetch. There may be secondary smoothing by longshore currents, generated by waves coming in on either side of the maximum fetch and causing longshore movements of beach sediment.

2.2.5 Wave energy

When orthogonal lines (rays) are drawn at equal intervals on wave refraction diagrams, at right angles to the alignment of waves in deep water. Projected shoreward they converge where they pass over a submerged bank or reef and diverge into where deeper water is traversed; in general, they converge towards headlands and diverge in embayments (Figure 2.2(b)). The spacing of these orthogonal lines shows how wave energy is distributed when the waves reach the coastline, a shoreward convergence indicating a concentration of wave energy, whereas a divergence indicates weakening. This can be expressed as a refraction coefficient (R), equivalent to the square root of the ratio between the distance between neighbouring orthogonal lines in deep water (S_o) and their spacing on arrival at the coastline (S). Calculated from measurements on a wave refraction diagram, this coefficient is an indication of the relative wave energy arriving on each part of the coastline. Such features as stream outlets or lagoon entrances are often found on coastal sectors of diminished wave energy (Bascom, 1954), while erosion is intensified where convergent orthogonal lines indicate a focussing of wave energy. However, there are many exceptions to the generalisation that erosion is concentrated on headlands and deposition in bays: on the Guyana coast, for example, there is deposition on convex parts of the coastline (Lakhan and Pepper, 1997).

On coasts sheltered from strong wave action, particularly where the nearshore zone is broad and shallow, wave energy is attenuated and breaking waves diminished. Wave energy may be dissipated to such an extent that waves do not reach the shore. The grassy marsh coastline of western Florida receives very little wave action (except during hurricanes), and in the broad, shallow Gulf of Bo Hai, in northern China, there is usually almost no wave action along the low-lying shores of the Hwang-ho delta. Marine vegetation, especially kelp, can diminish waves approaching the coast, as in the cool sea off California, while wave energy on humid tropical coasts may be diminished by greater viscosity in nearshore waters made turbid by large quantities of sediment in suspension.

2.2.6 Breaking waves

As waves move into shallow water ($d < L/2$) they are modified in several ways. Their velocity C_s diminishes according to the formula

$$C_s{}^2 = \frac{gL}{2\pi} \tan h \frac{2\pi d}{L}$$

The shallowing water also diminishes their length and period, and as they approach the shore their height increases and they steepen, the crests becoming narrower and sharper, and the intervening troughs wider and flatter. Orbital movements within each wave become increasingly elliptical, and shoreward velocity in the wave crest (used by surfboard riders) increases until it exceeds the wave velocity. When the orbital motion can no longer be completed, the oversteepened wave front collapses, forming a breaking wave (breaker). This sends forth a rush of water (swash or uprush) through the surf zone and up on to the shore, followed by a withdrawal, known as the backwash, by which water returns to the sea. These alternations can be documented photographically against a graduated scale, or measured with a dynamometer. In the nearshore zone the waves generate alterations of shoreward and seaward movement, the shoreward motion being generally stronger,

Figure 2.4 Plunging breakers on a shingle beach, with a strong backwash

except when breakers plunge and withdraw water and sediment seaward (Figure 2.4).

The pattern of breaking waves varies with wave height and nearshore gradient, so that low waves break closer to the shore than high waves, and waves of a specific height break closer inshore where the gradient is steep than where it is gentle. There are complications where the nearshore topography is rugged, with reefs that disrupt wave action, or where bars have been formed by wave action in the nearshore zone. Applications of this principle have been used to modify incident waveforms by re-shaping nearshore morphology (Bird, 1996a). Underwater structures have been built to diminish wave energy on an eroding coast near Niigata in Japan, and a 120 m long boomerang-shaped artificial reef has been constructed off Cottesloe, Western Australia, with the aim of creating waves suitable for surfing.

The ways in which waves shape and modify the shore depend on their incidence and dimensions as they break, and the resulting patterns of water flow. Breaking waves are influenced by several factors.

(a) Local winds (especially winds blowing offshore, which steepen wave fronts).

(b) Changes in nearshore water depth accompanying the rise and fall of tides or other short term sea level changes.

(c) Currents (including those resulting from wave motion).

(d) The gradient and topography of the sea floor.

(e) The configuration of the coastline.

Waves that arrive and break parallel to the coastline move water and sediment to and fro (onshore/offshore drifting) while those that arrive and break at an angle to the coastline move water and sediment along the shore (longshore drifting). Waves that break against hard rocky coasts or solid sea walls are reflected, and tend to move sediment seaward.

The energy of a breaking wave (E) depends largely on breaker height (H_b) and the density of the water (p), together with gravitational acceleration (g), according to the formula

$$E = \frac{1}{8} pg H^2$$

However, there is considerable variation in the dimensions of waves reaching the shore. The difficulty of estimating typical wave height can be overcome by using the concept of the significant wave ($H_{1/3}$), defined as the average height of the highest one-third of all waves observed over a period of 20 minutes, or the highest 33 in a train of 99 waves. The significant wave can be used to compare conditions in coastal waters at different times, or on different coastlines. On ocean coasts and those facing stormy seas significant wave height commonly exceeds 3 m. Dolan and Davis (1992) devised a storm intensity scale (duration × maximum $H_{1/3}$) for the Atlantic coast of the United States with five categories up to major hurricanes.

A breaker coefficient (B) was defined by Galvin (1972) as

$$B = \frac{H}{Gs\,T^2}$$

where H is wave height at break point, g is gravitational acceleration and T is wave period. Waves breaking on beaches show a range of forms, within which Galvin recognised four types.

1. Surging breakers, which are low and gentle waves until they sweep up a relatively steep beach.

2. Plunging breakers, with fronts that curve over and crash, producing little swash but a strong backwash.

3. Collapsing breakers that subside as they move towards the shore.

4. Spilling breakers, which are short and high, and produce foaming surf as the swash runs up a beach of gentle gradient.

Although four types of breaking wave have been distinguished, the important point is whether they are constructive (washing sediment up on to the beach: Galvin's 1, 3 and 4) or destructive (causing beach erosion: Galvin's 2). This depends on whether the swash and backwash generated by breakers (whether parallel or at an angle to the shore) achieves net shoreward or seaward movement of beach material. In general surging, spilling and collapsing breakers have strong swash (onshore flow) followed by a gentle backwash, producing net shoreward movement of sediment, whereas plunging breakers have a short swash and a stronger backwash, so that they withdraw sediment from a beach. This distinction takes us back to the work of Johnson (1956), who suggested that waves that have steepness ($H_o:L_o$) outside the breaker zone of less than 0.025 produce breakers that are constructive (moving sediment on to beaches), while those with higher ratios (especially >0.25) become destructive (withdrawing sediment from beaches). However, wave profiles also vary in relation to the nearshore gradient ($\tan \beta$), so that relatively gentle waves can be destructive where the nearshore slope is steep, and relatively steep waves may become constructive after traversing a broad nearshore slope of low gradient. Further discussion of interactions between wave characteristics and beach morphology is given in Chapter 5.

Water temperature variations have little effect on wave action in the tropics and temperate zones, but on cold arctic and antarctic coasts waves become ineffective when the nearshore water freezes. Shore ice formation is disrupted where the tide range is large, particularly where tidal bores form, as in Cobequid Bay and the Minas Basin at the head of the Bay of Fundy in Canada. Nearshore ice fringes protect the coastline in winter, but can become erosive when waves disrupt them in the summer thaw. The length of the cold season determines the period of frozen nearshore water, and when the ice thaws and disintegrates waves drive it up on to a beach, shore platform or salt marsh. On the shores of the St. Lawrence Gulf in Canada and around the Baltic Sea wave-piled ice fragments may accumulate on the shore as irregular aprons or ridges up to 5 m high.

2.2.7 Rip currents

Backwash is the seaward flow of water that had been carried shoreward by breaking waves, but the augmentation (wave set-up) of nearshore sea level by shoreward movement of water due to wave motion must also return seaward, typically as localised rip currents (Figure 2.5). These occur in distinct (though variable) patterns on many shores (particularly along beaches), related to longshore variations in wave set-up (see edge waves, Section 2.2.8). They are usually more or less evenly spaced along the shore at intervals of 50 m up to at least a kilometre): their spacing increases with breaker height (and so surf zone width) and decreasing beach gradient (Short, 1999). Within a rip current zone water flows back through the breaker line in sectors up to 30 m wide, attaining velocities of up to 8 km/hr before dispersing seaward. They are a hazard for swimmers, who can be swept out to sea, and who are advised to swim alongshore until they find water moving back towards the beach. The shoreward movement of breaking waves and the seaward return currents form the main components of the nearshore water circulation.

A light or moderate swell produces numerous small rip currents, and a heavy swell pro-

Figure 2.5 Rip currents (arrowed) in the surf on Woolamai Beach, Phillip Island, Australia. At the head of each rip current the beach has been lowered and cut back

duces a few more widely spaced and concentrated rips, fed by stronger lateral currents in the surf zone. Pulsations in outflow may occur in response to variations in the height of waves (or groups of waves) breaking on the adjacent shore. Rip currents may cut channels seaward through the nearshore zone (across any nearshore bars), and deposit fans of sediment as velocity falls at their seaward limits; in some places these channels have grown headward to form re-entrants in the beach. At low tide rip current velocity may increase as outflow is concentrated along these channels. When waves arrive at an angle to the beach the rip currents head away diagonally through the surf instead of straight out to sea, and cut oblique channels through the nearshore zone. Rising tides and onshore winds raise the water level along the shore, and thus intensify rip currents.

2.2.8 Wave currents

Waves that break parallel to the shore produce orthogonal swash and backwash on beaches, and mainly onshore–offshore movements of water and sediment, but there are usually lateral variations in breaker height, particularly where wave refraction has generated contrasts indicated by varying orthogonal spacing (Figure 2.2). Any such lateral variations in breaker energy are resolved by divergent longshore current flow in the nearshore zone. As a rule the natural variability of wave dimensions results in energy changes that result in pulsations in longshore and onshore–offshore current flow.

It has been suggested that waves breaking parallel to the beach generate orthogonal edge waves, standing oscillations that develop at right angles to the coastline as the result of resonance between waves approaching the shore and waves reflected from it (Hardisty, 1994). The existence of such edge waves is difficult to demonstrate, but it has been suggested that they produce longshore variations in breaker height as their crests augment incoming breakers. Augmented breakers form swash salients, sectors of higher water level from which there is longshore flow into intervening troughs, where outflowing rip currents develop. If edge waves are generated with the same (or doubled) period of incident waves, the two motions could combine to produce a regularity in nearshore water circulation that may explain the rhythmic nature of beach cusps (Figure 6.20) and certain multiple sand bars (Figure 8.14).

When waves arrive at an angle to the shore they deflect the nearshore water circulation, generating longshore currents. The effects of such wave-induced longshore currents are difficult to separate from the associated effects of oblique swash and orthogonal backwash produced when waves arrive at an angle to the shore and break on the beach. Both processes move sediment alongshore, the action of oblique wave swash causing beach drifting, while the wave-induced currents cause longshore drifting in the nearshore zone. Whereas current velocities of at least 15 cm/sec are needed to mobilise sand, agitation of sea floor sand by waves can lead to its entrainment by gentler currents.

2.2.9 Dominant waves

The direction, height and periodicity of waves approaching the shore have a strong influence on coastal outlines in plan and profile, particularly on cliffed coasts cut in soft formations or features formed by deposition. When wave conditions change there are corresponding adjustments in the morphology of depositional features such as beaches and spits. Wave conditions change frequently, and it is necessary to analyse several years' records to establish characteristic annual and seasonal wave regimes. Direct observations of wave conditions are made from lightships and coastal stations, and these can be supplemented by analyses of meteorological data, when the direction and strength of winds are correlated with the pattern and dimensions of locally generated waves, and by data on ocean swell regimes.

Dominant waves tend to determine such features as beach shape and alignment and the net direction of longshore drift and sorting on beaches. Variations in coastal aspect can result in differing directions of longshore drift in response to dominant waves.

2.2.10 High, moderate and low wave energy coasts

Coasts exposed to ocean swell and stormy seas (generally with deep water inshore) are known as high wave energy coasts, while those sheltered from strong wave action, bordering narrow straits, landlocked embayments or island or reef-fringed seas with limited fetch, or where wave energy has been reduced by intense refraction, are low wave energy coasts. It is useful to recognise an intermediate category of moderate wave energy coasts.

High wave energy coasts can be defined as those with mean annual significant wave height ($H_{1/3}$, measured as the waves break) exceeding 2 m, moderate wave energy coasts 1–2 m and low wave energy coasts less than a metre. High wave energy coasts typically have bold, rugged cliffs and long gently curving beaches, as on the stormy Atlantic shores of Europe and the ocean coasts of southernmost Africa and southern and western Australia. Coastlines become more irregular on moderate wave energy coasts, while on low wave energy coasts, beaches are typically shorter, with such features as cusps, lobes and spits, as well as deltas and marshlands diversifying the shore, numerous shoals and bars, few cliffs (except on weak formations such as clay) and a generally intricate configuration. Examples are found in the Danish archipelago, around Puget Sound, and on the shores of many estuaries and lagoons. The coasts of northern Sumatra and western Malaysia, bordering the Strait of Melaka, are typical of tropical low wave energy coasts, with extensive mangrove fringes, coastal plains formed by confluent deltas, occasional beach-fringed segments and very few cliffs. Similar features are seen where nearshore and fringing reefs diminish incident wave energy, as on the north coast of Viti Levu, in Fiji. The Gulf Coast of Florida, in the United States, has generally low wave energy, wave action be-

ing reduced by the broad, gently shelving offshore profile and a prevalence of winds blowing from land to sea. It shows an intricate coastline with only limited beaches interspersed with minor spits, deltas, and marshes and persistent offshore shoals. The sector between Tallahassee and Tampa was described by Tanner (1960) as a zero energy coast, with mean breaker height less than 4 cm and negligible longshore drifting (Section 2.2.5). However, the generally low wave energy coast is occasionally subject to drastic modification by strong wave action during hurricanes, when beaches are overwashed, and the shore profile may develop new erosional features that persist during subsequent calmer conditions.

2.3 Tides

Tides are movements of the oceans set up by the gravitational effects of the moon and the sun in relation to the earth. They are very long waves that travel across the oceans and are transmitted into bays, inlets, estuaries or lagoons around the world's coastline. Oceanic tides are indeed tidal waves, but this term has been widely misused as a synonym for tsunamis, which are large waves generated by tectonic events (Section 2.5).

The ebb and flow of tides produces regular changes in the level of the sea along the coast, and generates tidal currents. The lunar cycle produces semidiurnal tides (two high and two low tides in approximately 25 hours), well displayed around the Atlantic Ocean.

The solar cycle produces diurnal tides (one high and one low tide every 24 hours), as registered in the Caribbean, northern Java and the Philippines, and on the Antarctic coast. Elsewhere the two are mixed, yielding unequal high and low tides (e.g. high high, low low, low high and high low), as around much of the Pacific and Indian Ocean coasts. Where the effects of lunar

Figure 2.6 Tidal curves, tide ranges and shore zones. HAT – highest astronomical tide, MHST – mean high spring tide, MHNT – mean high neap tide, MIN HT – minimum high tide, MIN LT – minimum low tide, MLNT – mean low neap tide, MLSDT – mean low spring tide, LAT – lowest astronomical tide. The maximum shore width (AJ) is at maximum and minimum astronomical tides, diminishing to mean spring tides (BH), mean neap tides (CG) and minimum tides (DF)

gravity are stronger, high and low tides occur about 50 minutes later each day.

Variations in tidal type are significant in terms of durations of low tide drying (less than 12 hours with semidiurnal tides, but more than 12 hours with diurnal and mixed tides) which affect shore weathering and intertidal ecology. Tidal currents are also generally stronger with semidiurnal tides because of the greater rapidity of tidal fluctuations, but in general the most important aspect of tides in geomorphology is their vertical range.

The rise and fall of the tide (tide range) on a coast is measured by tide gauges, located chiefly at ports. The highest and lowest astronomical tides are those that occur at a particular point on the coast in calm weather over a period of at least a year. Tides recorded shortly after each new moon and full moon, when earth, sun and moon are in alignment, and have combined gravitational effects, are relatively large, and are known as spring tides (the term is a little confus-

ing, as these tides do not occur only during the spring season). The highest spring tides occur fortnightly (actually at intervals of about 14.6 days). Maximum spring tide ranges occur about the equinoxes (late March and late September), when the sun is overhead at the equator. At half-moon (first and last quarter) the sun and moon are at right angles in relation the earth, so that their gravitational effects are not combined; tide ranges recorded shortly after this are reduced, and are known as neap tides. The relationship between shore zones and tide ranges is shown in Figure 2.6.

There are also longer term variations of tide range in response to astronomical cycles of the relative positions of the sun, moon and earth. The moon, for example, returns to a similar position relative to the earth every 27.5 days, but its orbit is such that it returns to almost exactly the same position only once in 18.6 years. This long-term SAROS oscillation, based on the precession of the lunar orbit, is the nodal cycle (Pugh, 1987), which reached maxima in June 1950, February 1969 and October 1987, and caused exceptional flooding in low-lying Pacific Islands in 2006: the next is due in 2024. These maxima have been detected in sea level records in the Venice Lagoon (Section 3.10) and correlated with frequency and depth of tidal surges. Still longer astronomical tidal cycles have been identified, with maxima occurring in 1745 and 1922, and another expected in 2192, but these are of small dimensions (Cartwright, 1974).

Tidal forces also cause minor fluctuations in land levels (up to 30 cm) as the gravitational pull of the sun and the moon are exerted on the earth's crust: on the Atlantic coasts of Britain the land rises and falls by up to 10 cm twice daily. Tidal oscillations of sea level recorded on a coast are thus resultants of minor land movements as well as the upward and downward movements of sea level.

Tidal movements can also influence wave action. In general the transverse profile of the shore and nearshore zones is concave, and as the tide rises the water deepens, so that larger waves break upon the shore. Where tidal currents flow in one direction they can reduce the velocity and size of waves coming in from a contrary direction.

2.3.1 Tidal currents

Tidal currents, produced as tides rise and fall, alternate in direction in coastal waters, reversing as the tide ebbs; their effects may thus be temporary or cyclic, and can be measured using current meters (Morang, Larson and Gorman, 1997) or traced with the aid of floating drogues or seabed drifters. In the open ocean tidal currents rarely exceed 3 km/hr, but where the flow is channelled through gulfs, straits between islands, or entrances to estuaries and lagoons, tidal currents are strengthened, and may locally and briefly exceed 20 km/hr. Modifications of tides

transmitted into such basins are discussed in Chapter 11. Stronger tidal currents are generated by spring tides than neap tides because a larger volume of water is moved, and the strongest currents are generated where the tide range is large. Tidal currents attain 16 km/hr in the Raz Blanchart, between Alderney and Cap de la Hague in NW France, and are also strong through Apsley Strait, which separates Melville and Bathurst Islands off the north coast of Australia. Tidal waterfalls form between some of the islands in the Buccaneer Archipelago off the Kimberley coast in NW Australia, where tide ranges are very large (locally more than 10 m).

Tidal oscillations impinging on a coastline may set up longshore currents, as on the north Norfolk coast in England, where the longshore flow is westward during the 2–3 hr preceding high tide, then eastward for another 2–3 hr as the ebb sets in. Tidal currents have little direct effect on the morphology of coasts facing the open sea, but where they pass in and out of estuaries or through narrow straits they can scour the bordering shores. At the entrance to Deception Pass, on Whidbey Island in Washington (NW United States), strong tidal currents (up to 34 km/hr) have truncated the spit at West Point and deposited an elongated subtidal spit at right angles to the coastline. In a similar way, strong tidal currents have curtailed spit growth at Sandy Point in Westernport Bay, Australia (Figure 6.19). Spits that have grown into narrow inlets maintained by strong ebb and flow currents have been modified by current action, as at the entrances to Langstone and Chichester Harbours on the south coast of England.

In general, tidal currents do not cause beach erosion or deposition, but they carry sediment along the coast in the nearshore zone, and this may eventually be delivered to beaches alongshore. Longshore currents generated by tides may either augment or reduce those produced by oblique waves, depending on the direction of approach, but the effects of tidal currents are otherwise generally subordinate to the effects of waves in the nearshore zone. Transverse tidal currents flowing in and out of inlets and narrow bays can act like a breakwater, and so interrupt longshore drifting of beach sediments.

2.3.2 Tidal environments

Tidal environments can be classified in terms of tide ranges. These can be defined as the difference between mean high and mean low spring tides where the tides are semi-diurnal, or between mean higher high tide and mean lower low tide where the diurnal component introduces inequality in the levels of successive tides. Global data given in the Admiralty Tide Tables show that mean neap (or minimum) tides vary between 30 and 80 per cent of mean spring (or maximum) tide values and that astronomical tide ranges are 20–30 per cent higher than mean (or maximum) spring tide ranges. Coastwise comparisons can be conveniently made using mean spring (or maximum) tide range. Typically this shows more intricate variations along indented coastlines (as around the British Isles) than on open oceanic coasts.

Tide range in mid-ocean is small, of the order of 0.5 m, but is amplified where the tide is transmitted into shallow coastal waters, particularly in funnel-shaped gulfs and intricate embayments. Where the mean spring tide range is less than 2 m, the coast can be classified as microtidal, between 2 and 4 m mesotidal, 4–6 m macrotidal and more than 6 m megatidal. Large tide ranges are found on coasts where the tidal wave steepens across broad continental shelves and into gulfs and estuaries, especially where there is resonance between incoming and reflected tides, as in the Bay of Fundy, in eastern Canada. Microtidal ranges are typical of the more open coasts of the Atlantic, Pacific, Indian and Southern Oceans and landlocked seas such as the Baltic, Mediterranean, Black, Red and Caribbean. Because of local variations it is difficult to show global tide ranges on a textbook map, but the broad pattern is described in Panel 2.2.

A large tide range implies a broad intertidal zone, more than 20 km of sandflats and mudflats being exposed at low spring tides in the Bay of Mont Saint Michel, in France, and about 8 km off Broome in NW Australia. In Britain the intertidal zone is more than 3 km wide in Morecambe Bay, in The Wash and on Maplin Sands on the Essex coast. Wave energy is expended in traversing such a broad shore zone, and waves that reach the coast at high tide have been much diminished by friction in the shallow water passage. Zones of weathering and ecological processes become wide and dispersed. Strong tidal currents shape elongated parallel banks and troughs on the sea floor, particularly where they flow through narrow seas, as in the Straits of Dover and Torres Strait, between Australia and Papua New Guinea. Strong currents through Deception Pass in Washington have scoured the rocky floor, sweeping sediment out of either end of the narrow passage.

Movements of sea floor shoals close to the coast may modify wave patterns and result in coastline changes. Where the tide range is small, as around the Mediterranean and Baltic Seas and along the SW coast of Australia, wave action, intertidal weathering processes and biogenic activities are concentrated within a narrow vertical zone.

Changes in tide range may occur as the result of deepening or shallowing of the entrances to bays, inlets and estuaries. Tide range increases in a coastal lagoon when an entrance is cut or widened and deepened, allowing greater tidal ventilation. For example, the dredging of the approaches to Venice Lagoon has led to an increase in tide range within the lagoon, with stronger tidal currents and the dispersal of wave action over a larger vertical range than previously. A proposal to deepen the entrance to Port Phillip Bay, in SE Australia, to allow larger ships to enter

Panel 2.2 Global variations in tide range

Coasts where the mean spring tide range is less than 2 m can be classified as microtidal, between 2 and 4 m mesotidal, 4–6 m macrotidal and more than 6 m megatidal.

In Britain there is a microtidal sector in central southern England, between Portland Bill and Christchurch Bay, produced by a tidal circulation in the English Channel that yields large tides across on the north coast of France. Otherwise, macrotidal conditions predominate in England and Wales, with very high (megatidal) ranges in the Bristol Channel (Avonmouth up to 14.4 m), and from North Wales around the east coast of the Irish Sea (Liverpool 8.3 m). The coasts of western and northern Scotland are generally mesotidal, increasing to macrotidal south from Montrose. The east coast of England is largely macrotidal, tide ranges increasing southward from Berwick upon Tweed (4.1 m) to Boston (6.8 m), diminishing in Norfolk and Suffolk to mesotidal (microtidal 1.9 m between Caister and Lowestoft), then rising again to macrotidal southward from Clacton (4.7 m).

The west coast of the Atlantic Ocean is generally microtidal, except for sectors in southern Argentina (Puerto Gallegos 10.4 m), where there are strong currents in the Magellan Strait, the Amazon delta, northern New England and the Bay of Fundy (Parrsborough 11.6 m), the Gulf of St Lawrence (rising to 4.9 m at Quebec), Leaf Basin, a ria opening into Ungava Bay (11.5 m), and Frobisher Bay on Baffin Island to the north (9.0 m). Maximum tide ranges in Minas Bay at the head of the Bay of Fundy attain more than 18 m, and in Leaf Basin up to 20 m. The coasts of Greenland are mesotidal, as are those of western and northern Iceland, but SE Iceland is microtidal. Much of Norway is mesotidal but tide ranges diminish southward (Bergen 1.2 m) and are negligible around the Baltic Sea. The coasts of Denmark and NW Germany are generally microtidal, increasing to mesotidal in estuaries (Wilhelmshaven 4.0 m) and along the Dutch coast (Westkapelle 4.0 m), and continuing to increase through Belgium and northern France to the megatidal area of the Bay of Mont Saint Michel (Granville 11.5 m). They then diminish round Brittany (Brest 6.1 m) to mesotidal in the Bay of Biscay and Atlantic Iberia (Cadiz 2.7 m) and then microtidal (Gibraltar 0.9 m). The Mediterranean coasts are microtidal, tide ranges exceeding 1 m only near Sfax (1.3 m) in Tunisia. The coasts of West Africa are generally microtidal and mesotidal, the tide range diminishing south from Sierra Leone (2.6 m) and remaining microtidal around southern Africa until it becomes mesotidal in Mozambique (Beira 5.6 m) and the Strait of Madagascar. North from Kenya the Indian Ocean and Red Sea coasts are microtidal, as is the western side of the Arabian Gulf. Tide range increases to 2 m at Kuwait and the east coast of the Arabian Gulf is mesotidal, as is the Pakistan coast (Karachi 2.3 m). There is a small macrotidal sector in NW India (Bhavnagar 8.8 m), then a rapid diminution to microtidal conditions around India and Sri Lanka. Tide range increases in the Bay of Bengal, with macrotidal sectors on the Ganges delta and in Burma (Mergui 5.3 m). The coasts of Thailand and Peninsular Malaysia are generally microtidal, but increase locally to mesotidal and even macrotidal (Port Kelang 4.1 m). Microtidal and mesotidal conditions predominate in Indonesia, Papua New Guinea, the Philippines, Japan and east Asia, with a macrotidal zone in the Yellow Sea (Inch'on, South Korea, 8.2 m) and another exceeding 5 m in the northern part of the Sea of Okhotsk, where tides are augmented to 13 m at the head of the Zaliv Shelikhova gulf. The Arctic coast of Russia is microtidal westward to the Barents Sea, where it becomes mesotidal (Murmansk 3.2 m).

In the north Pacific Ocean tide ranges are small in the Bering Strait, but increase in the Gulf of Alaska, where Anchorage (9.4 m) and Port Essington (5.5 m) are macrotidal. Mesotidal conditions predominate in British Columbia but become generally microtidal on the southern shore of the Strait of Juan de Fuca (Port Angeles 1.8 m) and along the Pacific coasts of Washington, Oregon, California and Mexico. The Gulf of California has tide ranges of up to 9 m. They increase southward to macrotidal on the Panama and Colombian coasts (Balboa 5.0 m), then diminish to mesotidal in Ecuador (Guayaquil 3.6 m) and are microtidal through Peru and Chile to Cape Horn (Orange Bay 1.7 m), except for local augmentation in the gulfs of southern Chile (Puerto Montt 5.8 m).

On the Australian coast microtidal conditions prevail from Brisbane (1.8 m) south to Cape Howe, then along the whole of the southern and western coast with the exception of certain embayments. Tide range increases into Spencer Gulf (Port Augusta 2.0 m) and Gulf St Vincent (Port Wakefield 2.1 m), but the narrow entrance to Port Phillip Bay impedes tidal invasion, so that mean spring tide range diminishes from Point Lonsdale (1.1 m) at the entrance to about 0.6 m within the bay.

(Continued)

Panel 2.2 *(Continued)*

In NW Australia the tide range increases behind a wide continental shelf to 5.8 m at Port Hedland, 8.2 m at Broome and 10.5 m in Collier Bay on the Kimberley coast before diminishing to 5.5 m at Darwin. The Gulf of Carpentaria is microtidal, but in NE Queensland there is a mesotidal sector (Townsville 2.5 m) increasing to macrotidal in the vicinity of Mackay (4.9 m) then diminishing southward to Brisbane. New Zealand has generally microtidal coasts, increasing to mesotidal locally (Auckland 2.9 m, Westport 3.0 m), and the Pacific islands are microtidal. The coasts of Antarctica are also generally microtidal, with a few stations mesotidal (e.g. Breidbay 2.6 m).

---

and leave the Port of Melbourne raised concerns that it would increase tide range within this bay and raise high tide sea level, thereby accelerating shore erosion (Bird, 2006).

A reduction in tide range has occurred in estuaries impounded by barrages, as in the River Rance, NW France, and behind the dams on the Rhine delta. Such a reduction means that wave action becomes concentrated within a narrower vertical range than previously.

2.3.3 Tidal bores

Where tides drive water into funnel-shaped gulfs or estuaries the narrowing and shallowing configuration cause the front of the tidal wave to steepen and form a narrow wave that moves rapidly upstream as a tidal bore. The velocity (c) is equivalent to $\sqrt{g(h + H)}$, where g is gravity, h river depth and H wave height. During spring tides in the Bristol Channel the rising tide becomes a tidal bore up to 2 m high, moving up the Severn estuary at about 15 km/hr. It is preceded by forerunners, 20–30 m long low waves, and the sharp spilling bore front is followed by smaller waves and an upwelling of the river water. The bore penetrates upriver to beyond Gloucester.

Similar bores develop in estuaries and inlets on other macrotidal coasts, as in Turnagain Inlet in Alaska and towards the northern end of the Bay of Fundy in Canada. In the Amazon River the tidal bore (pororoca) moving upstream attains a height of up to 6 m. The large tides on the NW coast of Australia flow up several estuaries, but little is known of how far they run upstream and what dimensions they attain. The *mascaret*, a tidal bore that flowed upstream in the River Seine to Rouen, has been much reduced by dredging in the estuary downstream.

2.3.4 Tide-dominated coasts

The influence of tides on the coast depends largely on tide range, which determines the zone over which wave action can operate. On macrotidal and megatidal coasts the coastline is reached by waves only briefly and intermittently, wave action being withdrawn from the backshore for most of the tidal cycle. Beaches are poorly sorted because of the limited duration of effective wave action (Kidson, 1960). Brevity of wave attack also impedes cliffing, although cliffs may be cut in soft rock or unconsolidated deposits (such as glacial drift or salt marshes), as on parts of the shore of the Bay of Fundy and in estuaries with a large tide range. In such situations beaches are usually found at about high tide level, and the shaping of the beach profile, sorting of beach sediments and modification by longshore drift are all retarded by the vertical dispersal of wave energy as the tide rises and falls. Wave effects at any intertidal shoreline level are thus transitory: waves are probably more effective above mid-tide level, and most effective towards the high tide shoreline, particularly when storm waves arrive with a slightly raised sea level

(Trenhaile, 1987). Tidal currents play an important part in the shaping of intertidal topography exposed as the tide rises and falls: a morphology that can be termed tide-dominated. On muddy shores where wave action is weak the intertidal topography develops transverse profiles related to the tidal current regime, as in Westernport Bay, Australia (Section 10.1).

The distinction between wave-dominated and tide-dominated morphology raises problems. Beaches on microtidal coasts are almost entirely wave-dominated features, but wave action is usually present in areas classified as tide-dominated (even though its effects are modified as tides rise and fall). Salt marshes, mangrove and sea-grass banks and terraces are often considered tide-dominated, but waves certainly erode their margins and wash sediment into them as the tide rises. Strong currents within salt marshes and mangroves usually run transverse to the shore and are confined to the vicinity of tidal creeks as the tide falls (Figure 10.12). Purely tide-dominated morphology may exist on the floors of megatidal gulfs, such as the Bay of Fundy, although even there wave action contributes to the shaping of shoal and ripple forms as the tide rises and falls. Wave effects decline as the fetch diminishes, or where waves are impeded by reefs or shoals. The problem of defining tide-dominated environments has complicated efforts to use sedimentary sequences as indications of past changes in tide range and tidal current velocities (Hinton, 1998).

Extensive salt marshes have formed on low energy macrotidal and megatidal shores, as in Bridgwater Bay, in the Bristol Channel and on the shores of the Bay of Mont Saint Michel in NW France. Wide mangrove swamps occur in similar situations in low latitudes, as on the shores of bays and estuaries in northern Australia where the tide range is large.

On microtidal coasts, by contrast, wave energy is concentrated within a narrow vertical zone, facilitating cliff-base erosion and impeding the spread of salt marshes and mangrove swamps. Beach morphology, sediment transport and nearshore sea floor topography all become wave-dominated in these situations.

A change in sea level can modify tides. If sea level rises it is likely that there will be a slight increase in tide ranges around the world's coastline as the oceans deepen, the rise that actually occurs being modified as tidal amplitudes adapt to the changing coastal and nearshore configuration. If sea level falls tide ranges may diminish, except where the new coastal outlines result in increased local amplification.

2.4 Storm surges

Storm surges occur when strong onshore winds build up coastal water to an exceptionally high level for a few hours or days, and are most pronounced when they coincide with high spring tides. Strong onshore winds also generate large waves accompanying the raised sea level, overwashing beaches, flooding low-lying coastal areas and causing extensive changes in a short period. Beach erosion is usually severe during a storm surge, and if the coast consists of soft formations it may be cut back substantially. One of the best documented storm surges occurred in the North Sea on 31 January/1 February 1953, when a northerly gale raised the sea level up to 3 m along bordering coastlines, flooding extensive areas in eastern England and low-lying parts of the Dutch and German coasts. A similar surge accompanied the passage of a deep depression across the North Sea in 1978 (Steers et al., 1979). There have been many such episodes in this region in recent centuries, when coasts have been eroded and briefly submerged: the Hallig Islands off NW Germany were overwashed in 1758 and 1825. The limits reached by storm surges are commemorated on marker posts, as at Ribe in Denmark, or on the walls of buildings, as at Harbour Inn, near Southwold in Suffolk, and

Figure 2.7 A storm surge during a tropical cyclone produced large waves that dislodged and piled up beach rock on the shore near Port Hedland, NW Australia

in several Dutch coastal towns. Storm surges in the Adriatic Sea are responsible for the recurrent raising of sea level in the Venice Lagoon and marine flooding in Venice (acqua alta, Section 3.10).

Hurricanes raise sea level temporarily along the Gulf and Atlantic coasts of the United States by as much as 6 m, causing extensive erosion and damage to structures, and typhoons in the China Sea have similar impacts on the coasts of Japan and the Philippines. Major storm surges occur from time to time at the head of the Bay of Bengal, when large areas of the Sundarbans are devastated by flooding and wave overwash. In northern Australia tropical cyclones generate surges in a region where heavy rains accompany the summer monsoon, and cause widespread flooding in coastal districts. Erosion is severe during such episodes. Beaches are depleted, and such features as the thrown-up blocks of sand-

stone at Quobba in Western Australia and on Grand Cayman (Jones and Hunter, 1992), the piled-up beach rock near Port Hedland (Figure 2.7) or overwashed fans of beach sediment may persist for years or decades after such catastrophic events. On the Texas coast deep subparallel channels cut across low lying barrier islands by storm surges are called bogues, as on Andros Island.

2.5 Tsunamis

Apart from storm surges, exceptional disturbances of sea level occur during and after earthquakes, landslides or volcanic eruptions in and around the oceans. These produce tsunamis, very large waves that may attain heights of more than 30 m by the time they reach the coast. They are most common in the Pacific Ocean, which is bordered by zones of crustal instability, and they are responsible for occasional catastrophic flooding and beach erosion on Pacific coasts. The arrival of a tsunami is preceded by the withdrawal of the sea, an exceptional backwash exposing part of the nearshore area: a phenomenon that should be taken as a warning that very large waves are imminent.

Charles Darwin observed a tsunami, in the form of three large waves caused by an earthquake in Chile in 1835. In 1946 a tsunami was initiated by an earthquake off the Aleutian Islands, and waves travelling southward caused extensive devastation in the Hawaiian Islands, 3700 km away, 4.5 hr later. Large waves were recorded at many places around the Pacific Ocean, and at Scotch Cap, Alaska, the tsunami destroyed a lighthouse and radio mast on a cliff 30 m above sea level. An earthquake off the coast of Chile in 1960 also produced giant waves: a tsunami 11 m high devastated Hilo in Hawaii and a few hours later waves up to 40 m high broke on the east coasts of Hokkaido and Honshu in Japan.

In 1883 the explosive eruption of Krakatau, a volcanic island in Sunda Strait, Indonesia, generated a tsunami up to 30 m high on the nearby coasts of Java and Sumatra, sweeping away beaches, hurling coral boulders and blocks up on to fringing reefs and shore platforms and destroying the lighthouse at Anyer. In 1998 an earthquake 30 km off the north coast of New Guinea produced a 10 m tsunami that overwashed the sandy barrier at Sissano, west of Wewak, forming a coastal lagoon (Section 11.8.3).

A major tsunami occurred in the Indian Ocean on Boxing Day 2004, generated by a submarine earthquake (Richter Scale 9) off NW Sumatra. It caused extensive damage to coasts in Indonesia, Malaysia, Thailand, Burma, Bangladesh, India, Sri Lanka, Maldives and Somalia. There was extensive coastal flooding, particularly in sectors where a former mangrove fringe had been cleared, and large waves swept beach sand and gravel varying distances inland.

The deposits have distinctive characteristics that can be used to identify earlier tsunamis in coastal sedimentary sequences (Dawson and Shi, 2000).

Stratigraphic evidence of tsunamis generated by submarine earthquakes on a similar scale (Cascadian events) has been found on the Pacific coast of the United States. They have occurred at intervals of 300–1000 years, and the last was in 1700 A.D. (Atwater, 1987; Atwater and Hemphill-Haley, 1997).

The effects of a tsunami may persist long after the event (Dawson, 1994). In addition to changes effected on the coastline, very large waves may re-shape the nearshore profile and thereby change the pattern and dimensions of waves approaching the coast, resulting in subsequent erosion or deposition that would not otherwise have occurred. Like storm surges, they can throw large blocks and boulders up on to cliffs. Features possibly produced by earlier tsunamis have been found on various coasts, notably in SE Australia (Bryant, Young and Price, 1996).

2.5.1 Other giant waves

Giant waves can be generated in restricted areas by landslides and rock falls. A massive rock fall (30 million cubic metres of rocky debris) into Lituya Bay, an Alaskan fiord, in 1958 swept a wave 15 m high down the fiord at 160 km/hr, washing over a spit at the entrance and dispersing beach sediments seaward (Miller, 1960). Disturbances of a similar kind have occurred in the vicinity of ice coasts as the result of iceberg calving in Yakutat Bay, Alaska, in Greenland (where capsizing icebergs have generated waves 15 m high in fiords) and around Antarctica. Occasionally very large waves ('rogue waves') have been recorded on oil rigs and by ships at sea, some of which have been sunk. These seem to be of meteorological origin (Lawton, 2001). Possibly the so-called 'king waves' that break occasionally on the SW coast of Western Australia, forming swash that extends far up rocky slopes, sweeping them free of loose debris, are of this origin. Meteorites splashing into the sea could cause large waves, but examples have not been documented.

These are unusual phenomena. On most coasts waves influence the shaping of beaches within a zone a few metres above and below present mean sea level, the extent of this zone depending largely on tide range. Rapid changes occur on beaches during storms and occasional storm surges, but there are also more gradual gains and losses, leading to the re-shaping of the beach in plan and profile, during long intervening periods of less boisterous weather. There is sometimes difficulty in deciding whether particular coastal features are the outcome of brief catastrophic events or the product of gradual evolution in response to everyday processes.

2.6 Currents

See Panel 2.1. Reference has been made to rip currents that flow back into the sea through breaking waves at intervals along the shore (Section 2.2.7), wave-generated currents that flow alongshore when waves arrive at an angle to the shoreline (Section 2.2.8) and tidal currents generated by falling and rising tides (Section 2.3.1).

There are also ocean currents (Section 2.6.1), wind-driven currents (Section 2.6.2) and currents at the mouths of rivers (Section 2.6.3).

2.6.1 Ocean currents

Ocean currents are gentle movements of water in response to prevailing wind patterns and density variations in the oceans, or resulting from differences in the salinity and temperature of the water. They have little effect on beaches or nearshore morphology, except where they bring in warmer or colder water, which modifies ecological conditions and thereby influences the distribution of such features as coral reefs or kelp beds, the presence of which can affect beach forms and shore dynamics. The Gulf Stream in the North Atlantic, for example, brings relatively warm water to the Atlantic coasts of Britain and Norway, which could otherwise be subject to shore ice processes in winter.

2.6.2 Wind-driven currents

Wind-driven currents are produced where winds move the surface water, building up sea level to leeward and lowering it to windward as wave action proceeds. Wind-generated currents are not as regular as the alternating tidal currents, but their effects are cumulative in the direction of the prevailing wind. Strong currents are produced when winds drive surface water into gulfs, through narrow straits, or in and out of estuary and lagoon entrances. These may strengthen or oppose the currents produced by tides in similar situations, and it may be difficult to separate the effects of the two.

2.6.3 River-mouth currents

Currents are also produced by discharge from river mouths. In tidal estuaries and at inlets to lagoons fed by rivers fluvial discharge augments tidal ebb currents and diminishes tidal inflow (that may nevertheless be strong enough to move water in on a rising tide). River outflow may carry sediment into the sea, maintain or enlarge river outlets and form a seaward jet, which refracts approaching waves and can act as a breakwater impeding or interrupting longshore currents and sediment flow. Fluvial discharge currents are strong off streams fed by melting ice and snow from coastal mountains, as in Norway and Alaska during the summer months.

2.6.4 *Effects of currents*

Any of these currents can move fine to medium sand (grain diameter 0.1–0.5 mm) when their velocity exceeds about 15 cm/sec, stronger currents being required to move coarser material. Currents generated by winds and tides can be strong enough to move sand or even gravel on the sea floor, either contributing to longshore drift, supplying material to a beach, or carrying it away offshore. These effects are usually subordinate to the onshore–offshore and alongshore movements of beach sediment by wave action. Currents can prevent nearshore deposition, and erode channels or remove shoals, thereby deepening the nearshore water and permitting larger waves to break upon the shore. Sea floor shoals of sand or gravel may form adjacent to rocky headlands as the result of deposition by current flow, examples being the Skerries, off Start Point, and the Shambles, off Portland Bill, on the south coast of England. Such shoals persist where wave energy is too weak to move or disperse them. Sediment derived from them may in due course be washed on to a beach, but the more immediate effect is to modify wave refraction patterns in nearshore waters and reduce wave energy reaching the nearby shore.

In general, tidal and other marine currents are more effective in shaping sea floor morphology than in developing beach configuration. The early theory that long, gently curving beaches on oceanic coasts were produced by marine currents sweeping along the shore has given place to the view that these outlines are determined by refracted wave patterns. Nevertheless, changes in the topography of the sea floor, due to erosion by current scour or deposition from slackening currents, modify patterns of wave refraction and may thus indirectly affect beach outlines. Currents often play a part in removing material eroded by waves from the coast, or in supplying the sediment that is subsequently built into beaches by wave action.

2.7 Nearshore water circulation

The combined effects of wind-generated waves, astronomically generated tides, various forms of current flow and other disturbances of the sea produce a highly variable energy flux in nearshore waters. As has been noted, the several processes interact: a rising tide, for example, deepens nearshore water (the nearshore zone being generally concave upward in profile), thereby increasing the height and energy of waves that reach the shore. A tidal current flowing in one direction can reduce the velocity and dimensions of waves moving in the opposite direction. Marine currents in the nearshore zone are the resultant of potential flows generated by winds, waves, tides and other forces, and there is much variation in current direction and velocity. In addition, wave variability results from the arrival of waves of differing height and length, generated from differing distances and directions, and there are often irregular wave patterns arriving in the nearshore zone. The outcome is a complex nearshore hydrodynamic system that moves sediment on the sea floor and onshore, offshore and alongshore, and influences the shaping of the coastline, including beaches, and the nearshore sea floor.

It is possible to deduce the coastal features that would be shaped by a set of wind, wave and current processes continuing uniformly over a period of several hours or days, but such an adjustment is rarely attained before there are changes in one or more of the driving components and the partially adjusted features begin to be modified. Stability is thus an ephemeral concept in terms of depositional coast morphology, where any attempt to define a stable coastline must be in terms of a stated timescale. On hard rock coasts there may be prolonged phases of stability when there is no response to nearshore processes, while on soft rock coasts, beaches, marshes and deltas there is almost continuous instability, as nearshore processes frequently mobilise sediments.

2.8 Wind action

In addition to the effects of waves and currents (which may be at least partly generated by wind action), coastal landforms may be shaped or modified by the wind. Strong winds deflate fine grained sediment (sand, silt and clay) from beaches and tidal flats, lowering their surfaces, and causing the movement of rock particles onshore, alongshore or offshore. Sand blown from the beach or foreshore is transferred to the zone above high tide level, and deposited as dunes (Chapter 7), which may remain in position or be swept inland or along the coast by wind action. Wind-drifted silt and clay may be deposited down-wind from source areas such as intertidal mudflats that dry out sufficiently for the wind to mobilise this fine grained sediment. Weathering on rocky shores (Section 5.1.2) may produce sediment fine enough to be carried away by the wind, which thus contributes to the lowering of shore rock surfaces.

Wind-blown rock particles may become airborne, or may bounce or roll as they travel downwind. Impacts with rock surfaces result in the rounding and attrition of such particles, while rock surfaces are scoured or smoothed by abrasion as the wind drives sand particles at and across them. Wind action also enhances evaporation, drying out wet rock outcrops on cliff faces and shore platforms, especially at low tide.

2.9 Other processes

Other processes influencing coastal evolution include runoff after heavy rain or from the melting of snow or ice, which causes gravur-

ing of cliff faces and gulleying of coastal slopes, forming downwashed fans, and the outwashing of sand from beaches. Weathering processes that have influenced the shaping of coastal landforms include physical weathering (by insolation, freeze-and-thaw or wetting and drying), chemical weathering (by solution, salt crystallisation, mineral decomposition or base exchange) and biological weathering (by shore organisms, burrowing animals or root penetration).

Some shore processes are conditioned by the salinity of seawater. Salinity can be estimated from water density, but is usually measured either as total dissolved solids or instrumentally from the concentration of chloride ions (chlorinity), and is expressed in parts per thousand (ppt). In the oceans salinity averages 35 ppt salt concentration (mainly chlorides), but there are variations in salinity from almost fresh in parts of the Baltic Sea to hypersaline (>35 ppt) in the Red Sea and Shark Bay in Western Australia. Sea salinity increases in warm and windy environments, especially along arid coasts where the salt concentration is augmented by high evaporation. Salinity diminishes in cold areas, particularly near the mouths of rivers and along coasts where melting glaciers and ice sheets supply fresh water. Salt water and sea spray have corrosive effects on shore rock outcrops (Section 5.1.2) and also produce distinctive habitats for marine and estuarine flora and fauna, which may influence weathering, erosion, transportation and deposition of sediment in the coastal environment.

2.10 Modelling coastal processes

Laboratory simulation of these various processes has been attempted using scale models such as water tanks in which waves, tides and currents can be generated and their combined effects assessed. The aim is to test hypotheses concerning the ways in which these processes cause erosion, move sediment and promote deposition on the sea floor and along the coast. Such physical models have limitations because of the difficulty of scaling down materials and processes without modifying their physical properties (e.g. coherence, friability, expansion and contraction of sediments; viscosity and surface tension in water), but they have been useful in exploring potential responses to marine and nearshore processes (Silvester, 1974).

Measurements of nearshore processes, using wave recorders, tide gauges and current meters, have been made at coastal laboratories, such as the Scripps Oceanographic Institution at La Jolla in California. The United States Army Corps of Engineers Research Facility at Duck in North Carolina and the Black Sea Laboratory at Sochi on the Russian Black Sea coast have piers equipped for instrumental surveys and computerised recording of accompanying sediment movements and changes in coastal and nearshore morphology. Conclusions from such monitoring apply only to the nearby coastline, and there is a risk that insertion of numerous instrument-bearing structures may modify the natural processes, and cause changes that would not otherwise have occurred.

Mathematical modelling has been much used by engineers as a basis for computer simulations of coastal processes. Such modelling can be used to study the effects of integrated process systems (waves, tides and currents) on nearshore sediment flow, and the ways in which these processes and responses will be modified by the introduction of structures such as groynes or breakwaters. It is important to be sure that the information used is accurate and comprehensive, and to test predictions against what actually happens,

in order to refine the model and improve subsequent forecasts.

2.11 Summary

Processes in coastal waters include waves generated by wind action locally (including storm waves) and remotely (ocean swell transmitted from distant storms), tides generated by astronomical forces and related to coastal and nearshore configuration, disturbance by storm surges and tsunamis, and associated currents. They act in combination, but are conveniently treated separately.

Coastal outlines are produced largely by waves, which also generate longshore drifting of sediment. Breaking waves may be constructive (moving sediment shoreward) or destructive (causing erosion), and variations in their energy may be measured by such parameters as significant wave height (the height of the highest one-third of a set of waves). Coasts may have high, moderate or low wave energy.

Coasts may be microtidal, mesotidal, macrotidal or megatidal, according to the vertical tide range. Tides generate currents that can shape sea floor and intertidal topography and influence the form of estuaries and lagoons. Some coasts may be classified as tide dominated rather than wave dominated.

Storm surges, tsunamis and other giant waves may cause major and persistent changes in coastal landforms. Currents, generated in various ways, may indirectly affect coasts by modifying nearshore topography and therefore incident wave regimes. Other processes shaping coastal features include wind action, water runoff and weathering by physical, chemical or biological agents. These various processes combine in a nearshore water circulation that causes movement of sediment along the coast and on the sea floor, particularly in shallow water.

Many coasts show features that developed when the sea stood at different levels in the past, or when it was rising or falling. The operation of waves, tides, currents and the other processes that have been discussed has sometimes been at a higher level than it is now, and sometimes lower.

There have been phases of still-stand when the relative levels of land and sea remained constant at particular altitudes, and phases when the sea was rising or falling relative to the coastal land. It is now necessary to examine the history of changing levels of land and sea.

3

Land and sea level changes

3.1 Introduction

Around the world's coastline there are features that formed when the sea stood higher or lower relative to the land, especially during Quaternary times. Beach deposits or marine shell beds stranded above present high tide level are indications of an emerged coastline on which the sea was formerly at a higher level, while drowned valley mouths indicate a submerged coastline where the valleys were excavated by their rivers when the sea stood at a lower level. The terms emerged and submerged are used in coastal geomorphology in preference to emergent and submergent. It is sometimes difficult to determine whether a change in relative sea level (RSL) has been the result of upward or downward movements of the land, or an actual rise or fall in the level of the sea, or some combination of the two. Changes resulting from uplift or subsidence of the land are known as tectonic movements, while those due to an actual rise or fall of sea level are eustatic movements, and relative sea level is the resultant of these land and sea level changes.

A marine transgression occurs when sea level rises to invade the land, and a marine regression when sea level falls, exposing the former sea floor as a land area. Relative sea level has risen and fallen frequently through geological time. Few coastal areas have been tectonically stable for long, and many show obvious indications of continuing instability, notably earthquakes and volcanic activity. On the other hand, there have been phases of still-stand, when the sea has remained at or close to its level relative to the land long enough for recognisable coastline features to have developed. Some of these are above present sea level (emerged), others below (submerged), and on much of the world's coastline the past 6000 years have been a phase of still-stand, with the sea at its present relative level.

Sea level depends partly on the volume of water in the oceans, which is determined by the balance of evaporation and precipitation produced by the hydrological cycle, and partly on the size and shape of the crustal depressions that contain seawater. Mean sea level provides a datum from which upward or downward movements of relative sea level can be measured.

3.2 Mean sea level

A definition of mean sea level must exclude short term variations such as those caused by tides and waves, described in the previous chapter. Tides (Section 2.3) have a vertical range varying

Coastal Geomorphology: An Introduction, Second Edition. Eric Bird.
© 2008 by John Wiley & Sons, Ltd

from almost zero to about 20 m around the world's coastline, and tidal cycles vary from about 12 hours to 18.6 years, with some even longer, though minor, astronomical cycles. Tidal oscillations are in theory symmetrical about a long term mean sea level, conventionally defined as the arithmetic mean of the height of calm sea surface (i.e. excluding waves and oscillations related to winds and atmospheric pressure variations) measured at hourly intervals over at least 18.6 years. In practice mean sea level at points on the coast is related to a national datum such as American Sea Level Datum, or Ordnance Datum in the United Kingdom (Kidson, 1986).

There are also short term fluctuations of mean sea level (several days to weeks) in relation to weather conditions, notably the variations in atmospheric pressure that accompany the passage of depressions and anticyclones, together with the effects of associated winds. Strong wind action produces storm surges (Section 2.4), which drive water shoreward and raise sea level temporarily in the nearshore zone by up to several metres, as well as producing very high waves that break upon the coast. Persistent onshore winds can maintain high water levels, especially in gulfs and semi-enclosed bays areas.

Sea level also varies seasonally in relation to temperature, pressure, and wind regimes. It rises by about a centimetre with each millibar fall in atmospheric pressure. Analyses of monthly or seasonal data show that maximum annual sea levels occur at different times around the world's coastline: in September in the eastern North Atlantic, for example, and in April along the eastern seaboard of Australia. In the South China Sea mean sea level is about 40 cm higher during the NE monsoon (November to March) than during the SW monsoon.

There are sea level changes related to longer term atmospheric pressure cycles such as the North Atlantic Oscillation, which is correlated with sunspot cycles, and the 2–7 year El Niño

Southern Oscillation, which produces high pressure and low sea level over the SE Pacific Ocean and low pressure and high sea level in the Indian Ocean, then reverses these. Komar (1986) showed that sea level rose temporarily by about 20–30 cm along the Pacific coast of North America during the 1982–83 El Niño Southern Oscillation.

It is necessary to exclude these variations and so determine mean sea level before deciding the extent of sea level changes over particular periods.

3.3 Causes of sea level change

Apart from these short term variations in sea level, upward or downward movements of sea level can result from several causes, as indicated in the following sections.

3.3.1 Eustatic movements of sea level

Sea level rises when the volume of water in the ocean basins increases and falls when it is reduced. These changes are world-wide because the oceans are interconnected. They have been termed eustatic, a term that was introduced (Suess, 1906) when it was assumed that such changes were equivalent throughout the world's oceans, but it is now realised that there have been regional discrepancies in the amount of sea level rise or fall on particular coastline coastlines (see below).

Through geological time the volume of water in the oceans has been gradually augmented by the arrival of small quantities of juvenile water supplied from the Earth's interior, primarily from volcanic eruptions, implying a long term rise in sea level, but it is unlikely that these accessions have had much influence on sea level changes during the Quaternary.

3.3.2 Steric changes

An increase in atmospheric temperature results in warming and expansion of the oceans, and sea level rises, whereas if the oceans cool they contract, and sea level falls. The volume of sea water also diminishes as salinity increases, and rises as it freshens. Ocean volumes also vary with the density of sea water, related to temperature, salinity and atmospheric pressure. These are known as steric changes.

It has been calculated that a rise of 1°C in the mean temperature of the oceans would increase their volume so that sea level would rise by about 2 m. Estimates of Pleistocene variations of mean ocean temperature (based on palaeo-temperature measurements on fossils from ocean floor deposits) are within 5°C of the present temperature, so this could only account for sea level oscillations of up to 10 m.

3.3.3 Sedimentation

Sea level can also rise because of a gradual reduction in the capacity of the ocean basins resulting from the accumulation of sediment carried from the land to the sea, whether by rivers, melting glaciers, slope runoff, landslides, wind action or coastal erosion. This is a very slow process, termed sedimento-eustatic. Transference of all the land above present sea level into the ocean basins would raise the level of the oceans by more than 250 m, but present estimates of denudation rates account for a sea level rise of only about 3 mm per century.

3.3.4 Tectonic movements

Relative sea level can change because of tectonic movements, upward or downward, of the Earth's crust, changing the shape of ocean basins and raising, lowering or deforming the continents. These may be epeirogenic, orogenic or isostatic movements. Epeirogenic movements are broad, rather uniform tectonic uplift or depression, which have tended to raise continents and depress the floors of ocean basins, with warping restricted to a marginal hinge-line. Some coasts lie close to this zone of marginal warping, with uplift on the continental side and depression on the oceanic side, as on much of the coast of southern Africa.

Changes in the ocean basins

Epeirogenic movements have changed the shape and capacity of the ocean basins. An increase in the capacity of an ocean basin (measured below sea level) causes a lowering of sea level, and a reduction causes sea level to rise. It has been suggested that the floors of the ocean basins, notably the Pacific Ocean basin, have been subsiding intermittently, particularly during Quaternary times, increasing their capacity and resulting in the successively lower sea levels marked by emerged stairways on bordering coasts. Depression of ocean floors of the oceans has been accompanied by lateral movements of the continental plates at rates of up to 5 cm/year, which have also modified the configuration of the ocean basins. As tectonic movements within the ocean basins raise or lower sea levels all over the world, this kind of change is termed tectono-eustatic.

Epeirogenic land movements

Epeirogenic movements have raised each of the continents during and since Mesozoic times, but the uplift has been punctuated by phases of stability during which planation surfaces have been formed, with land areas reduced to base level by long term weathering and erosion. A succession of such planation surfaces, separated by bold scarps initiated during each episode of epeirogenic uplift, was recognised in Africa by King (1962), who suggested that they were present on each of the continents. Some of the planation surfaces may have been cut by marine processes during phases of higher relative sea level. Intermittent epeirogenic uplift has produced stairways of marine terraces in which the oldest is the highest and farthest inland, separated by ancient cliffs cut when the sea stood at each level. Such stairways could also be produced by intermittent eustatic lowering of sea level, or some combination of eustatic and epeirogenic movements.

Orogenic movements

Orogenic movements are more complicated deformations of the Earth's crust associated with mountain-building, notably at the convergence of tectonic plates. They are active around the margins of the Pacific Basin, notably on the west coast of the United States, in Japan, the Philippines, Indonesia, Papua New Guinea and New Zealand. These are termed neotectonic regions, characterised by frequent earthquakes and volcanic activity, and their coastlines show evidence of irregular displacement by uplift, lowering, tilting, folding and faulting. Sea level changes on tectonically active coasts have been at least partly due to the rising or sinking of the land margin, either gradually or as the result of sudden earthquakes. Such displacements are much in evidence around the Mediterranean, where there has been recent tectonic uplift in the Oman region of Algeria, the southern Peloponnese and western Crete, and the Izmir region of southern Turkey. Submergence of ports such as Carthage in Tunisia, built in the classical Greek and Roman era, about 2000 years ago, indicates coastal subsidence. On the southern coast of North Island, New Zealand, orogenic movements around Wellington have led to a juxtaposition of uplifted and emerged coasts with downwarped and submerged coasts, parts having been displaced by faulting. The sequence of uplifted and tilted marine terraces culminates in the raised beaches on Turakirae Head, to the east.

Some tectonic displacements have been recorded historically. In the vicinity of Wellington parts of the coastline were uplifted by up to a metre during the 1855 earthquake, forming emerged shore platforms. An earthquake at Hawke Bay, New Zealand, in 1931 raised the coastal plain near Napier by up to 2 m and drained a lagoon. The 1964 Alaskan earthquake raised intertidal shore platforms out of the sea, as on Montague Island, where a sudden uplift of 11 m led to an advance of the coastline of 400 m while parts of Homer Spit, in Kachemak Bay, sank by nearly 2 m (Shepard and Wanless, 1971). Successive earthquakes have raised coasts around Tokyo Bay, producing a sequence of emerged Holocene terraces, with shore platforms uplifted by the 1703 and 1923 movements. During the Colombian earthquake of 1979, the coast around Tumaco subsided by up to 1.6 m (Herd *et al.*, 1981). In the Rann of Kutch, on the border between Pakistan and India, subsidence caused by the 1819 earthquake resulted in the sea rising to submerge an area of about 500 sq km.

Similar neotectonic features have been reported from sectors of the Antarctic coastline, and radiocarbon dating of sediments indicating former sea levels along the coast of Argentina has revealed alternating sectors of orogenic uplift and depression (Codignotto, Kokot and Marcomini, 1992).

Isostatic movements

Isostatic movements are adjustments in the earth's crust resulting from loading or unloading of the surface. Areas heavily laden by the accumulation of lava or sedimentary deposits show crustal subsidence, as in the vicinity of large deltas, where the load consists of sedimentary deposits accumulating at the mouth of a river. Parts of a delta not maintained by active sedimentation become submerged as this subsidence proceeds (Section 12.4).

Isostatic movements have also occurred in regions that were glaciated during Pleistocene times, as in northern North America and northern Europe. As ice accumulated the Earth's crust was loaded and depressed, and when it melted the unloaded crust gradually rebounded. On the shores of the Gulf of Bothnia, sea level has been falling at rates of up to a centimetre per year because of isostatic land uplift due to the melting of former glaciers and ice sheets. Recovery of depressed crustal areas continues for some time after the ice has gone, until a physical equilibrium is restored.

Sea level changes have also resulted from another form of isostatic movement, the loading of continental shelves during a marine transgression. The weight of the deepening water depresses the continental shelf, and the adjacent coast is downwarped as a result, a response known as hydro-isostatic subsidence. Sea level rise is thus augmented, the extent of hydro-isostatic subsidence varying in relation to the width and slope of the continental shelf (which determines the volume and weight of water load gained during a sea level rise) as well as to the structure and strength of geological formations in the area affected. A marine transgression may thus be greater on steep coasts with deep water close inshore than on gently shelving coasts. Sea level rise can be similarly augmented where soft sediments, such as peat, are being compressed beneath the sea floor by the gathering weight of water.

Volcanic movements

Volcanic activity has also resulted in a relative rise or fall of sea level. In the 19th century the geologist Charles Lyell observed that the limestone pillars in the Roman Market of Serapis, in the town of Pozzuoli near Naples in Italy, showed a zone about 2 m above present sea level that had been drilled by molluscs when the sea rose to flood the structure (Figure 3.1). Sea level evidently rose as the land subsided after the pillars were built in Roman times, and fell sharply when the area was uplifted, an oscillation thought to be due to evacuation and re-filling of a subterranean lava chamber associated with the nearby volcano of Vesuvius.

Glacio-eustatic movements

Major oscillations of sea level accompanied the waxing and waning of the Earth's ice cover in response to alternations of cold (glacial) and mild (interglacial) climate during the Quaternary Epoch. These sea level changes are termed glacio-eustatic. During glacial phases the Earth's hydrological cycle was interrupted when the climate cooled sufficiently for precipitation to fall as snow, which accumulated as glacial ice and persistent snowfields in polar and mountain regions. Retention of large amounts of water frozen on land depleted the oceans, and there was world-wide lowering of sea level. In the milder interglacial phases the trend was reversed, water released from melting snow and ice flowing back into the ocean basins to produce a world-wide sea level rise (marine transgression), the sea at times extending above its present level.

During the Last Glacial phase, late in Pleistocene times, sea level fell about 140 m (Donn et al., 1962), but about 18 000 years ago the polar ice sheets and mountain glaciers began to melt, initiating a world-wide sea level rise. This was called the Flandrian transgression when it was first recognised by Dubois (1924), working in Flanders and along the north and NW coast of France, but it was subsequently found to be world-wide. It continued into Holocene times (10 000 years ago), and came to an end about 6000 years ago when the sea attained

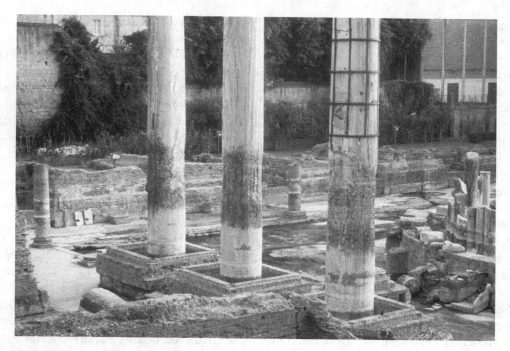

Figure 3.1 Dark zones formed on columns by marine organisms during a mediaeval phase of higher sea level in the Roman Temple of Serapis at Pozzuoli, near Naples, Italy. The bare zone below the marine organisms may indicate that a sudden uplift occurred, or it may have been protected by a sedimentary deposit that was subsequently removed

approximately its present level. Although it is strictly a Late Quaternary marine transgression, it is convenient to refer to it as the Holocene marine transgression.

The Earth is still in an Ice Age, in contrast to the ice-free conditions that have prevailed through most of geological time, apart from earlier glaciations in the Pre-Cambrian and Permian. If the remaining land-borne ice sheets, glaciers and snowfields on the Earth were to melt they would release sufficient water to raise the level of the oceans by at least 60 m, causing widespread coastal submergence and the loss of existing coastal lowlands (including most of the world's major cities). Melting of floating ice, as in the Arctic Ocean and the ice shelves bordering Antarctica, will not increase the volume of the oceans, and will have no effect on sea level.

Human impacts on sea level

Sea level has risen on coasts where the land has been subsiding as the result of human activities such as groundwater extraction, which depletes the aquifers under and around coastal urban and industrial centres. As underground water is withdrawn, the sediments of the aquifers are consolidated and compressed by the weight of overlying rock formations (and buildings, if any), and the loss in volume results in subsidence of the land surface. This has contributed to a relative sea level rise in the Venice region, and around Bangkok in Thailand.

Relative sea level has risen where oil or natural gas has been pumped from underground strata, as in southern California and the Ravenna region in Italy. Similar submergence has followed the loading of coastal land with artificial structures, and some port and land reclamation schemes, including the construction of artificial islands, have caused subsidence and changes in local tide regimes, raising relative sea level, especially in bays and estuaries. A local rise in relative sea level may follow the deepening of nearshore areas as the result of dredging, with increased tidal penetration. Shore walls and tidal barrages built in the Thames and Medway estuaries have resulted in higher high tides and penetration further upstream of waves driven by storm surges, effectively raising relative sea level.

Geoidal changes

Sea level changes have also been influenced by shifts in ocean surface topography that result from geoidal changes. Evidence from satellite sensing has shown that the oceans have an undulating surface configuration, with domes and troughs rising and falling up to 90 m above and below a smooth geoid because of gravitational, hydrological and meteorological forces. These domes and troughs are related partly to gravity patterns and other geophysical phenomena, including tides and the Earth's rotation, and partly to climatic patterns and ocean circulations. In the North Atlantic, tide gauge records show that, when the effects of coastal tectonic movements were excluded, mean sea level rose between 1920 and 1950, the rise being faster on the American coast than in Europe, probably because of variations in sea surface topography (Pirazzoli, 1989).

Similar changes have occurred as the result of variations in the distribution of oceanic domes and troughs, especially during the Quaternary. Shoreward movement of domes and troughs results in a sea level rise where high areas move coastward and a sea level fall as low areas move in.

Changes related to ocean currents

Variations in sea level also result from changes in transverse gradients that develop where ocean currents are strong. Off the east coast of the United States and in the China Sea, the ocean is up to 2 m above its general level in areas adjacent to the poleward currents of the Gulf Stream and Kuroshio, and any variation in the velocity or extent of these currents will lead to changes of sea level on nearby coasts. Moreover, the ocean envelope is related to the axis and velocity of the earth's rotation, and any variation in these will cause differential changes of sea level around the world's coasts.

3.4 Measuring changes of level

Having outlined the reasons why relative sea level has changed, the evidence of emerged and submerged coastlines can be considered. It is necessary to relate the levels of coastline features that now stand above or below present sea level to a specific datum. Some have reported the levels of emerged or submerged coastlines above or below mean sea level, the datum used on topographic maps in many countries. Others have used low water spring tide level, the datum used on nautical charts of coastal regions and others high water spring tide level, that is more easily determined and more readily accessible for surveying work on the coast. Each of these has advantages for particular purposes, the first fitting in with surveyed contours and benchmarks on topographic maps, the second enabling chart soundings and submarine contours to be used without modification and the third being the most practical in field work. The discrepancies between them increase with tide range, and it is necessary to adjust levels to the same datum before comparisons are made. The most convenient method is to convert all measurements of emerged and submerged coastline features to mean sea level datum, and state the local spring tide range above and below this level.

Emerged or submerged shore features usually consist of benches or terraces, backed by former cliffs, and sometimes bearing beach deposits or marine shell beds. Their levels can be measured at the base of the abandoned cliff, but this is frequently obscured by beach deposits, dunes or sediment that has been washed down from adjacent slopes, and it is rarely possible to determine it accurately. For this reason, the levels of emerged or submerged coastlines are usually given approximately, e.g. 5–8 m above or below mean sea level.

3.5 Correlation and dating of former coastlines

Where marine deposits are associated with emerged coastlines, correlation may be possible in terms of distinctive fossils. Specific assemblages of minerals may occur in old beach sands, but lateral variations in mineral composition of present-day beaches are such that correlation in terms of this evidence can only be tentative.

A more reliable means of correlation is based on radiocarbon dating of samples of wood, peat, shells, bone or coral obtained from deposits associated with former coastlines. Nuts are useful because they formed in a particular year, whereas the others cover life spans or longer periods. Corals and shells give less reliable dates because they can ingest older carbon. Radiocarbon analysis permits estimates of the age of samples less than about 50 000 years old, and is a means of distinguishing Holocene deposits (less than 10 000 years old) from those formed during Pleistocene times. Dates obtained from radiocarbon measurements are stated in years BP (before the present, the present being defined as the year AD 1950). The margin of error is usually indicated as ± the standard deviation of the measurements made (e.g. 5580 ± 200 years BP).

Other geochronological methods include potassium–argon dating, notably of Quaternary volcanic rocks, and uranium–thorium dating, which has been used to date corals, other marine carbonates and dune calcarenites. Uranium–thorium dating of emerged coral reefs has enabled a Last Interglacial high sea level to be dated at 125 000 years BP. Use has also been made of fission track (ionised trail) dating of minerals (notably zircon) in volcanic ash deposits. Radioactivity trapped within quartz and feldspar crystals is cumulative, and can be released for dating by thermoluminescence (TL), based on glow graphs relating intensity of light emission to temperature. It has been used to estimate the age of sand grains in coastal deposits (Bryant, Young and Price, 1996), and is applicable to sand deposits originally formed in daylight, such as dunes and loess, more than 50 years old. Laser generated optically stimulated luminescence (OSL) has similar applications. Lead isotope (210Pb) ratios have been used to date salt marsh deposits within the past century (French, 1996). These and other methods of dating are being developed to determine the age of emerged and submerged coastal deposits and of associated landforms (Duller, 2003).

3.6 Emerged coastlines

On many coasts there are stranded beach deposits, marine shell beds and platforms backed by steep clifflike slopes, all marking former coastlines that now stand at a higher level than when they originally formed. Many are well above present high tide level. These are emerged coastlines, and they owe their present position either to uplift of the land, or a fall in sea level, or some combination of movements of land and sea that has left the coast higher, relative to sea level, than it was before. Most are of Quaternary age, but some of the older and higher coastline features may have originated in Tertiary times, along with shore deposits that mark the limits of Tertiary and earlier marine transgressions. As well as providing data for the geological sequence of land and sea level changes they demonstrate landforms and associated deposits developed by marine action in the past.

Sequences of terraces or emerged beaches measured on one sector of coast may not be recognisable on adjacent sectors because terrace preservation requires a particular relationship between the rock resistance, the intensity and duration of marine erosion at the higher coastline phase and the degree of subsequent degradation and dissection. As a rule, permeable sandstones, limestones (including coral) and gravelly formations retain terraces better than impermeable rocks that are more resistant (restricted terrace development at the higher coastline phase) or less resistant (greater destruction of terrace by subsequent denudation). Coastal terraces are rarely well preserved on soft or rapidly weathering rock formations.

Emerged coastlines have been reported at various levels around the world's coastline. Some are essentially platforms or benches, thought to have been formed by marine action, and backed by a rising slope interpreted as a former cliff that has been degraded by subaerial processes. There are good examples in western Scotland, notably in the Oban district, and on the Californian coast at Palos Verdes, where the modern cliffs intersect 13 marine terraces at various levels up to 380 m above the sea. In Western Australia emerged shore platforms cut in Pleistocene dune calcarenite, notably at Point Peron, stand 2–3 m above sea level on a coast where tide range is negligible. There is a sharp drop (sometimes with a notch below a visor) to the modern low tide shore platform, which is generally awash.

Solution notches have been cut in cliffs on limestone coasts above present sea level, notably on emerged coral islands in the Pacific Ocean. In Phang-nga Bay, southern Thailand, emerged notches in a limestone cliff indicate a sudden uplift or sea level fall (Figure 3.2): a gradual change would have produced a taller single enclave as the basal ledge was cut down.

In southern Britain the best-preserved emerged beach is between 5 and 8 m above mean sea level (typically 1–3 m above high spring tide level). There are lateral variations in level that may indicate that several stages of higher sea level are represented, but the junction between modern shore platforms and the cliff base also shows lateral variations of several metres in relation to present sea level, and the emerged beach may have shown a similar height range. In the early literature these were sometimes called raised beaches, when it was assumed they owed their higher elevation to the uplift of the coastal land, but with the realisation that there have been substantial variations in sea level the term emerged beach is preferred.

The emerged beaches of southern England consist of sand, rounded pebbles and cobbles and occasional boulders, often with some shelly material, and sometimes cemented into sandstone or conglomerate by carbonates or ferruginous precipitates. There are good examples on the coasts of Cornwall, notably in Gerrans Bay and Falmouth Bay (Figure 3.3), and along the south coast at Portland Bill and Black Rock, near Brighton. They are overlain by Late Pleistocene periglacial deposits and so probably date from a Last Interglacial phase of higher sea level. On the limestone coasts of the Gower Peninsula in South Wales similar emerged beaches are overlain by glacial drift deposits, and are therefore also of Pleistocene age.

On some coasts there are emerged features with associated deposits that have yielded a Holocene radiocarbon age. These may result from one or more episodes of higher Holocene

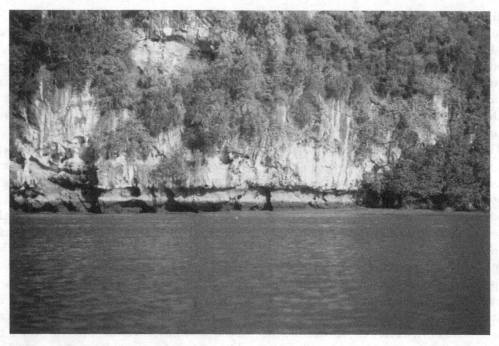

Figure 3.2 Notches on a limestone cliff in Phang-nga Bay, Thailand, the upper one formed during a higher (1–2 m) relative sea level phase, the lower one at present mean sea level

sea level or they may be due to tectonic uplift within the past 6000 years. In places there are indications that the higher sea level was also warmer, such as the fossil oysters on a Holocene terrace 10 m above sea level at Vestervøy in SE Norway. There are also low level emerged terraces of Holocene age on uplifted coral reefs, notably in the Pacific Ocean.

Beach ridges (Section 6.18) that show a seaward decline in the levels of ridge crests (and intervening swales) have been cited as evidence that sea level has fallen, but as the dimensions of beach ridges are also related to the heights of the waves that built them a seaward decline of ridge crests could result from diminishing wave heights. The size of beach ridges is also influenced by the rate of wind-blown sand supply, and a seaward decline in ridge crest and swale levels could result from tilting by tectonic movements. There are emerged beach ridges in west-

ern Scotland that have certainly been raised by isostatic uplift. On the island of Islay in western Scotland beach ridges are found at various levels up to 37 m above sea level, stranded on a wide Pleistocene emerged shore platform (McCann, 1964).

Stairways of emerged features are found on several coasts. In North America, the Atlantic coastal plain bears depositional terraces that have been correlated with successive alternations of sea level since Miocene times, the sea attaining levels of 13.5, 6, 7.5, 4.5 and 0 m respectively, with intervening low sea level phases (Oaks and Coch, 1963). On the Pacific coast similar terraces have been dislocated by the San Andreas fault near San Francisco, and warped and uplifted terraces are found near Los Angeles and on Santa Catalina Island (Emery, 1960). In Chile, coastal terraces have been displaced tectonically in a region subject to earthquakes, and as the Chilean

Figure 3.3 Emerged gravelly 'raised beach' on the coast of Falmouth Bay in Cornwall. The beach lies upon an emerged shore platform (arrowed in the distance) and has been exhumed from a mantle of periglacial Head deposits

coast runs parallel to the main axes of Andean uplift there is a possibility that terraces found at the same height above sea level have been raised epeirogenically (Fuenzalida *et al.*, 1965).

In the SE of South Australia a series of stranded beach and dune ridges runs roughly parallel to the coast at successive levels up to more than 60 m above present sea level. They were tilted transversely when the Mount Gambier region rose and the Murray-mouth area subsided during Quaternary times (Figure 3.4). The levels of the successive beach ridges result from land movements as well as sea level oscillations.

On the NE coast of New Guinea (Huon Peninsula), a remarkable series of emerged Quaternary coralline terraces gains elevation and multiplies across an axis of uplift in the vicinity of the Tewai gorge, where the highest attains more

than 600 m above sea level. Radiocarbon dating of the younger terraces has been supplemented by uranium–thorium dating of the earlier and higher terraces to indicate that the sequence was initiated about 220 000 years ago, and that the terraces formed during successive still-stands, the last of which was about 6000 years ago. Similar results were obtained from studies of the emerged coralline terraces that form a stairway on the tilted coral cap of Barbados (Chappell, 1983). On the north coast of New Guinea, localised uplift has raised a mangrove swamp deposit 50 m above sea level near Aitape.

Coasts bordering recently deglaciated areas have also been subject to continuing (isostatic) uplift of the land, and show sequences of differentially elevated coastlines, as on the southern shores of Hudson Bay in Canada, where Holocene beaches have been raised by up to 315 m.

(a)

(b)

Figure 3.4 Emerged dune calcarenite ridges in the SE of South Australia and western Victoria rise to an axis of uplift near Mount Gambier. A cross-section SW to NE through Naracoorte shows the ridges and intervening swales (former coastal lagoons) that occur at a higher level on the Mount Gambier axis and a lower level to the north, towards the Murray valley

On Skuleberget in NE Sweden a hilltop beach on a coastline 8600 years old has been raised 286 m above the present level of the Gulf of Bothnia by the Holocene isostatic rebound. Other Holocene emerged coastlines have been traced around the Baltic Sea, the most notable being the Litorina Sea coastline, formed between 7000 and 4000 years ago and prominent as far west as Denmark. An equivalent former coastline has been identified locally in NE Scotland (Walton, 1956) and may extend into NE England. Much of Scotland is still rising isostatically, but the Outer Hebrides are undergoing submergence. Northern Ireland has a low postglacial (Holocene) terrace that diminishes in altitude and disappears southward, indicating tilting related to isostatic recovery. Similar features occur around Puget Sound in the NW United States.

Planation surfaces cut across tilted and folded geological formations have been found at various levels on each of the continents, some of them dating from Tertiary or Mesozoic times. Some are emerged marine terraces with associated beach deposits, formed during Pleistocene and earlier phases of higher sea level. In SW England, for example, there is an extensive planation surface (known as the Trevena Platform) between 90 and 130 m above sea level, rising to a break of slope that is thought to be a former coastline. St. Agnes Beacon, in Cornwall, rises sharply above this platform, and has old beach and dune deposits at about 130 m, indicating that this hill was once an island in the sea.

Planation surfaces at various levels have been described from Wales (Brown, 1960). St David's Peninsula in Pembrokeshire is dominated by a broad plateau 45–75 m above sea level, considered to be a marine planation surface with residual higher hills that may once have been islands when the sea stood at a higher level. Pen Beri and Carn Llidid are hills of dolerite rising above this plateau, but if they were formerly islands, any beach sediments on their coasts have been removed, possibly by glacial or periglacial processes after marine planation came to an end. It is possible that marine planation occurred under periglacial conditions, when rock weathering is rapid, in which case the plateau is a kind of strandflat with residual islands (hutberge) as on the SW coast of Norway (Section 5.2.5). A more extreme view is that the planation was achieved entirely by periglaciation rather than marine processes, and that the residual hills were not islands.

In SE England a planation surface rising to a possible former coastline about 210 m above present sea level bears marine deposits that were assigned an early Pleistocene (Calabrian) age on palaeontological grounds, their contained fossil assemblage matching marine deposits of this age elsewhere. Similar platforms have been described at various levels (notably rising to breaks of slope about 130, 60, 30 and 15 m above sea level) in the British Isles, but confirmatory evidence of contemporary beach deposits is generally lacking.

While some planation surfaces are emerged sea floors, shaped by marine processes when the sea stood at higher levels, others may have been formed by subaerial processes, especially rivers, and some may be the outcome of periglacial or desert weathering and erosion. Whatever their origin, they have little influence on coastal geomorphology beyond determining the form and altitude of cliff crests. Thus in SW England there are stretches of even-crested cliffs cut into platforms at various levels above the sea. Along the north coast of Cornwall cliff recession has cut into the 90–130 m platform, whereas along the south coast the flat-topped cliffs of the Lizard Peninsula border a platform about 60 m above sea level.

3.7 Submerged coastlines

Submerged coastlines formed when the sea stood at various lower levels, during still-stands that punctuated falling sea levels and ensuing marine transgressions. Sea floor morphology has been charted using depth soundings and radar, and evidence of submerged coastlines has been found in the form of sharp breaks of slope, thought to be submerged cliffs and shore platforms, at various levels offshore. Stairways of submerged terraces have been mapped off various coasts, one of the clearest being the paired sequence of seven terraces at successive levels down to 110 m below present sea level, bordering Tsugaru Strait, between Hokkaido and Honshu, Japan (Emery, 1961). A submerged coastline at a depth of about 18 m has been reported off the coasts of the United States (Shepard, 1973).

Some submerged coastlines maintain consistent levels, and could have been produced when sea level was lowered by eustatic movements, or as the result of tectonic subsidence. Stearns (1974) described a sequence of submerged coastlines around the Hawaiian Islands down to a depth of 1100 m, regarding those down to 140 m as having formed during the Late Quaternary glacio-eustatic oscillation of sea level, while those at greater depths were the outcome of earlier tectonic subsidence. Submerged beaches have been found at various levels around the Australian coastline. In Bass Strait a well defined break of slope indicates a submerged coastline at a depth of 60 m below present sea level, possibly formed during the Last Glacial low sea level phase in Pleistocene times (Jennings, 1959).

Other submerged coastlines have been tilted, folded or faulted by tectonic movements that have disrupted the sea floor, as off southern California (Emery, 1960). Local tectonic subsidence has led to partial submergence of prehistoric structures on the shores of the Golfe du Morbihan in Brittany, and Cork Harbour in Ireland.

Beach deposits have been found along some submerged coastlines, as on the floor of the Gulf of Mexico. In Australia evidence of a submerged coastline came from sediment with nearshore fossils, dredged from a depth of 130 m off the New South Wales coast. Dune calcarenite ridges that formed behind Pleistocene coastlines now submerged can be traced off the coasts of Western Australia and in Encounter Bay, South Australia, while sea floor contours near Flinders Island show the outlines of submerged parabolic dune topography.

Solution notches have been found at various levels below present sea level off limestone coasts on Crete and elsewhere in the Mediterranean, formed low still-stands of sea level followed by sudden subsidence or a rapid sea level rise. Coastal submergence in these areas is also indicated by the drowning of port structures built about 2000 years ago.

Submerged forests are found in the intertidal and nearshore zones on many coasts (particularly in the British Isles), and are indicative of a relative sea level rise in Holocene times. At Cliff End, near Hastings on the south coast of England, the submerged forest exposed at low spring tides consists of black logs on a clay deposit that has been eroded and partly overrun by inwashed sand. At Formby in Lancashire there are the remains of a submerged forest in a former dune swale. Submergence is also indicated where Holocene beach ridges pass landward under swamps, as on the coast of Sarawak.

Coastal submergence has occurred in the SE North Sea, where field patterns seen on the sea floor near Pellworm off Schleswig-Holstein in northern Germany are considered to be part of a landscape inundated by a storm surge in 1362. The notion that a storm surge can result in permanent submergence of a coastal landscape needs refinement, for storm surges cause only temporary inundation. Permanent submergence is more likely to be due to land subsidence here.

Much evidence of submerged coastlines has doubtless been lost because smoothing of the sea floor by wave action during successive episodes of marine transgression has obliterated or concealed them. Submerged beaches may be hidden beneath a veneer of finer deep-water sediments. Moreover, the emerged sea floors during low sea level phases became landscapes shaped by terrestrial processes (rain and rivers, frost and ice, and wind action), and coastline features that had formed during the preceding sea level fall may have been removed or buried by subaerial erosion and deposition. Because of these effects, remnants of the older submerged coastlines are unlikely to persist on the present sea floor. Those that have been found probably date from the Late Quaternary.

3.8 Sea level variations

When the levels of emerged and submerged coastlines are graphed against their geological age, relative sea level fluctuations of sea level relative to the land are shown over a specified period. Attempts to draw graphs showing absolute sea level movements (i.e. the pattern of sea level rise and fall that would be registered on a tectonically stable coastline) on a global scale have been abandoned because this is now seen as a theoretical abstraction. Where coastal submergence or emergence has taken place it is difficult, if not impossible, to separate land movements from a rise or fall in sea level, but it is useful to produce graphs showing the relative levels of land and sea over time on particular sectors of the coast.

Some sea level graphs show the elevation of former coastlines against their age, determined by radiocarbon or other methods of dating. Others have been drawn on the assumption that the large-scale oscillations of sea level and climate during Quaternary times were related to variations in solar radiation. Estimates of these variations, based on astronomical theory, are projected back through time to show a sequence of warmer and colder periods during the Quaternary, which are correlated with interglacial and glacial phases to give a timescale for sea level fluctuations. An alternative approach used oxygen isotope ratios measured from analyses of fossil foraminifera in sedimentary cores obtained from the floors of the oceans, where sedimentation has been very slow, and deposits marking the whole of the Quaternary era are only a few metres thick. The ratio of the oxygen isotopes O^{16} and O^{18} in fossil foraminifera is an indication of the temperature of the environment at the time they formed and thus, compared with present ocean temperatures, an indication of the scale of warmer or colder climates in the past. Foraminifera obtained from successive levels in stratified ocean floor sediments have yielded evidence of the sequence of past changes of climate, with warmer and colder phases that can be correlated with interglacial and glacial phases in the Pleistocene period. It is assumed that each climatic oscillation was matched by a glacio-eustatic oscillation of sea level, a warmer ocean indicating a higher sea level, and a cooler ocean a lower sea level, and this can be adapted to a graph of changing sea levels calibrated with vertical dimensions.

Shackleton and Opdyke (1973) analysed oxygen isotopes and palaeo-magnetism in ocean-floor cores from the Equatorial Pacific, representing a sedimentation sequence over the past 870 000 years. Palaeo-temperatures within the upper 2.2 m (deposited in the past 128 000 years) were used to deduce a sea level graph that showed maxima correlative with those derived from studies of terrace sequences in Barbados and New Guinea. This indicates a high interglacial sea level (the Eemian or Ipswichian stage) about 80–120 thousand years ago, a Last Glacial minimum about 20–25 thousand years ago and the ensuing Holocene marine transgression.

3.9 Late Quaternary sea level changes

Sea level graphs indicate that about 80 000 years ago sea level around much of the world's coastline was slightly higher than it is now. It then began to fall, and remained low during the Last Glacial phase, between 80 000 and 6000 years ago (known as the Devensian in Britain, the Würm in Europe and the Wisconsinan in North America). The extent of lowering of sea level during this Last Glacial phase has been estimated from the volume of water abstracted to form the late Pleistocene glaciers and ice sheets, which indicates that the sea fell more than 100 m below its present level. Extrapolation of the pre-Holocene floor of the Mississippi valley (traced beneath a thick alluvial fill) out to the edge of the continental shelf suggests a lowering of about 140 m.

As a result of sea level lowering during the Last Glacial phase the world's continental shelves emerged as wide coastal plains, and coastlines advanced towards their outer edges. The British Isles became a peninsula of Western Europe, there were wide coastal plains off what are now the Gulf and Atlantic coasts of the United States, and Australia was enlarged and linked to Tasmania and New Guinea (Figure 3.5). Rivers extended their courses to the lowered coastlines, incising valleys across the continental shelves and dissecting earlier terraces. Cliffs that had formed on earlier coastlines became degraded to bluffs, segments of which have been preserved where deposition has protected them from Holocene marine erosion. Examples can be seen on the Sussex coast at Cliff End, east of Hastings, and on the Bass Strait coast of Victoria, Australia, at Two Mile Bay, near Port Campbell.

About 18 000 years ago the Earth's climate was still very cold, glaciers and ice sheets were close to their maximum extent and the sea stood about 140 m below its present level. Then the Earth's climate became warmer, and the ice cover started to melt and water returned to the oceans, producing the Holocene marine transgression. The Last Glacial landscapes of the continental shelves were submerged as the sea rose to roughly its present level about 6000 years ago (Section 3.3.4.6).

Evidence of stages in Holocene marine submergence has been obtained from borings in coastal plains and deltas, notably on the Gulf Coast of the United States and in the Rhine delta. Freshwater peat and relics of land vegetation have been encountered well below present sea level, overlain by marine deposits formed as the sea rose. These stages have been determined by dating (mainly by radiocarbon) such materials as shells, wood or peat associated with these deposits found at specific depths below present sea level. In attempting to relate dated materials to contemporary sea levels it is necessary to take account of the extent to which stratigraphic horizons have been lowered, relative to present sea level, by subsequent compaction of underlying sediments (especially peat and clay) and by the crustal subsidence (an isostatic response to sedimentary loading) that has taken place in many deltaic regions.

When the dates are plotted against former sea levels a graph is produced that traces the sequence of sea level changes relative to the land for particular coastal regions. The resulting graphs show that the sea level rise was rapid, averaging just over a metre per century. Most studies have deduced an oscillating rise, with pauses and occasional slight regressions, particularly where there are Holocene stratigraphic sequences of alternating peats and marine sediments. A spasmodic sea level rise, with minor advances and retreats, would have aided shoreward sweeping of sea floor sediment (Section 6.4.3). However, some studies have indicated a smooth and steady increase accompanying the warming of global climate, and have judged the supposed oscillations as statistical aberrations. Further research is needed to decide which is the correct view.

There are certainly discrepancies between Holocene sea level graphs from different parts of the world's coastline (Figure 3.6), some of which result from the complicating effects of

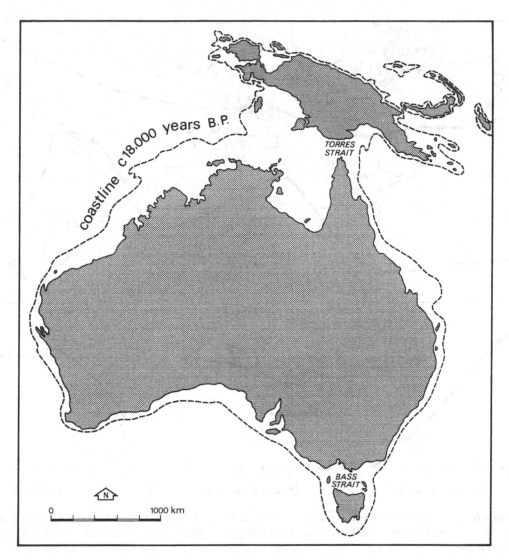

Figure 3.5 The coastline of Australia and New Guinea about 18 000 years ago, before the Late Pleistocene–Holocene marine transgression rose to submerge the continental shelf and separate Australia from Tasmania by the formation of Bass Strait (9000 years ago) and from New Guinea by the formation of Torres Strait (8000 years ago)

land uplift or depression in coastal regions, and some from regional variations in the scale and sequence of sea level changes. The contrast between Late Quaternary sea level history on the Atlantic coast of the United States, where the marine transgression is still proceeding slowly, and SE Australia, where the marine transgression slackened or came to a halt about 6000 years ago, is well known, as is the fact that post-glacial isostatic recovery has produced a falling sea level in Scandinavia and northern Canada.

Figure 3.6 Major variations in relative sea level change during the Holocene (after Pirazzoli, 1996): (a) in high northern latitudes, where land uplift has resulted in a falling sea level; (b) in middle latitudes, where the sea has risen at a slackening rate to attain (but not exceed) its present level; (c) in much of the southern hemisphere and SE Asia, where the sea rose above its present level between 3000 and 6000 years ago and has since fallen back. There are many other local variations related to local land uplift or subsidence, and to geodetic changes in ocean level

Submergence of Late Pleistocene landscapes as this marine transgression proceeded may be the basis for the story of Atlantis, a land area lost beneath the sea. Archaeologists have suggested that the submergence of Atlantis, mentioned by Plato as a lost land off the west coast of Europe but now thought to be near Crete in the Mediterranean, was due to the tsunami generated by the explosive eruption of the Thera volcano (Santorini) about 3500 years ago. Although this would account for the widespread damage of coastal features in the eastern Mediterranean, it would not explain the permanent submergence of a former land area (James, 1995).

Similar legends recur in the folklore of coastal people. In SW England the lost land of Lyonesse is thought to have lain in the area of the Isles of Scilly, where there are submerged archaeological features between Bryher and Tresco. On the Welsh coast stories of drowned cities and palaces probably originate from imaginative interpretations of the sarns, bouldery ridges of overwashed Pleistocene glacial drift that extend out from the coast and across the sea floor, submerged by the marine transgression. Coastal tribes of Australian aborigines tell stories suggesting that their ancestors retreated from lands now submerged on the sea floor.

Islands have disappeared in Chesapeake Bay as the result of submergence: Sharp's Island in Maryland had an area of 700 acres in the late 17th century, but was reduced by erosion to about 600 acres by 1850. Even in the first decade of the 20th century it had farms and a hotel. By about 1950 there was only a small grassy islet with just room for half a dozen men to stand, and now only the Sharp's Point Light shows where it was (Douglas, Kearney and Leatherman, 2000).

Evidence of landforms that existed during the Last Glacial low sea level phase on what is now the sea floor is scanty, because most of these were destroyed by wave action as the sea rose during the ensuing marine transgression. Relics of submerged cliffs, beaches and dunes have been found on the sea floor, and there were deposits of sand and gravel that later drifted shoreward to form beaches and barriers (Figure 6.5). It is sometimes possible to trace the courses of valleys incised by rivers that extended their courses across the sea floor during the period of lowered sea level, and some submarine canyons may have been thus initiated.

Processes of marine erosion and deposition began to shape existing coastlines as the sea approached its present level. Areas that had been uplands on the emerged coastal plain during the Last Glacial low sea level phase became islands or reefs offshore, their outlines modified by wave erosion and deposition. Valleys that had been incised across the sea floor were submerged and largely filled with sediment. It is sometimes possible to trace their offshore alignments in existing sea floor morphology, and they can be located by seismic surveys, which detect the sediment fill. Along the present coastline the landforms described in later chapters took shape as low lying areas were submerged to form embayments and inlets, valley mouths became estuaries and higher ground persisted as steep coasts and promontories. Cliffs were cut as the rising sea encountered rock outcrops of varying resistance, sand and gravel were deposited to form beaches, spits and barriers, wind-blown sand accumulated as coastal dunes, river sediments began to fill estuaries, and in some places to build protruding deltas, and corals and other marine organisms built reef structures.

Much remains to be done to elucidate the sequence and effects of sea level changes on coastal landforms. An example of a problem that remains to be solved is the presence of glacial erratics on the shores (and in some low level emerged beaches) of south and SW England and Brittany, beyond the limits of Pleistocene glaciation. These are large shore rocks that came from distant outcrops and were delivered to their present position by icebergs or ice rafts, possibly thrown on to the shore by storm surges or tsunamis (Figure 3.7).

Another problem is posed by accordant valley mouths (i.e. where streams descend to present sea level), which may result from uplift of coastal land offsetting the sea level rise that would otherwise have submerged the valley mouth. Examples of this occur on the coasts of southern Norway and eastern Scotland, where land uplift has occurred as a result of isostatic recovery following glaciation. Alternatively, as in the chines

Figure 3.7   Giant's Rock, near Porthleven in Cornwall, a large boulder (1.8 m high, 2 m wide and 3.6 m long) thought to have been deposited on the coast from a melting iceberg in Late Pleistocene times

of the Isle of Wight, stream downcutting may have been matched by cliff recession with the sea at its present level (Figure 4.26).

3.10 Modern sea level changes

Around much of the world's coastline the sea level has been relatively stable during the past 6000 years, apart from minor oscillations, of the order of a metre. This has been a period of Holocene still-stand, when the relationship between land and sea level has been more or less stable. There are, nevertheless, sectors where the coastal land has continued to rise or fall within the modern period (which may be approximately defined as the past century).

Emerging coasts (as around the Gulf of Bothnia) border shallowing seas, and often show progradation as the result of sea floor sediments being carried shoreward by wave action, whereas submerging coasts border deepening seas, and often show increasing erosion as waves grow larger and sediment is lost to the sea floor. Table 3.1 summarises the various features of emerging coasts.

Continuing submergence is obvious in the Venice region, where sea level has risen by more than 30 cm since 1890, largely because of land subsidence following groundwater extraction, although the dredging of channels to give larger ships access to the Port of Marghera has also contributed. The frequency of city flooding by high tides (known as acqua alta) has increased from four times a year in the 1900s to more than 100 times in the 2000s. Bangkok is one of several other coastal cities where sea level has risen because of land subsidence related to groundwater extraction. Near Port Adelaide an annual relative sea level rise of up to a centimetre

Table 3.1 Features characteristic of emerging coasts

1. Progradation of beaches (beach ridge plains, declining seaward). Any cliffs behind them are cut off from marine erosion and become subaerially degraded to bluffs (Figure 6.11).

2. Widening of coastal salt marshes or mangrove swamps.

3. River-mouth rapids, as in the Fotlandsvatnet in SW Finland (Ristaniemi *et al.*, 1997) and the Selfloss River in Iceland.

4. Incision of streams in deltas and coastal plains.

5. Drainage and dissection of salt marshes and mangrove swamps as tidal creeks become incised, as on the shores of Solway Firth, NE England and SW Scotland.

6. Abandonment and subaerial degradation of cliffs behind emerged shore platforms that were originally cut when the cliff was receding, and have been maintained by occasional storm waves.

7. Shoreward drifting of sea floor sediment to widen beaches and dunes, as at Kalajoki on the Gulf of Bothnia coast of Finland.

8. Shallowing and shrinkage of coastal lagoons and enlargement of bordering spits, as on the Punta do Aceira, which projects into Lagoa de Araruama, north of Rio de Janeiro, Brazil.

has been attributed partly to extraction of groundwater and partly to wetland drainage (Belperio, 1993). Coastal subsidence south of Los Angeles, California, has resulted from the pumping of oil from underground strata, and there have been similar effects near Galveston in Texas and on the Bolivar coast in Venezuela. In the Ravenna region a sea level rise of up to 1.3 m occurred between 1950 and 1986, due to subsidence resulting from subsurface compaction following the extraction of natural gas as well as groundwater. This drawdown has been most severe over the Ravenna Terra gas field NE of the city, and there has been increased sea flooding along the Adriatic coastline where it intersects the subsidence bowl. Table 3.2 summarises the various features of submerging coasts.

Evidence of modern sea level changes has been obtained from analyses of tide gauge records,

Table 3.2 Features characteristic of submerging coasts

1. Beach erosion initiated or accelerated, and more rapid erosion of cliffs, with increased losses of sediment, particularly to the sea floor.

2. Increased erosion of seaward margins of coastal salt marshes or mangrove swamps.

3. Higher water levels (higher high tides) in estuaries, lagoons and deltas, with increased salinity penetration and associated ecological changes (freshwater ecosystems replaced by brackish water ecosystems).

4. Higher water tables in coastal regions, leading to formation or the deepening and expansion of wetlands or lakes on low lying parts of coastal plains.

5. Dissection of salt marshes and mangrove swamps as tidal creeks are enlarged.

6. Inundation of shore platforms.

from geodetic and altimetric surveys and from certain biological indicators.

3.10.1 Tide gauge records

Some measurements of changes in sea level have been based on data from tide gauges, using calculations of annual and decadal means and running means to determine changes in mean sea level during the period for which the records are available. Most tide gauges are instruments installed for the purposes of navigation and harbour operations rather than to provide scientific information. They register actual fluctuations of the sea surface, including the effects of wind stress, waves and changes in atmospheric pressure, and also register local effects, such as current swirl. Many are located at ports, on structures where there may be local sea level anomalies resulting from wave reflection and ponding, and where there is a risk of damage and disturbance in the course of ship movements and port activities. Their datum levels are some-

times poorly maintained, and if a disturbance or accident results in the subsidence of a harbour structure the tide gauge records this spuriously as a rise in mean sea level.

Some caution is therefore necessary in measuring sea level changes from available tide gauge records. More scientific instrumentation is being developed around the world's coastline, and this will in due course provide improved monitoring of sea level changes in the future.

3.10.2 Geodetic surveys

An alternative source of information on sea level changes in recent decades has been provided by repeated geodetic surveys of land areas, based on precise levelling. In Sweden and Finland, for example, successive geodetic surveys have confirmed the fall in sea level indicated by tide gauges along the shores of the Gulf of Bothnia, which results from continuing isostatic uplift of Scandinavia following deglaciation (Figure 3.8). Substantial areas of land have emerged from

Figure 3.8 Rates of land uplift around the Gulf of Bothnia, based on Swedish and Finnish geodetic surveys

■ Progradation 1773-1967.

GULF OF BOTHNIA

N

☐NYKARLEBY

Lappo a.

0   5   10km.

Figure 3.9 A sector of the Finnish coast (arrowed in Figure 3.8) of the Gulf of Bothnia showing the extent of coastline advance (progradation) due largely to emergence (land uplift) during the period 1773–1967

the sea floor in recent centuries (Figure 3.9). In the United Kingdom there is evidence from tide gauges that mean sea level has risen faster in the south and SE (Sheerness and Newlyn) than in the north (North Shields and Aberdeen) (Woodworth, 1987), but geodetic confirmation of transverse tilting is still awaited.

Geodetic surveys are elaborate and expensive, and it will be some time before they can contribute much to knowledge of global sea level changes. Monitoring from satellites is providing information on the rise and fall of land and sea levels around the world, using measurements based on the Earth's centre as a datum. This will eventually provide an integrated global tidal facility, and it may then be possible to distinguish the effects of land uplift or subsidence and sea surface rise or fall in vertical changes in sea level relative to the land (Goldsmith and Hieber, 1991).

3.10.3 Biological indicators

Evidence of sea level changes may also be obtained from repeated surveys of the levels of marine organisms such as oysters, barnacles, mussels, algae and kelp, where these are found at specific tidal levels in vertical zones encrusting cliffs, rocky shores, sea walls and pier supports. Biological zones are generally correlated with the depth and duration of marine submergence with the sea at its present level, and can therefore be expected to move up or down in relation to a rise or fall of sea level. A sea level rise should be indicated by upward movement of such zones on cliff faces, stacks and rocky protrusions, as well as on artificial structures, accompanied by landward movements across shore platforms and intertidal outcrops.

Measurements of the upper and lower levels of oysters and barnacles on concrete piling at Miami Beach, on the subsiding coast of SE Florida, showed that these horizons moved 15 cm upward between 1949 and 1981, consistent with the relative rise of mean sea level registered on nearby tide gauges (Wanless, 1982). It may be possible to detect such migrations with reference to historical photographs where mean sea levels can be determined in relation to fixed features such as steps or decking, and it would be useful to document existing levels of zoned organisms as a basis for future measurements. The most suitable organisms are readily identifiable plants or animals that occupy particular parts of the intertidal zone with consistent, well defined upper or lower boundaries that can be correlated with particular stages of the tide, preferably at or close to mid-tide level. Indicator organisms should be able to migrate upwards as sea level rises by rapidly colonising higher levels that are either untenanted, or occupied by other plants or animals that can be displaced or overrun.

An example from SE Australia is the calcareous tubeworm *Galeolaria caespitosa*, which occupies a well defined intertidal zone and is

Figure 3.10   Cushions of *Galeolaria caespitosa* on pier supports at Dromana, Port Phillip Bay, Australia

best developed on sites sheltered from strong wave action and abrasive sand movements (Bird, 1988). It forms either a thin layer, or cauliflower-like encrustations similar to the trottoirs, accretionary ledges of coralline algae that protrude from some Mediterranean rocky shores (Section 13.10). The vertical range of *Galeolaria* is typically 20–40 cm, with an upper limit close to mid-tide level, irregular on sites exposed to strong waves and variable swash and spray, but horizontal on sheltered sites such as the inner sides of harbour walls and on piers (Figure 3.10). Monitoring of the upper limit of *Galeolaria* in Port Phillip Bay, initiated in 1988, indicated an average rise of about 5 mm during the past 18 years, possibly in response to a sea level rise. This seems to be independent of the long term SAROS tidal oscillation, which peaked in 1987 and 2006.

Sea level changes may also be indicated by studies of changes in zoned patterns of vegetation in salt marshes and mangrove swamps, but it is difficult to obtain precise measurements of sea level changes from such evidence.

3.11 Recent changes of land and sea level

Evidence of recent changes of land and sea level from tide gauge records was discussed by Pirazzoli (1986), who analysed tide gauge records from 229 coastal stations that had been maintained for at least 30 years. He found that 63 (28.5 per cent) showed a mean sea level rise exceeding 2 mm per year, 52 (22.5 per cent) between 1 and 2 mm per year and 47 (20.6 per cent) less than 1 mm per year, the remaining 65 (28.5 per cent) having shown a mean sea level fall. As over 70 per cent of the records showed a positive trend a global sea level rise seemed likely, but the geographical distribution of the

229 stations was uneven, with strong northern hemisphere mid-latitude clustering, and only six in the southern hemisphere. There is the difficulty that most tide gauges are located at ports, and may not be reliable indicators of mean sea level changes.

Emery and Aubrey (1991) extended this review, using 664 tide gauge stations, 65 per cent of which had at least 30 years' records. Analysis of 98 key stations showed that sea level had fallen only on sectors where the coast has been rising tectonically or isostatically. The Global Sea Level Observing System (GLOSS) is developing a much more representative network of 287 tidal stations around the world's coastline, and will provide more accurate information in the next few decades (Woodworth, 1991), both globally and regionally (Pugh, 1991). The Intergovernmental Panel on Climate Change has used evidence from satellite altimetry to measure sea level changes since 1993 (IPCC, 2007).

Sea level changes over the past few decades should be considered against the background of factors (listed previously) that are known to have influenced Holocene sea level trends. Climatic fluctuations are thought to be responsible for the major oscillations of water level that have taken place in the Great Lakes in North America during the past century. Climatic changes have contributed (along with dam construction in tributary rivers) to the variations in level recorded on tide gauges around the Caspian Sea: it fell by 2.67 m between 1930 and 1975, and has since been rising (Figure 3.11), with the onset of more humid conditions in the surrounding area (Kaplin and Selivanov, 1995). The rising sea level has formed lagoons bordered by barrier beaches, both of which are transgressing on to bordering land.

Figure 3.11 Changes in the level of the Caspian Sea between 1930 and 2000, as indicated on the Baku tide gauge. Until 1977 there was an intermittent lowering of sea level, but subsequently there has been a rise of about 2 m, probably because of a trend towards a more humid climate in the region. By 2006 the sea level had risen to 26.3 m below Mediterranean sea level. Information supplied by Dr. S. Lukyanova of Moscow State University and Dr. Nasser Saddedin, Geological Survey of Iran

Tectonic movements have certainly influenced changes of land and sea level in recent decades. Sea level has continued to fall on coasts where isostatic land uplift has followed deglaciation, as in parts of Scandinavia, Northern Canada and Alaska, and where the land is rising tectonically, as in northern New Guinea, some Indonesian and Philippine islands and parts of the Japanese coast. In Alaska isostatic uplift has proceeded at the rate of at least 3.5 cm/year, as indicated by Holocene glaciomarine sediments 230 m above sea level near Juneau. Similar uplift has followed deglaciation on the mountainous coasts of southern Chile and Argentina.

An emerged terrace implies a phase of stillstand when a shore platform was cut at a particular level followed by uplift to raise it above present sea level. Emerged terraces at various levels on the coasts of Scandinavia and Scotland have been seen as the result of postglacial isostatic recovery, but their separation implies an intermittent rather than steady uplift history.

Sea level has been rising on coasts where the land margin is subsiding, as on the Gulf and Atlantic seaboards of the United States, the south and SE coasts of Britain, the Netherlands and north Germany, NE Italy and several other areas (Milliman and Haq, 1996). The floor levels of Roman London are now about 4 m below high-tide level in the Thames estuary, indicating an average rate of submergence of about 2 mm per year in this area over the past two millennia. At Newlyn in Cornwall, mean tide level in the 1990s was about 20 cm higher than that recorded for 1915–21, when the mean sea level at this site was calculated to establish a datum for British topographical surveys: a rise of just over 2 mm/year. It is possible that the records from this tide gauge station have been affected by hydro-isostatic subsidence of the adjacent Land's End Peninsula. Continuing subsidence in southern and eastern England may be responsible for the persistence of broad open es-

tuaries in Suffolk and Essex, the wide mouth of the Thames estuary, and embayments such as Chichester Harbour and Southampton Water: in each case submergence has exceeded the rate of sedimentary infilling. By contrast, mean tide level at Dundee in Scotland fell just over 2 mm per year in this period, any sea level rise being more than offset by the effects of the ongoing isostatic uplift following deglaciation. If a sea level rise is taking place around Britain, it has probably been in progress since the Little Ice Age phase of colder climate recorded in the mid-18th century. The possibility that tectonic deformation has continued in Britain since Late Tertiary times (the past 5 million years) cannot be ruled out.

Figure 3.12 shows the location of sectors of the world's coastline that have been subsiding in recent decades, as indicated by evidence of tectonic movements, increasing marine flooding, geodetic surveys and tide gauges recording a rise of mean sea level greater than 2 mm/year over the past 30 years. Some are in areas of isostatic downwarping around major deltas; others are at least partly the outcome of human activities, notably groundwater extraction, as in the Venice and Bangkok regions, and oil extraction, as in Southern California.

The balance of the evidence suggests that a global sea level rise is in progress (Pirazzoli, 1996), but more extensive monitoring is required to confirm its dimensions. It is possible that the so-called contemporary world-wide marine transgression of between 1 and 2 mm/year has been over-estimated, or that more complete global studies will show that there have been geographical variations in the nature and scale of sea level change during the past century similar to those indicated by Holocene sea level graph discrepancies. Certainly any statement that global sea level has risen by a specific amount in recent decades is provisional until these geographical variations have been assessed.

Figure 3.12 Sectors of the world's coastline that have been subsiding in recent decades, as indicated by evidence of tectonic movements, increasing marine flooding, geomorphological and ecological indications, geodetic surveys and groups of tide gauges recording a rise of mean sea level greater than 2 mm per year between 1970 and 2000 **Key:** 1, Long Beach area, southern California; 2, Colorado River delta, head of Gulf of California; 3, Gulf of La Plata, Argentina; 4, Amazon delta; 5, Orinoco delta; 6, Gulf and Atlantic coast, Mexico and United States; 7, Southern and Eastern England; 8, the southern Baltic from Estonia to Poland; 9, North Germany, the Netherlands, Belgium and northern France; 10, Hebrides, Scotland; 11, Loire estuary and the Vendée, western France; 12, Lisbon region, Portugal; 13, Guadalquavir delta, Spain; 14, Ebro delta, Spain; 15, Rhône delta, France; 16, Northern Adriatic from Rimini to Venice and Grado; 17, Danube delta, Rumania; 18, Eastern Sea of Azov; 19, Poti Swamp, Soviet Black Sea coast; 20, SE Turkey; 21, Nile delta to Libya; 22, NE Tunisia; 23, Nigerian coast, especially the Niger delta; 24, Zambezi delta; 25, Tigris–Euphrates delta; 26, Rann of Kutch; 27, SE India; 28, Ganges–Brahamputra delta; 29, Irrawaddy delta; 30, Bangkok coastal region; 31, Mekong delta; 32, Eastern Sumatra; 33, Northern Java deltaic coast; 34, Sepik delta; 35, Port Adelaide region; 36, Corner Inlet region; 37, Hwang-ho delta; 38, Head of Tokyo Bay; 39, Niigata, Japan; 40, Maizuru, Japan; 41, Manila, Philippines; 42, Red River delta, North Vietnam; 43, Northern Taiwan; 44, East Siberian coastal lowlands; 45, Maritime Provinces, Canada; 46, Mackenzie delta and northern Alaska; 47, Yukon delta, Alaska.

The coasts of the Caspian Sea are included as sea level rose by more than 2 m there between 1977 and 2000 (Figure 3.11).

3.12 Future sea level changes

In 2007 the Intergovernmental Panel on Climate Change (IPCC, 2007) reported on global warming, related primarily to the 'greenhouse effect' of increasing concentrations of carbon dioxide, methane and nitrous oxide generated by human activities (notably agriculture, industry and the burning of fossil fuels) since 1750. The Panel tabled data showing that global average sea level had risen by 1.3–2.3 mm/year between 1961 and 2003, and 2.4–3.8 mm/year between 1993 and 2003. Estimates of sea level rise between 1980–99 and 2090–99 were between 0.18 and 0.59 m.

During the next few decades monitoring of sea level changes will clarify the pattern of upward and downward movements of the sea relative to the land around the world's coastlines. Predictions of changes on particular coasts should take account of the known variability of sea level in relation to upward or downward movements of the land, the migrations of troughs and domes on the ocean surface and other factors that could increase or reduce sea level rise (Section 3.3). The nature and effects of a worldwide sea level rise will be discussed in Chapter 14.

3.13 Summary

Mean sea level shows short term variations related to meteorological factors and long term variations related to tidal cycles, but has been relatively stable over the past 6000 years except on coasts where there has been continuing land uplift or subsidence. Eustatic movements are caused by changes in the volume of the oceans, but there are also changes related to ocean temperature and salinity, current patterns, sedimentation, gravity variations and tectonic movements, which may be epeirogenic (continental uplift or subsidence), orogenic (related to mountain building) or isostatic (related to loading or unloading of the Earth's crust, notably by the accumulation and melting of ice sheets). There are also changes of sea level related to volcanic activity. Human activities affecting sea level include the extraction of groundwater, oil and natural gas, dredging, land reclamation and loading with heavy structures. There are former coastlines, both emerged and submerged (above and below present sea level), that can be dated and used to compile graphs of sea level changes over time. The most important of these in coastal geomorphology is the worldwide Holocene marine transgression, which began about 18 000 years ago and ended (on stable coasts) about 6000 years ago. Modern changes of sea level have been determined from tide gauge records, geodetic surveys and biological indicators. Global warming due to increasing 'greenhouse gases' in the atmosphere is likely to cause sea level to rise globally.

4
Cliffs

4.1 Introduction

Cliffs are steep (usually more than 40°, but often vertical and sometimes overhanging) coastal slopes cut into rock formations. They are generally receding as the result of marine erosion at their base, accompanied by subaerial erosion of the cliff face. Many cliffs are fronted by shore platforms exposed as the tide falls (Figure 4.1).

Three-quarters of the world's coastline is cliffed and rocky. The geomorphology of cliffs and rocky shores has been discussed in textbooks by Trenhaile (1987) and Sunamura (1992), and summarised briefly by Griggs and Trenhaile (1994). Cliffs are strongly influenced by the geology of coastal regions, particularly the structure and lithology of rock formations that outcrop on the coast and their response to weathering and erosion processes. Rock formations of varying age, from Pre-Cambrian (more than 560 million years old) to Holocene, outcrop on the world's coastlines, but most cliffs have been shaped during Pleistocene and Holocene times, mainly the past 6000 years, when the sea has stood at or close to its present level.

Some cliffs have been produced by uplift of the land margin as the result of faulting; others follow fault lines, but are partly or wholly the outcome of differential erosion, where faulting has placed weak rock formations alongside resistant rocks. The first category, a steep or cliffed coast produced by faulting, where the seaward slope coincides with the plane of the fault, along which the land has been raised, is termed a fault coast: a tectonic feature known as a fault scarp. Some coasts were initiated as fault coasts, but have been cut back by marine erosion and now stand landward of the fault. Alternatively, a steep or cliffed coast that follows a fault line, where the seaward slope has been formed by differential erosion of rock formations juxtaposed by the prior faulting, is a fault-line coast.

Climate has been an important influence on the weathering of coastal rock outcrops, which results from physical, chemical and biological processes, related partly to subaerial conditions and partly to the presence or proximity of the sea. Rocks are decomposed or disintegrated by such processes as repeated wetting and drying, solution by rainwater, thermal expansion and contraction, freeze–thaw alternations and shore ice effects, all related to temperature, precipitation and evaporation regimes in the coastal environment. Rock debris falls to the cliff base as talus, which must be consumed or removed by wave action if cliff recession is to continue; if it persists as a wide protective beach, and

Coastal Geomorphology: An Introduction, Second Edition. Eric Bird.
© 2008 by John Wiley & Sons, Ltd

Figure 4.1 The intertidal shore platform cut in chalk west of Birling Gap, Sussex, looking towards Seaford Head. The platform is strewn with flint cobbles that have come from layers of flint that can be seen in the cliff to the right. At high tide the narrow beach of flint shingle at the cliff base can be mobilised by breaking waves and used to abrade the lower part of the cliff, where undercutting has resulted in the collapse of a sector of cliff, forming a basal apron of chalk talus (arrowed). The knobbly shore platform is also being abraded as waves move the flint cobbles to and fro. On the left the lower part of the shore platform is covered with marine wrack, and abrasion is less active

marine erosion at the cliff base is halted, the cliff becomes degraded by subaerial processes.

Where the rocks are very resistant (e.g. massive granites), steep and high coastlines have changed little, if at all, as the result of marine erosion over the past 6000 years. Less resistant formations have been cut back as cliffs, some bordered by irregular rocky shores, others with smoother shore platforms at least partly exposed at low tide. Elsewhere, there are cliffs or coastal slopes that plunge into deep water. A distinction is sometimes made between hard rock cliffs and soft rock cliffs, also known as earth cliffs (May, 1972).

The simplest cliffs are found where marine erosion has attacked the margins of a stable land mass of coherent rocks, removing a wedge of material to form a steep (usually more than 30°, and often vertical) slope fronted by a shore platform (Figure 4.2). Early stages in this evolution can be seen on the shores of reservoirs after they fill with water, and wave action begins on a bordering slope. On many cliffed coasts there are complications introduced by the lithology and structure of outcropping rock formations, the degree of exposure to wave attack, the effects of subaerial weathering (physical, chemical and biological) on the coast and the history of changing land and sea levels (Emery and Kuhn, 1982). Even on relatively weak rock formations, the cutting of cliffs and shore platforms takes time, and implies that the sea has remained at or close to the same level in relation to the land for a prolonged period (up to 6000 years on many coasts: Figure 3.6).

Figure 4.2 An abrasion notch cut in steeply dipping Devonian sandstone at Cape Liptrap, Victoria, Australia. Notebook scale width 20 cm

Cliffs can form and recede quickly on very soft materials, such as salt marshes and mangrove swamps (Chapter 10), where low cliffs (clifflets or microcliffs) are cut along the seaward margin. The same is true of cliffs cut into peat bog deposits, as in NW Ireland where black cliffs up to 2 m high, cut into blanket bogs that descend to sea level around Broad Haven and Blacksod Bay, are receding rapidly. Cliffs cut in freshwater swamp deposits are seen at Lang Lang on the NE coast of Westernport Bay in Victoria, Australia (Section 4.6), at Owenga on the east coast of the Chatham Islands (Section 9.14) and near Budir in NW Iceland.

4.2 Cliff evolution

Cliffs are cut back mainly during storms, when the cliff base is undercut by the hydraulic pressure of wave impact and the abrasive action of water laden with rock fragments (sand and gravel) that during storms are hurled repeatedly at the cliff base. A basal abrasion notch is thus formed, typically 1–2 m high and recessed by up to 3 m. As it grows the cliff face becomes unstable, and an overhanging rock mass eventually collapses (a rockfall). Cliff-base notches are better developed on relatively resistant rock formations because a weak rock outcrop will not sustain them. Often there is a sloping abrasion ramp at the cliff base, declining to a shore platform.

After a storm the backshore is littered with debris that has fallen from the cliff. Agitated by wave action, this becomes broken and worn down (a process known as attrition), and is either retained as a beach (which may protect the cliff base from further abrasion), or carried away along the shore or out to sea by the action of waves and currents.

The vertical cliffs of chalk in SE England and northern France have been formed and

Figure 4.3 Vertical grooves and buttresses on the Chalk cliff near the mouth of Cuckmere River in Sussex

maintained by marine erosion in this way. May and Heeps (1985) noted that marine erosion cut basal notches and marine processes removed basal talus, but found that cliff recession was largely due to rockfalls caused by saturation after heavy rain, and particularly to freeze–thaw action in winter, especially when mild weather follows a cold winter, as in 1940, 1947, 1963 and 1979. The massive rockfall on Beachy Head in Sussex early in 1999 was attributed to the expansion of saturated chalk during a spell of very cold weather (Section 4.8). Freezing and thawing can cause toppling and slumping, producing half-conical basal talus fans of broken rock, and leaving a white scar on the cliff face (which is otherwise grey, as the result of weathering and colonisation by algae). The fallen rock is gradually consumed by weathering and corrosion (the chalk being dissolved by rainwater and aerated spray and surf) and removed or dispersed by wave action. Off the chalk coasts bordering the English Channel the sea is typically grey–green or opaline, with chalk particles in suspension. The vertical cliff profile thus restored is then further undermined by basal wave erosion until it collapses again.

Some cliffs show vertical grooves and buttresses cut out along joint planes. Examples are seen on the chalk cliffs near the Seven Sisters in Sussex (Figure 4.3) and on Triassic sandstone at Peak Hill on the south coast of Devon.

On soft formations, such as clays, unconsolidated sands, deeply weathered rocks or unconsolidated glacial or periglacial drift deposits, cliffs and steep coastal slopes recede by recurrent slumping, particularly after wet weather or the thawing of a snow cover. Intermittent recession of this kind is seen on the soft Tertiary sands and clays of the Bournemouth coast and the northern shores of the Isle of Wight, on Jurassic clays and shales on the Dorset and Yorkshire coasts, on Coal Measures clays and sands in Pembrokeshire and on glacial drift deposits, as in eastern England, the Danish archipelago, New England or the islands of Puget Sound. Cliff recession is dramatised where buildings are undermined and lost, and little is left of coastal churches at Dunwich in East Anglia and Trzesacz in Poland, both damaged by recession of cliffs cut in glacial drift.

Recession of the cliff crest often takes the form of irregular breakaways, when lumps of rock become dislodged and fall away from the cliff. On the Dorset coast slumping has occurred where cliff-top paths have become breakaways (Figure 4.4). Cracks develop behind, and parallel to, receding cliffs as a prelude to calving or slumping, and deformation and fracturing parallel to the

Figure 4.4 Breakaway on the cliff crest at Black Ven, near Lyme Regis in Dorset (cf. A in Figure 4.5)

coastline is seen on many cliff-top roads and damaged buildings close to cliff edges, as on the Holderness coast. Slumping coastal slopes are irregular, and basal debris fans are undercut by the waves, forming a slope-over-wall profile. The vertical cliff grows in height as it is cut back, but there is soon further slumping. The cliffs thus recede as the result of alternating marine and subaerial erosion. Figure 4.5 shows such a sequence on the coast of Lyme Bay.

Coastal slopes and cliffs in soft formations are dissected by gullies cut by runoff after heavy rain or melting snow, as on the Wealden clays and sands on the SW coast of the Isle of Wight. A cliff of soft Tertiary sandstone at Black Rock Point, on the coast of Port Phillip Bay, Australia,

has been dissected and cut back by gullies cut by occasional runoff after heavy rain, accompanied by the exudation of fine sediment by groundwater seepage. The cliff became steeper and smoother in profile after cliff-top stabilisation diverted runoff (Bird and Rosengren, 1987). Such features do not persist on cliffs that are being cut back rapidly by marine erosion.

Cliff weathering includes physical processes that disintegrate the rock surface, chemical processes that decompose rock outcrops and biological processes that may attack or protect the rock. Solution processes are particularly active on limestone coasts (including dune calcarenites, which are calcareous sandstones formed by the lithification of coastal dune deposits (Section 9.10). Seeping groundwater dissolves carbonates and washes out fine particles, forming cracks and crevices in the cliff face, and precipitation of carbonates hardens the surface, which is often darkened by algal colonisation. Some limestone coasts show a contrast between basal abrasion, where waves armed with sand and gravel have scoured a smooth rock surface and excavated fissures and caves, passing upward to a cliff face pitted by solution in rainwater and sea spray. On fine-grained rock outcrops weathering by repeated wetting and drying, possibly accompanied by salt crystallisation, produces large cavities (tafoni) and more intricate honeycombing (Figure 4.6), particularly where a harder surface crust is penetrated to underlying softer rock. Caves may be excavated on cliff faces scoured by wind action.

Strong onshore winds can sweep sediment from the cliff up and over the cliff crest, where it may be deposited as a cliff-top dune. On the Port Campbell coast in Victoria, Australia, fine-grained sediment winnowed from cliffs of calcareous siltstone has been swept up and over the cliff crest by the wind and deposited as a cliff-top levee, sloping down landward.

At lower levels the cliff face may become intricately pitted (muricate) as the result of

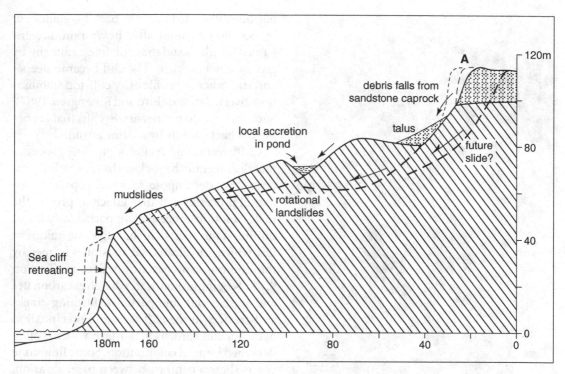

Figure 4.5 Recurrent landslides and basal cliffing on the coast near Lyme Regis in Dorset. The cliff crest (A) has receded about the same distance as the basal cliff (B) during the past century (after Brunsden and Jones, 1980)

weathering, which includes the effects of recurrent wetting and drying, salt crystallisation, corrosion by rainfall and sea spray, and bioerosion by marine organisms. Salt weathering is more intensive in the upper intertidal zone (between mean high tide and the limit of swash and splash at high spring tide) on arid coasts, where salinity rises to hypersaline levels, than on coasts with high rainfall, where salinity is reduced. Commonly rock surfaces above high tide level are whitened by the deposition of a thin covering of salt during dry periods, and this disappears when the rocks are splashed by rain or spray.

Coastal vegetation and fauna often include species the growth and metabolism of which lead to decomposition or dissolution of rock outcrops, especially limestones. Bioerosion can also be achieved by terrestrial organisms, notably penetration of and widening of joints and

fissures by the roots of plants growing on the cliff face. Rabbit burrows and nesting holes excavated by birds, particularly sand martins, contribute to the erosion of soft sandstone cliffs, while burrowing mortar bees have dissected the sandy cliff near Redend Point in Dorset.

The effects of some organisms may be protective or even constructional. A dense growth of kelp or barnacles, or a firm encrustation of algae, may protect a rocky shore from abrasion.

Superficial induration of the cliff face by the precipitation of cementing materials washed down the cliff or brought to the surface by seepage can slow erosion, at least until the hardened crust is breached and dissected. The vertical cliff face at Demons Bluff on the coast of Victoria, Australia, has been hardened in this way, and shows slumping in sectors where the indurated crust has been breached (Figure 4.7).

Figure 4.6 Honeycomb weathering on Cretaceous sandstone in the spray zone (just above high tide level) on the Otway coast of Victoria, Australia

Figure 4.7 Cracks in the mudrock cliff at Demons Bluff, Victoria, Australia, which recedes as masses of rock fall from the cliff face

It is convenient to consider cliff outlines in plan, then in profile, noting that cliff morphology is three dimensional and that cliff features need to be explained in terms of processes that affect both plan and profile.

4.3 Cliff morphology

Cliff morphology is three dimensional, and includes cliff profiles shaped at right angles to the coastline and cliff outlines in plan, i.e. coastlines as seen on maps or vertical air photographs. It is convenient to consider cliff profiles first (Sections 4.4–4.6.7) and then cliff outlines in plan (Section 4.7), but some overlap is inevitable because both are shaped by similar processes. Cliff morphology is much influenced by hinterland topography and the varied resistance of rock formations outcropping along the shore. Ridges that intersect the coast end in headlands and valleys in bays or inlets, and there are generally headlands on hard rocks and bays cut out in soft rocks. It is possible to recognise categories of hardness of rocks attacked by the physical forces of marine erosion (Clayton and Shamoon, 1998).

(a) Very hard quartzites and sandstones, massive granite and indurated metamorphic rocks.

(b) Moderately hard slates, shales, grits and basalts.

(c) Weak limestones (including chalk) and sandstones.

(d) Very weak mudrocks (defined as rocks containing at least 90 per cent silt and clay, and formerly known as siltstones, shales and clays) and unconsolidated sands.

Cliff profiles on resistant rocks are generally bold and steep, becoming gentler on weaker outcrops. Many of the bold headlands on the coast of Cornwall are on outcrops of dolerite (greenstone), in contrast with gentler coastal slopes on slates or phyllites. Cliff morphology is also influenced by geological structure, including dip, joints, faults and folds, and the disposition of harder and softer rock outcrops.

4.3.1 Cliff profiles

Cliff profiles are related not only to the structure and resistance of outcropping rock formations to abrasion, but also to their durability in the face of physical, chemical and biological weathering processes.

Vertical cliffs are best developed on homogeneous or well stratified rock formations, notably sandstones and limestones (including emerged coral reefs). Usually, the cliff faces have formed by breakaways along vertical joint planes, as in the Lower Cretaceous sand-rocks east of Hastings in Sussex. Coherent silty sediments such as brick-earth (loess) stand in vertical cliffs on the shores of Pegwell Bay in Kent, and along the SE shore of the Sea of Galilee, and there are vertical cliffs cut in glacial drift deposits along the Holderness coast in NE England. The simple association of vertical cliffs fronted by wide seaward-sloping shore platforms is well illustrated on the chalk coasts bordering the English Channel (Figure. 4.1).

On coasts formed by faulting (Section 4.1) the cliff face may be the plane produced by faulting or (more often) a scarp where a fault plane has been exposed by differential erosion. Cliff profiles are related to variations in lithology and structure, picked out by erosion as the cliff base is cut back. The more resistant parts of coastal rock formations protrude as ledges, or persist as caprock on rocky stacks and islands offshore, whereas the weaker elements are cut back as cliffs and caves.

Rock resistance depends on several factors. Massive rocks are generally more resistant to

Figure 4.8　Escarpment cliff in landward-dipping chalk at Ballard Down in Dorset

erosion than rock formations divided by many joints, bedding planes, cleavage planes and fractured zones, which facilitate cliff dissection. Solid and massive formations are generally eroded more slowly than formations that disintegrate readily, such as friable sandstones, rocks with closely spaced joints and bedding planes or rock formations shattered by faulting. On volcanic islands (such as Surtsey, off the south coast of Iceland) the softer sediment (volcanic ash or tuff) is soon washed away, but where hard lava outcrops at or above sea level, cliffs and platforms predominate.

Most rock formations have planes of division that are weakened by weathering processes and penetrated by marine erosion, influencing the outline in plan of a cliffed coast. These include bedding planes, cleavage planes, joints and faults, overthrusts and zones of soft or shattered rock. These can be excavated by weathering processes and wave scour to form crevices, clefts, inlets and caves along the coast. Shore rocks are attacked by quarrying, the hydraulic pressure of breaking waves forcing air and water into fissures, which are gradually enlarged to form clefts and gullies. Intricate dissection by wave action along planes of weakness (including joints and faults) has produced the irregular outlines of the Tintagel cliffs in North Cornwall, with many steep sided coves, inlets, caves and gullies, all

closely related to the intricacies of local geology (Wilson, 1952).

On the coasts of Kent and Sussex vertical cliffs have been cut in chalk where the strata dip gently seaward (Figure 4.1). In the Isle of Wight and on the Dorset coast, where the chalk outcrop dips steeply and is often strongly folded, there are more irregular features related to minor variations in lithology, such as ledges and reefs on the more resistant layers (e.g. the Melbourn Rock) and gentler slopes on softer marly horizons within the chalk. Where the dip is landward, as at Ballard Down in Dorset (Figure 4.8) and at the eastern (Culver Cliff) and western (Tennyson Down) ends of the Isle of Wight, the cliffs undercut the chalk escarpment, and may be described as escarpment cliffs (Bird, 1995, 1997). The influence of dipping rocks on cliff profiles is illustrated in Figures 4.8 and 4.9.

Contrasts related to lithology and structure are also well displayed in the profiles of cliffs cut in Jurassic formations along the Lyme Bay coast in Dorset. On Golden Cap (Figure 4.10) the harder sandstone outcrops form ledges and the softer sands and clays gentler (sometimes vegetated) slopes and areas of subsidence.

Figure 4.11 shows structural ledges as steps on horizontal sandstones separated by weak clays. On the north coast of Devon the cliffs near Hartland Point are generally vertical, cut back

Figure 4.9 Influence of seaward dip on the profile of a sandstone cliff near Boat Harbour on the north coast of Tasmania

by stormy seas across Carboniferous sandstones and shales compressed into tight zigzag folds along vertical axes running at right angles to the coastline. In detail the harder sandstones protrude and the weaker shales have been cut out, while on a larger scale there are ridges on anticlinal zones (upfolds) that have proved slightly more resistant, and coves and inlets along the transverse synclines (downfolds).

More resistant formations, such as the granites of Cornwall, have been dissected by wave action along horizontal, inclined and vertical joint planes to form headlands and clefts, which in places have a castellated appearance related to the cuboid jointing, well known at Land's End. Distinctive cliffs have also developed on columnar basalts, as in the Giant's Causeway in Northern Ireland, and there are vertical cliffs and stepped rocky shores on columnar dolerites at Pillar Point in SE Tasmania. The Old Red Sandstone is another resistant formation that forms

high cliffs on the north coast of Scotland between Dunscansby and Skirza Head and on the west coast of the Orkney Islands. Here cliffs up to 60 m high are retreating by way of frequent rockfalls and removal of the fallen debris by wave action. Also in Scotland the rugged cliffs of St Abb's Head show an impressive array of crags, clefts, caves, gullies, reefs, stacks and skerries, formed by dissection along numerous planes of weakness in Silurian slates and Devonian volcanic rocks. Rapid variations in cliff profiles are related to intricate outcrops of sandstone, limestone and volcanic rock at Dunbar.

On the SW coast of the Isle of Wight sandstone horizons in the clay-dominated Wealden Beds produce relatively steep cliffs where they outcrop at the cliff crest, as at Barnes High, and cliff-face ledges where they outcrop in the cliff profile, as at Sudmoor Point (Bird, 1997). In general, bolder cliffs occur on coasts exposed to strong wave action from the open sea, and

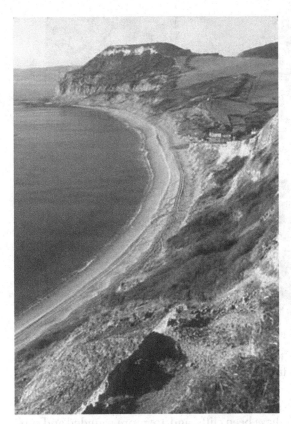

Figure 4.10 Slope-over-wall profile at Golden Cap, Dorset. There is a capping of Upper Greensand, the slope is cut in Middle Lias sands and clays and the basal cliff in Lower Lias shales and marls

more subdued slopes on relatively sheltered sectors. On the SW and SE coasts of Australia cliffs have been cut into Pleistocene dune calcarenite (Section 9.10), and show features related to the various components of these lithified dunes. These include dune-bedded sandstone, layers of hard calcrete (caliche) cemented by carbonate precipitation, soil horizons (palaeosols), often with plant structures preserved in carbonate, calcrete gravels (calcirudites) and soft unconsolidated dune sand.

On the high wave energy coast near Port Campbell in Australia vertical cliffs have been cut by marine erosion in horizontally stratified Miocene calcareous siltstones. The huge waves that break against these during storms have cut out ledges along the bedding planes at various levels up to 60 m above high tide (Baker, 1958). These structural ledges are rarely more than 5 m wide, and are the product of present-day storm wave erosion; they are not emerged shore platforms. The power of storm waves is illustrated where large boulders have been thrown up and over cliffs, as at Quobba in Western Australia.

Cliffs on more sheltered sections of the coast, where strong wave action is intercepted by headlands, islands or offshore reefs, or attenuated by a gentle offshore slope, may show profiles partly formed by subaerial weathering and erosion as well as those shaped by marine attack. Cliffs in these situations may develop slope-over-wall profiles (Section 4.3.3).

The profiles of steep coasts on similar rock formations show intricate variations related to exposure to wave attack. The bold profiles of cliffs of massive sandstone facing the ocean in the Sydney district contrast with the gentler, often vegetated, slopes on the same geological formations on the sheltered shores of Sydney Harbour and Broken Bay. The boldness of cliffing developed in these situations is closely related to the local fetch, which limits the strength of attack by local wind-generated waves.

Cliffs rising to between 100 and 500 m above sea level are termed high cliffs, while those exceeding 500 m are termed megacliffs (Guilcher, 1966). They are found on the Pacific coasts of Peru and Chile where coastal land has been uplifted tectonically, and on volcanic islands such as the Canary Islands, Madeira and Tenerife in the Atlantic Ocean. The cliff at Enniberg on the north coast of Vidoy (Faerö Islands) is 725 m high. Megacliffs on the SW coast of Lord Howe Island are cut into almost horizontal lava flows, and rise more than 500 m above sea level They show scars where masses of rock have fallen away, and sectors of angular boulders that have fallen to the underlying shore. These contrast with

Figure 4.11 Structural ledges on the cliff at Wonboyn, New South Wales, Australia, corresponding to the upper surfaces of flat lying Devonian sandstones separated by layers of soft clay

rounded boulders on sectors where there has been no recent fall. The cliffs have thus shown intermittent and local retreat.

High vertical cliffs border lava flows on the Scottish island of Skye, eroded by the sea along joint planes, but their morphology is similar to (and passes laterally into) that of lava cliffs inland, produced by glacial and periglacial processes inland, where they are fronted by grassy colluvial aprons. Many megacliffs are also plunging cliffs, passing steeply down into deep nearshore water: these will be considered after dealing with shore platforms.

4.3.2 Coastal bluffs

A distinction is made between cliffs (steep to vertical coastal slopes that expose hard or soft rock outcrops) and bluffs (which may formerly

have been cliffs, and are more rounded and subdued in profile, with rock outcrops concealed or largely concealed by a weathered mantle, soils and vegetation). When marine erosion ceases the cliff becomes degraded to a gentler slope, and fallen and downwashed sediment accumulates as a basal talus apron. Bluffs have slopes determined by the geotechnical properties of the rock outcrop: they are usually between 8 and 10° on soft clays and steeper on more resistant formations. Bluffs are extensive on coasts where former cliffs have been subaerially degraded because emergence has withdrawn wave action, as in NE Scotland and in the Wellington district, New Zealand.

Alternatively, coastal bluffs have formed where deposition (e.g. beaches, beach ridges and dunes) has protected them, as on the South Wales coast eastward from Pendine. Here Savigear (1952) traced stages in subaerial

Figure 4.12 Incipient cliff degradation is seen behind the reclaimed area at Samphire Hoe, west of Dover, formed by dumping chalk excavated from the Channel Tunnel in 1987–92. The vertical cliff has become a steep but partly vegetated escarpment

degradation (including periglaciation) of former cliffs cut in Old Red Sandstone to rounded slopes because of the growth of a large protective sandy foreland. On the North Island of New Zealand at Matata cliffs have become vegetated bluffs behind a prograded beach ridge plain. On the north Norfolk coast a line of bluffs, formerly Pleistocene cliffs, was cut off by the development of spits, barrier islands and marshlands formed during and since the Holocene marine transgression. In Australia, enclosure of former embayments by the growth of coastal barriers has been followed by the degradation of the former sea cliffs on the enclosed coast, as behind the Gippsland Lakes region in Victoria (Bird, 1978a).

Coastal bluffs may also form where cliffs become degraded following artificial protection by sea wall construction, as at Byobuguara on the coast of Japan, or the chalk coast at Samphire Hoe, near Dover in Kent (Figure 4.12).

Cliffs are often (but not always) actively receding, their gradient depending on relative rates of cliff crest and cliff base recession. Bluffs are usually more stable, although some are retreating by way of intermittent and local slumping, and could be termed wasting bluffs. Examples are seen on the Pacific coasts of Oregon and Washington, where there are steep bluff sectors carrying scrub and forest that appear fairly stable, but are in fact receding as the result of intermittent and localised slumping. Landslides generally occur here in the wetter winter months, during or after occasional storm surges or in response to a long term (e.g. 300 year) tsunami cycle (Komar and Shih, 1993) (Section 2.5). The slump scars are quickly revegetated in this cool and moist environment as the debris fans are consumed by

the sea. Similar processes are effective on vegetated bluffs on humid tropical coasts, such as
the Ivory Coast in West Africa.

On the NE coast of Port Phillip Bay, Australia,
vegetated bluffs have become unstable when vegetation is cleared from the cliff-top area. This
is because the cliff-top vegetation canopy intercepted some rainfall, and its root systems
drew upon water percolating down through the
soil. When the cliff-top vegetation was cleared
runoff and seepage increased, causing erosion of
the bluff. Coastal bluffs can also be rejuvenated
as cliffs when marine erosion is intensified by
coastal submergence, or the loss of a protective
beach.

Steep coastal slopes that are not actively receding, but descend steeply to the shore, are seen
on the North Devon coast bordering Exmoor
(Figure 4.13), and the coast of the French and
Italian Riviera. Much of the New Zealand coast
is similarly bold (Cotton, 1974). The building of
roads along steep coasts often results in instability, both during their construction when the
slope is excavated and debris spills down, perhaps to the shore, and when the road is disrupted
by slumping or rockfalls.

In addition to cliffs and bluffs there are other
steep coasts consisting of slopes that have been
shaped primarily by subaerial processes, and are
dissected by short, steep stream valleys. Some
descend to basal cliffs and perhaps shore platforms, as on the steep coast of the Otway Ranges
in SE Australia, others to rocky or boulder-
strewn shores, as on the steep coast of Macalister Range in NE Australia. Both are relatively
straight steep slopes, possibly upfaulted along
the coastline. Off glaciated coasts the intertidal
zone is often bouldery, with coarse residual

Figure 4.13 The steep coast of North Devon near Lynmouth, where vegetated slopes descend to a rocky and
bouldery shore. The shore widens off the mouth of the River Lyn across a gravelly intertidal delta, which was
enlarged during a severe flood in 1952

material where moraines or boulder clay have been washed over by the sea, and waves have swept away the finer sediment.

Cliffs were extensive on many coasts before the Last Glacial phase of low sea level, but during that phase, in the absence of marine attack, they were degraded by subaerial denudation. As a rule, the bluffs thus formed were undercut as the sea attained its present level and so rejuvenated as receding cliffs, but it is possible to find sectors where the Pleistocene bluffs have been preserved, usually behind areas of Holocene deposition. Thus the sandstone cliffs east of Hastings in Sussex pass into grassy bluffs at Cliff End: these can be followed along the northern side of Romney Marsh, where they mark the cliffed coastline that existed prior to the formation of the Dungeness foreland, until they pass into the active Lower Greensand cliffs of the Folkestone coast. The receding vertical cliffs of the Port Campbell coast in Victoria, Australia, are interrupted at Two Mile Bay by a sector of Pleistocene bluff standing behind an emerged shore platform that has not yet been removed by the sea (Bird, 1993a). This is a small remnant of the bluffs that bordered Bass Strait during the Late Pleistocene low sea level phase, the rest having been revived as cliffs after the return of the sea to its present level (Section 3.9).

4.3.3 Slope-over-wall profiles

Some steep coasts consist of slopes that descend to basal cliffs. Such slope-over-wall profiles may develop in various ways. On soft homogeneous formations a coastal slope formed by subaerial processes (runoff, slumping, soil creep) may descend to a basal cliff kept fresh by wave attack during occasional storms. Such coasts may show alternations in profile, being degraded by subaerial weathering and runoff during occasional downpours and cut back and steepened after dispersal of basal fans by wave action when storms

accompany high tides. The removal of basal sediment produces instability, which is transmitted up the cliff face until the cliff crest begins to recede (Figure 4.5).

Slope-over-wall profiles are seen where the upper part of a cliff is a slope cut in weathered rock while the lower part is a vertical cliff in more coherent unweathered rock. On the Dorset coast between St Albans Head and Durlston Head a slope in the soft Purbeck beds descends to a cliff cut in harder Portland limestone. On the limestone cliffs of south Pembrokeshire thinly bedded strata have weathered into a convex upper slope over a vertical wall of underlying more massive limestone, as at Arnold's Slade and Mount Sion. On Flamborough Head in Yorkshire a grassy upper slope (gradient 20–30°), cut in soft glacial drift deposits, declines to a vertical cliff cut in firmer chalk, with a basal talus apron (Figure 4.14).

Alternatively, the upper slope could follow a seaward-dipping bedding plane, joint or fault plane that has been undercut and cliffed at the base by wave action, as on Porth Island near Newquay and Glebe Cliff south of Tintagel in Cornwall.

Where coastal bluffs have formed as the result of subaerial weathering and erosion on former cliffs protected by a wide beach, a coastal barrier or an artificial structure such as a harbour breakwater, the removal of the protective feature will result in such bluffs being undercut by cliffs, forming a slope-over-wall profile until cliff recession consumes the whole of the formerly degraded slope.

Slope-over-wall profiles have also formed on coasts in high latitudes where cliffs cut in relatively resistant rock were degraded by periglacial freeze–thaw and solifluction (the downslope movement of frost-shattered rubble) during cold phases of the Pleistocene when sea level was lower. The Pleistocene cliffs then became bluffs mantled by an earthy solifluction deposit with angular gravel (known as Head deposits), that

Figure 4.14 Composite cliff profile at Flamborough Head, NE England. The upper convex slope is on a capping of glacial drift, the cliff is cut in chalk and there is a grassy basal talus slope behind a beach of flint and chalk gravel

extended out on to what is now the sea floor in a broad, diminishing apron (Figure 4.15).

This process is still active on arctic and antarctic coastal cliffs (see below). During and since the Holocene marine transgression these periglaciated coastal slopes became vegetated, and were undercut by marine erosion, forming a steep or vertical basal cliff, exposing solid rock formations and producing a slope-over-wall profile. The vegetated coastal slope, mantled by frost-shattered debris, is thus a legacy of past periglacial conditions, for it cannot be explained in terms of processes now at work (Figure 4.16).

Coastal landforms of this type are found on the west and SW coasts of Britain, where vegetated slopes (typically 20–30° but locally up to 45°) descend to steeper, rugged, rocky cliffs. The coastal slope may be a bevel of almost uniform gradient (especially where it follows seaward-dipping bedding, cleavage, joint or fault planes), but more often it is convex in form, like a hog's back, and occasionally it is concave, where the lower slopes of the head deposit are preserved.

(a)

(b)

(c)

(d)

▦ Head

Figure 4.15 The evolution of a slope-over-wall coast profile where a Pleistocene cliff (a) became degraded by periglacial solifluction to form a slope mantled with head deposits (b) during a low sea level stage. When the sea rose to its present level the base of the Head apron was undercut (c), and eventually a bedrock cliff was exposed below a coastal slope of Head deposits (d)

Hog's back coasts are well developed on the Exmoor coast in north Devon, where the geological strike is parallel to the coast, forming coastal escarpments where the dip is landward (Figure 4.13).

The proportion of relict periglaciated slope to actively receding cliff depends on the degree of exposure of the coast to wave attack. On sheltered sectors of the south coast of Cornwall (as on the eastern side of the Lizard Peninsula near Coverack) the relict slope is well preserved down almost to high tide level. On the more exposed north coast of Cornwall, open to Atlantic storm waves, the relict slope has been largely, and in places completely, destroyed by Holocene marine erosion. Little if any of the coastal slope remains on the coast of Watergate Bay in north Cornwall, which has high, almost vertical receding cliffs. The periglaciated coastal slope is better preserved on the sheltered eastern sides of major headlands, as on Dodman Point (where the slope of solifluction deposits becomes concave behind Vault Beach), than on their more exposed western shores.

There were probably several alternations of periglaciation and marine cliffing on these coasts as climate varied and sea level oscillated during Pleistocene times. In places the coastal slope may include several facets, the gradient of which diminishes upward between successive breaks of slope. Savigear (1962) suggested that each facet could have been initiated as a vertical marine cliff, which became degraded to a coastal slope, the upper facets being older and therefore more degraded than those at lower level, the latest being the active (Holocene) marine cliff undercutting the composite slope. Multi-faceted cliff profiles on the coast of Cornwall (as on Beeny Cliff, near Tintagel) could be interpreted as the outcome of such a sequence, but generally the evidence for several stages of slope evolution has been destroyed by the intense Late Pleistocene periglaciation, which shaped the coastal slopes that have been subsequently undercut by the sea.

Figure 4.16   Slope-over-wall cliff coast profile at Dodman Point, Cornwall, with a steep slope formed by Pleistocene periglacial solifluction undercut by Holocene marine erosion, as shown in Figure 4.15

Periglacial disintegration of rock outcrops and the formation of Head deposits by solifluction on coastal slopes left some residual massive rock formations (mainly quartzite), which survived frost-shattering, protruding from the debris-mantled coastal slope as tors or buttresses. These are seen as tors near the top of coastal slopes, and buttresses where they jut out from the slope at lower levels. In Cornwall there are good examples in the Dodman district (Figure 4.17). There has also been slumping of the periglacial drift mantle on coastal slopes, forming hummocky vegetated topography, as at Dizzard Point on the north coast of Cornwall (Bird 1998).

On the north coast of Cornwall where islets and stacks such as those at Bedruthan Steps and Tobban Horse near Porthtowan retain segments of the Head-mantled slope it is possible to estimate the extent of cliff-base recession during the last 6000 years, since the sea rose to its present level: typically at least 100 m, and locally as much as 200 m.

Periglaciated coastal slopes mantled with Head deposits are found between 60°N and 39°N on the coasts of Scotland, western and southern Ireland, Wales, Brittany, western France and northern Spain (as at Cabo Finisterra) and south to Cap Mondego in Portugal. They are not well developed in equivalent latitudes of the Pacific coast of North America, partly because of the rarity of suitably resistant coastal rock formations, but more because the periglacial zone narrows and fades out south of the limit

Figure 4.17 A buttress of resistant quartzite protruding from the Head-mantled slope on the coast at Boswinger, near Dodman Point in Cornwall

of Pleistocene glaciation on the west coast of North America. The contrast in coastal landforms is thus the outcome of differing Late Quaternary environmental histories (Williams *et al.*, 1991). Periglaciated slope-over-wall profiles are also found on the coasts of Korea and eastern Russia, and on the shores of subantarctic islands, such as Macquarie Island (55°S) and the Auckland Islands (51°S).

4.4 Cliff weathering in cold climates

Where periglacial processes are still active on high coasts, rocky outcrops of cliffs are frost-shattered, producing basal fans of angular, unsorted debris. At Cape Lisburne in Alaska an unvegetated talus slope over 300 m high descends into the sea, and similar features are seen on Baffin Island in northern Canada, in Chilean Patagonia and around Antarctica. Frost-shattering has led to rapid slope recession on the Pacific coast of Sakhalin in eastern Russia (Zenkovich, 1967).

On the shores of the St Lawrence estuary sectors of rocky cliff and shore platform that have developed alongside narrow straits where the fetch is limited are the outcome of frost-shattering and shore ice-plucking, the waves washing away the products of cold climate rock disintegration. Similar ice-plucked features have been described from the rocky coast of Disko, in Greenland. Rocky cliffs in Greenland and Antarctica show scoured features resulting from etching by strong winds, especially where they mobilise sandy material. Rocky coasts around the Baltic Sea are weathered by freezing salt spray. On the island of Rügen on the German Baltic coast there are chalk cliffs, fronted by flint shingle and glacial boulders. Aprons of frost-shattered rubble form each winter, and are undercut and washed away by the summer sea.

Permanently frozen ground (permafrost) influences coastal evolution in northern Alaska and the arctic coasts of Canada and Russia, where the summer thawing of interstitial ice contributes to rapid degeneration of tundra cliffs, consisting of peat and morainic sediments, which slump into the sea and are swept away by wave action. This thermoabrasion on tundra cliffs can cause them to retreat tens of metres in a few weeks.

4.4.1 Ice cliffs

A special kind of cliff is found where glaciers or ice sheets have advanced beyond the edge of the land and end in a wall of ice descending to, and below, the sea. These ice cliffs are well developed in Antarctica, where the Ross Ice Shelf terminates in cliffs up to 35 m high. In summer icebergs calve from these cliffs and in winter the adjacent sea freezes, a fringe of sea ice persisting until winds and the spring thaw break it into floes. Melting in late spring is hastened by the growth of diatoms that darken the ice surface and increase its solar heat absorption. Protruding snouts of glaciers that reach the sea form floating ice tongues, somewhat richer in morainic debris than ice shelves. This debris is released and deposited on the adjacent sea floor, and some of it may subsequently be washed up on beaches. Wave action is weakened by the predominance of winds blowing offshore from the ice cap and impeded when sea ice forms.

Ice cliffs can be eroded by iceberg collisions. In April 2005 a 120 km long, 2500 sq km iceberg that had broken away from the Ross Ice Shelf collided with the Drygalski Ice Tongue, breaking off a large fringe and setting the ice cliff coastline back by up to 5 km.

Ice cliffs are also seen behind Icy Bay, Alaska, and where glaciers still occupy the heads of some fiords, notably in Greenland, northern Canada and southern Chile. A concealed rocky coastline is protected from wave attack as long as the ice fringe persists. The effectiveness of marine processes depends on the seasonal duration of shore ice, which exists for 100–150 days per year in the St Lawrence estuary, and for longer and shorter periods in lower and higher latitudes.

4.5 Cliffs and bluffs on humid tropical coasts

Coastal slopes with a vegetation cover are extensive in humid tropical regions, where they may show little if any basal cliffing. These profiles result from the intensive subaerial weathering and denudation characteristic of the humid tropics, such that many coastal rock formations have been weakened as the result of rapid and deep decomposition by chemical weathering, and do not form cliffs. Instead there are forested bluffs, steep coastal slopes with a soil and vegetation mantle as in north Queensland, where the Great Barrier Reef prevents strong wave action reaching the mainland coast. Within the Indonesian archipelago weak waves across limited fetches achieve little more than the removal of material that moves down the slope and would otherwise accumulate on the shore. Vegetation extends down to a trim line marking the upper limit of wave swash (Figure 4.18).

Cliffed coasts are thus comparatively rare on humid tropical coasts in SE Asia and India, on the east and west coasts of tropical Africa, in Brazil and on the Pacific coast of Central America. Cliffing occurs where wave action is relatively strong, on exposed promontories, and is more extensive on coasts receiving ocean swell, especially where the coastal rock formations are

Figure 4.18 Tropical bluffs at Etty Bay, near Innisfail on the NE coast of Australia, showing rain forest descending almost to the high tide line

weak, as in Paraiba, NE Brazil, where 21 per cent of the coastline is retreating cliffs cut in sandy clays (Guilcher, 1985). Yampi Sound in northern Australia is bordered by low crumbling cliffs of deeply weathered metamorphic rock, from which protrude bolder promontories of quartzite, a type of resistant rock that has been little modified by chemical weathering. Marine erosion therefore works upon coastal rock formations, the resistance of which is related to prior weathering under humid tropical conditions rather than their original lithology. An example is the persistence of a dolerite headland at Mamba Point in Liberia, where adjacent outcrops of thoroughly weathered granite and gneiss have formed vegetated bluffs. Coastlines with humid tropical bluffs are generally fairly stable, but may be receding where there are recurrent landslides, which become rapidly revegetated.

4.6 Cliff dissection

As cliffs recede, weathering and erosion penetrate zones of weakness such as faults, joints or outcrops of less resistant rock, cutting clefts and crevices that may develop into caves and blowholes or deep, narrow inlets, with stacks the outcome of irregular cliff retreat. Cliffed coasts showing an array of such features are found on Carboniferous limestone in south Pembrokeshire and on Old Red Sandstone on the island of Jura in Scotland. Caves, archways and stacks have been formed by the dissection of chalk cliffs in Freshwater Bay, in the SW of the Isle of Wight, and the Port Campbell limestone (a calcareous siltstone) coast in Australia has spectacular examples of these dissection features.

The low cliffs cut in peaty clay at Lang Lang on the NE shore of Westernport Bay, Australia, have a crenulate outline that may have originated when numerous bays were cut by waterfalls draining from swamp woodland on a coastal plain where floodwater runoff was not confined to stream channels.

An unusual form of erosion occurred at Fairhaven, Washington State, late in the 19th century, when hydraulic sluicing was used to wash sand and gravel from Post Hill Point, a bluff of glacial drift, to provide material to fill and reclaim Commercial Point, an adjacent bay. The bluff is now stable and vegetated behind the reclaimed area of shipbuilding yards, a ferry terminal and factories.

4.6.1 Caves

Caves form where marine erosion has penetrated zones of weakness, and are common in rock formations that have numerous joints or faults, or segments of weaker rock. Cathedral Caves on the Catlin Coast of South Island, New Zealand, are tall caves that have been excavated between vertical joints in sandstone beneath a roof of harder rock. Fingal's Cave on the Scottish island of Staffa is 20 m high and 70 m long, cut out along the vertical planes that border columns of Tertiary basalt. On limestone coasts the sea may penetrate and widen fissures that originated as subterranean solution caverns prior to the Holocene marine transgression. The caves of Bonifacio in southern Corsica are believed to have formed in this way. On the Port Campbell coast in Victoria, subterranean caves excavated along vertical joints in Miocene calcareous siltstone have been etched out by marine erosion to form deep inlets in the cliffed coast, with large caves at their heads, as at Lochard Gorge (Bird, 1993a). Caves are less likely to develop in weak rock formations, where any penetration by the sea is likely to be followed quickly by collapse of the undercut rock.

Unusual small caves have been cut in low cliffs of periglacial drift (head deposits) at Porth Hellick on St Mary's, Isles of Scilly, where breaking waves have rotated granite boulders to form a horizontal pothole.

4.6.2 Blowholes

Where the hydraulic action of incoming waves and the compression of trapped air within a cave puncture the roof a blowhole develops, and water and spray may be driven up through it as fountains of spray. Blowholes formed in this way are seen on Porth Island, near Newquay in Cornwall and on the Old Red Sandstone coast near Arbroath in eastern Scotland. Australian examples occur at Kiama in New South Wales at the head of an elongated cave cut out along joints in basalt, and at Quobba in Western Australia. The forno on the Sagres Peninsula near Cape St Vincent, southern Portugal, are blowholes rising from caves where air and spray are expressed noisily, and on Downpatrick Head in Mayo, NW Ireland, blowholes formed in layered Carboniferous sandstone are known locally as puffing holes.

On the coast of Cornwall some blowholes have grown into large circular cavities linked to the sea through archways, as at the Devil's Frying Pan near Cadgwith and the Lion's Den on the Lizard Peninsula (Figure 4.19).

Sinkholes are similar to blowholes, but originate as solution hollows in limestone. Rounded cauldrons or coves are formed where receding cliffs intersect them, as at The Grotto on the Port Campbell coast in Victoria, Australia.

4.6.3 Coastal gorges

Coastal gorges are spectacular steep sided clefts, chasms or inlets formed at an angle to the shore in coastal rock formations. They are found

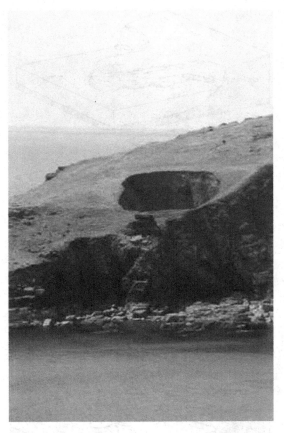

Figure 4.19 The Lions Den, a round hole formed by the collapse of a cave roof on the coast near Lizard Point, Cornwall

where the roofs of caves have collapsed or where rock has been excavated by wave action along or between lines of weakness such as joints, faults, cleavage planes, steeply dipping bedding planes or intruded dykes. Examples can be found on many cliffed coasts. On the Isle of Skye in Scotland some intruded dykes of Tertiary basalt have been weakened by fracturing or weathering, and have been excavated as clefts. Such features are known as geos or yawns on the NE coast of Scotland, where Moista Sound is a chasm 120 m high, 180 m long and only a few metres wide, cut out by waves along a weathered dyke exposed in cliffs of Old Red Sandstone. Wife Geo, near Skirza Head,

is a steep sided inlet 250 m long with an adjacent tunnel. Similar gorges are called zawns on the granitic Land's End peninsula in Cornwall, where some have been at least partly shaped by tin mining. Elongated inlets have been excavated between joint planes on the rocky coast of Le Puits d'Enfer in western France and in granite–gneiss on the south coast of Western Australia, where the Gap, near Albany, is a 24 m deep chasm cut out between parallel joints. Deep gorges of this kind are incised between vertical joints in the bold sandstone cliffs of the Jervis Bay district in New South Wales, and the spectacular calcareous siltstone cliffs of Port Campbell in Victoria. On the South Pembrokeshire coast flat-topped cliffs cut into Carboniferous limestone have been dissected along joints and fault planes to form narrow, steep sided rocky inlets such as Huntsman's Leap (Figure 4.20). A distinction is made between gorges cut out entirely by marine erosion and stream-cut valleys that can also form steep sided inlets where their mouths have been submerged by the sea.

4.6.4 Natural arches

Where powerful wave action has excavated caves along joints and bedding planes on a cliffed coast, some may extend through headlands to form a natural arch (Figure 4.21).

An example is the Green Bridge of Wales on the Pembrokeshire coast, an arch of massive Carboniferous limestone beneath which thinly bedded strata have been cut out along joint planes (Figure 4.22).

The Porte d'Aval is a slender natural arch on the cliffs of hard chalk at Etretat on the Normandy coast of France. Several natural arches have been called London Bridge, notably at Torquay in Devon and Portsea on the Victorian coast in Australia. On the Port Campbell cliffs in Australia a natural arch was lost when a storm removed the trunk from Elephant Rock, and

Figure 4.20 Huntsmans Leap, a cleft cut out between joint planes in Carboniferous Limestone on the cliffed coast of Pembrokeshire in SW Wales

another when the inner arch of London Bridge collapsed in 1990 (Figure 4.23).

Tunnels are elongated natural arches. On the north coast of Cornwall Merlin's Cave is a tunnel 100 m long excavated by marine erosion along a thrust plane between slates and volcanic rocks on Tintagel Island. On limestone coasts some long tunnels may have been excavated by underground streams before they were invaded by the sea.

4.6.5 Stacks

Dissection of promontories on a cliffed coast may also isolate stacks (residual islets), formed

Figure 4.21 The dissection of a rocky promontory by caves to form natural arches, which collapse leaving stacks that are gradually reduced by erosion

either where a natural arch has collapsed, or where a transverse inlet has been cut along a zone of weakness through the headland. They are gradually reduced by marine erosion, and may

Figure 4.22 The Green Bridge of Wales, a natural arch cut in Carboniferous Limestone on the Pembrokeshire coast in SW Wales

be undercut until they collapse, leaving only a basement platform. Stacks occur off many cliffed coasts: the Twelve Apostles near Port Campbell, Victoria, Australia, are well known examples (Figure 4.24). Off the Oregon coast many stacks are residual hard rock elements isolated from coastal rock formations as the softer matrix was removed in the course of cliff recession (Figure 4.25).

4.6.6 Coastal valleys

Cliffs are interrupted where valleys come down to the coast. Some end in estuaries where the sea has risen to invade them (Chapter 11), or alluvial plains and deltas formed by sedimentation (Chapter 12). Others end in hanging valleys truncated by cliff recession, their floors well above present sea level, with rivers pouring out as waterfalls cascading down to the shore,

as at Ecclesbourne Glen, on the Sussex coast. On the NW Devon coast there are opportunities to study the relationships between cliff recession and valley incision, which has produced a series of coastal waterfalls, as at Speke's Mill Mouth. At Tintagel in Cornwall, Trevena Brook pours out over the cliff from a valley truncated by cliff recession, and on the Dorset coast there are several waterfalls, including one at St Gabriel's Mouth near Golden Cap. Numerous waterfalls pour down the steep slopes of the tectonically uplifted coast south of Jackson Bay in Fiordland, in the South Island of New Zealand.

Deep and narrow coastal ravines known as chines on the SW coast of the Isle of Wight and in Bournemouth Bay have been cut by small streams in valleys truncated by rapid cliff recession (Figure 4.26). There are waterfalls along the chines and down the cliff face. Incised notches in cliffs are usually

Figure 4.23   The inner natural arch at London Bridge, near Port Campbell in SE Australia (above), a month before it collapsed on 15 January 1990, leaving an outer arch (below)

Figure 4.24 The Twelve Apostles are stacks in calcareous siltstone off the cliffed coast of Port Campbell in Victoria, Australia. Photograph by Jenifer Bird

Figure 4.25 Stacks on the Oregon coast at Pistol River. These are residuals of more resistant rock that persisted when the coast was cut back by marine erosion

Figure 4.26 Whale Chine, on the south coast of the Isle of Wight, formed where a valley cut into soft Cretaceous sands and clays has been truncated by cliff recession

truncated valley mouths, but one near Milford-on-Sea on the Hampshire coast is a beheaded valley.

Accordant valley mouths, with unaggraded stream channels opening to the shore, are rare, occurring only where cliff recession has matched the rate of stream incision or where coastal uplift has prevented valley-mouth submergence during the Holocene marine transgression. Gullies cut into recently deposited volcanic ash and agglomerate (solidified volcanic gravel) on the shores of Anak Krakatau, an Indonesian volcano (Figure 4.27), are also accordant, downcutting having proceeded at a similar rate to cliff recession in these unconsolidated sediments (Bird and Rosengren, 1984).

Dry valleys (cut by rivers when the water table was higher or the ground frozen and impermeable) have been truncated by the recession of chalk cliffs on the Sussex coast to produce the undulating crest of the Seven Sisters along an almost straight coastline (Figure 1.2). The undulating cliff crest crosses successive valleys lined with deposits of frost-shattered chalk (Coombe Rock) produced by Pleistocene periglacial action. These dry hanging valleys, truncated by cliff recession, decline seaward, and in places intersect the shore platform, producing either a re-entrant or a sector where the chalk passes beneath a sand and gravel fill, as at Birling Gap, near Beachy Head (Figure 1.2).

Steep-sided incised valleys cut by glacial melt water, notably water overflowing from ice-dammed lakes, have been identified on the coast of Cardigan Bay in Wales, where the Irish Sea ice sheet abutted the coast in late Pleistocene times and enclosed lakes in the valleys of the Teifi and Nevern Rivers. These lakes rose in level and overflowed, cutting valleys across coastal promontories, as on Cemmaes Head and Dinas Head. Similar valleys have been truncated by cliff recession on the Atlantic coast of North Devon, which

Figure 4.27   Gullied slopes of volcanic ash below the rim (R) of the crater of Anak Krakatau, Indonesia, have been truncated by cliffs (C) cut by marine erosion. A more resistant lava flow (L) protrudes into the sea on the left. Photograph by Neville Rosengren

was also close to the ice margin at one stage in Pleistocene times, but it is uncertain whether these were cut by water overflowing from ice-dammed lakes.

4.6.7 *Hinterland topography*

Cliffs are also influenced by the geomorphology of the immediate hinterland, in particular the topography intersected as the cliff recedes. Flat topped cliffs have been cut into a coastal plateau, as on the Atlantic coast of Cornwall, particularly south of Portreath, where the even crested cliffs of Reskajeage Downs rise about 100 m above sea level. Cliffs cut into high ground recede more slowly than cliffs eroding into low lying topography on similar rock formations, so that interfluves tend to become promontories between valley-mouth embayments. This is illustrated on the Yorkshire coast north of Flamborough Head, where cliffs have been cut into a landscape of plateaux and incised valleys on Jurassic formations, forming headlands and bays. Similar features are seen on the northern Oregon coast, and on the Australian coast near San Remo in Victoria a crenulate outline of ridge headlands and valley bays has developed on the margins of dissected hilly country on Cretaceous sandstones. Where the hinterland slopes away from the coast cliffs diminish in altitude as they recede, as on Beachy Head in Sussex and in the vicinity of Childers Cove near Warrnambool in Australia.

4.7 Outlines in plan of cliffed coasts

The influence of lithology and structure on cliff profiles is also expressed in coastwise variations.

The Pembrokeshire coast between St David's Head and Strumble Head has a succession of headlands on igneous intrusions and intervening bays and coves cut out in relatively weak Ordovician sedimentary outcrops. Old Castle Head in south Pembrokeshire protrudes because hard rocks outcrop at the cliff base, and is bordered by bays cut out in softer formations. Coastlines have also been influenced by the topography of the hinterland, so that ridges tend to end in headlands and valley in bays.

A distinction may be made between discordant cliffs, cut across the strike of several geological formations, as on the Lyme Bay coast in Dorset, and concordant cliffs, that run parallel to the strike along a single formation, such as the rugged cliffs on Whin Sill dolerite on the Northumbrian coast. Lateral transitions in lithology within an outcropping formation can result in morphological variations. Point Addis, on the coast of Victoria, Australia, is a bold cliffy headland with a structural shore platform at its base, both cut in hard, stratified Oligocene limestone, but this formation grades laterally eastward into softer calcareous clay and marl at Jan Juc. As it does so the cliffs become less bold and subject to recurrent slumping and the shore platform disappears (Bird, 1993a).

Where a relatively resistant coastal rock formation is backed by parallel weaker outcrops, penetration of the outer wall by marine erosion is followed by the excavation of coves and embayments in the softer rock. In such situations there has often been prior incision of valleys parallel to the coastline, cut in the softer outcrops behind a coastal ridge on harder rock. A well known example of this is seen on the Dorset coast east of Weymouth, where several stages of dissection can be seen. Stair Hole, near Lulworth, is an early stage, only a narrow breach in the outer wall of steeply dipping Jurassic limestone backed by a hollow with slumping Weald clay. Close by, Lulworth Cove (Figure 4.28) has a wider entrance through the limestone wall, and an almost circular bay has been carved out between converging valleys cut in soft Cretaceous sands and clays, backed by a high ridge of chalk.

East of Lulworth Cove a much broader embayment has developed along the clay lowland that fronts the chalk ridge at Worbarrow Bay, while to the west the Jurassic and Wealden rocks have been removed almost completely, leaving only the small residual peninsula on the harder Jurassic limestone at Durdle Door. This coastline is the outcome of partial Holocene marine submergence of limestone ridges and clay vales produced by subaerial denudation of the Jurassic formations, followed by marine cliffing.

The outline in plan of a cliffed coast tends to become simplified and smoothed with the passage of time, except where there are marked contrasts in the structure and lithology of coastal rock formations, which may perpetuate irregular outlines as the cliff recedes. Where the recession of cliffs cut in a relatively weak formation such as glacial drift or dune calcarenite uncovers outcrops of harder basement rock at or near sea level, the latter emerge as headlands while the softer rocks are cut back as embayments. Such coasts, termed contraposed coasts (Clapp, 1913), are exemplified by the west coast of Eyre Peninsula in South Australia, where receding cliffs cut in Pleistocene dune calcarenite extend behind and between exhumed promontories of harder Pre-Cambrian rock. In NW Brittany granite has been exposed to form headlands as the Pleistocene periglacial deposits (known as head deposits) that formerly concealed them were cut back by the sea, leaving intervening embayments backed by cliffs and bluffs cut into head deposits. Similar features are seen on coasts where glacial drift deposits rest on solid rock formations that have been exhumed to form rocky headlands that protrude seaward between bays backed by cliffs or bluffs in the drift

Figure 4.28 Lulworth Cove was formed behind a gap cut through an outer rampart of Portland Limestone, and excavated in softer parallel outcrops of Purbeck Beds, Weald Clay, Lower Greensand and Gault with a backing Chalk escarpment. The cove is not simply the product of marine erosion, for subaerial denudation had already excavated converging valleys in the softer formations, forming a lowland that was then invaded by the sea. The curvature of the shingle beach and cliff base backing Lulworth Cove has been shaped by the refraction of waves coming in through the entrance. Published by courtesy of Aerofilms Limited

deposits, as on the Atlantic coast of the Outer Hebrides.

Where coastal outcrops are comparatively uniform in relatively soft rock, a receding cliffed coast develops an outline in plan that is related to the prevailing wave patterns. An example is the cliffed coast of the Great Australian Bight east of Eucla, which appears relatively straight on small scale maps, although some sectors present a serrated outline (Figure 4.29). These cliffs are difficult to access except at widely separated lookouts, but their outlines in plan can be seen on Google Earth (Internet), which shows that there have been numerous landslides, leaving under-

cliffs of tumbled rock in front of the vertical cliffs.

Where the waves moving towards the coast are refracted by offshore topography or adjacent headlands cliffed coasts develop gently curved outlines in plan, much like those on depositional coasts. A curved cliffy coast may pass smoothly into a curved depositional coast, as in the asymmetrical Waratah Bay on the SE coast of Australia (Figure 6.18) and in Drakes Bay, California, where the curved outline of cliffs cut in soft Pliocene sedimentary rocks is continued by the curvature of a sandy beach (Figure 11.8). Similar curved outlines have developed

Figure 4.29 Cliffs bordering the southern edge of the Nullarbor Plain, a remarkable plateau that is essentially the depositional surface at the top of the Miocene Nullarbor Limestone. This sedimentary plain has been uplifted and tilted gently southward and eastward, and in a semi-arid to arid climate it has been preserved without any dissecting river valleys, although there are underground caves. The cliffed coastline along its southern margin does not coincide with a fault line, and Jennings (1963) suggested that the cliffs were the outcome of evenly distributed marine erosion of a coast of uniform height and uniform rock resistance. The origin of the serrated outline has not been explained

on the Tertiary cliffs of Bournemouth Bay in southern England, shaped by refracted waves approaching from the SW.

4.8 Coastal landslides

Coastal landslides are movements of large masses of rock, earth or debris down a coastal slope: they are also called landslips, particularly in Britain (Lyell, 1833). Mass movements occur on weak and weathered rock outcrops and in unconsolidated sediment on steep and cliffed coasts. Instability develops when an increase in

shear stress (e.g. by greater loading) or a decrease in shear strength (i.e. weakening of the rock formation) reaches a threshold level and is relieved by movement downslope.

Such downslope movements can take place in various ways: falls, slides and flows (Brunsden and Prior, 1984). These include the following.

(a) Rockfalls and toppling, where masses of rock collapse from the cliff face (Figure 4.30).

(b) Translational slides, where rock masses slip down a seaward dipping plane.

Figure 4.30 Subsiding and toppling of columns of rock eroded from the cliff of soft calcareous siltstone of Miocene age near Port Campbell in SE Australia

(c) Rotational slides, which collapse seaward down a curved plane to form back-tilted rock masses.

(d) Mudslides, where masses of coherent silty or sandy clay move irregularly down across a sheared surface.

(e) Mudflows, where highly lubricated fine grained sediment moves downslope as a slurry.

Mudslides and mudflows form lobate tongues that spread out across the shore and into the sea. They occur especially in the winter or the rainy season, when they may be accompanied by surface runoff, including streams that cut gullies and deposit basal sediment fans. An example of a coast disrupted by landslides is seen on the coast of St Bride's Bay in Pembrokeshire, where an outcrop of Carboniferous sandstones and clays has subsided in broken and tilted blocks separated by chasms, as at Settling Nose and Haraldston Chins.

Groundwater seepage from permeable formations has also contributed to instability, coastal landslides being most active during wet periods when the rock formations become saturated (shear stress increased by loading), or when frozen rock thaws and becomes incoherent after a cold spell (shear strength diminished). Saturation loading of weak or porous coastal formations with groundwater has led to cliff collapse in the head deposits on the Cornish coast at Downderry and Gunwalloe.

By contrast, desiccation can lead to flaking and falling of particles from cliffs. In the dry summer of 1976 basal fans of collapsed clay particles formed beneath cliffs cut in Jurassic at Kimmeridge on the Dorset coast. Slumping may also occur as a consequence of changes in volume as ground temperatures rise and fall, and can be initiated by the widening of crevices in the cliff face as the result of penetration by the roots of cliff-top vegetation and the pressure exerted as such roots grow larger. Burrowing animals, notably rabbits, can contribute to slumping and erosion of cliffs and bluffs cut in soft rocks.

Near Newport, Oregon, the loading of a cliff by buildings has been followed by landslides that have damaged or destroyed many houses and led to the demolition of a condominium (Viles and Spencer, 1995). The weight of cliff-top buildings was partly responsible for the collapse of Holbeck Hall Hotel on a steep coastal slope at Scarborough in Yorkshire in 1993.

Coastal landslides are common where geological formations dip seaward. On the coast of Lyme Bay in Dorset several large coastal landslides, backed by receding cliffs, have occurred where masses of gravel-capped Upper Greensand, dipping seaward, have subsided over slumping Jurassic clays at Black Ven and Cains Folly (Figure 4.4). The town of Lyme Regis is threatened by the recession of upper cliffs behind enlarging coastal landslides. Measurements on a mudslide in Worbarrow Bay, Dorset, using erosion pins and electronic recording, showed various types of movement, some in frequent brief (slip–stop) episodes, some gradual, some in sudden surges (e.g. 3 m in 20 minutes) and some apparently random (Allison and Brunsden, 1990). This is one of several places on the south coast of England where pebbles from capping gravel deposits move gradually down through the landslide and are eventually deposited on the beach. Further east the profiles of Gad Cliff, Hounstout Cliff and St Alban's Head show massive cappings of Portland Limestone that have disintegrated, toppled and subsided over Kimmeridge Clay to form hummocky topography, undercut by basal cliffs.

A seaward dip in the coastal rock formations facilitates landslides and rockfalls, the undercut rock sliding down bedding planes into the sea, leaving the exhumed bedding planes as a coastal slope. On the south coast of England coastal landslides are common where the permeable Chalk and Upper Greensand formations dip seaward, resting on impermeable Gault. Water seeping through the permeable rocks moves down the inclined clay surface, and if marine erosion has exposed the junction at or above sea level, lubrication of the interface leads to slipping of the overlying rocks. A spectacular landslide occurred in this situation on the Devon coast near Axmouth in 1839. Similar landslides have occurred on the south coast of the Isle of Wight, and between Folkestone and Dover in Kent, where a railway built along the undercliff

is damaged from time to time by falling rock. The physical effects of wetting and lubrication of the clay surface are accompanied by chemical processes, for when the seeping water, rich in dissolved calcium carbonate, reaches the glauconitic gault, base exchange occurs, calcium ions displacing potassium ions so that alkalinity increases and the clay is deflocculated. The upper layers of the clay become a soft wet slurry that flows out at the base of the cliff and hastens the undermining of the chalk and Upper Greensand.

On the south coast of the Isle of Wight masses of Chalk and Upper Greensand have broken away from an upper escarpment and subsided to form the Ventnor Undercliff between St Catherine's Point and Luccombe. To the west successive landslides have led to rapid recession of a cliff cut in Lower Greensand, truncating and largely consuming a valley at Blackgang Chine, behind a slumping undercliff terrace on soft clays and sands (Figure 4.31). On the SW coast of the Isle of Wight there is a series of broad amphitheatres cut in the Wealden Beds, backed by cliffs that overlook hummocky subsided terraces (Hutchinson et al., 1991).

On the Oregon coast there have been major landslides at Humbug Mountain, forming lobate slumped promontories that temporarily prograded the coastline, before being consumed by marine erosion. There was massive slumping on the bluffs at Anchorage, Alaska, as a result of the 1964 earthquake. Cones of slumped talus that extended up to 600 m into the sea have been cliffed and cut back subsequently by wave action. The high cliffs on the coast of Hawke Bay north of Napier in New Zealand were shaken by earthquakes in 1931 and 1990, and have a vegetated talus apron that is being undercut by the sea.

West of Moonlight Head in Victoria, Australia, a series of amphitheatres backed by cliffs in Tertiary sandstone is fronted by tumbled rocks, where landslides have occurred on the underlying Palaeocene clay. Similar such lobes at Fairlight in Sussex form natural breakwaters

Figure 4.31 Undercliff terrace formed by landslides at Blackgang, Isle of Wight. The cliff on the right, cut in Lower Cretaceous sands and clays, has receded as the result of recurrent landslides

that intercept eastward drifting shingle, and this has also occurred on the shores of Lyme Bay, where shingle has accreted updrift of landslide lobes instead of moving eastward to Chesil Beach (Figure 6.24). Below the Lyme Bay cliffs are several arcuate festoons of boulders, which commemorate the seaward limits of former lobes from which finer sediment has been dispersed by wave action (Figure 4.32).

Attempts to stabilise landslide-prone cliffs include artificial slope grading, the insertion of drainage systems and the planting of vegetation, the most satisfactory results being where a combination of these procedures is used.

4.8.1 Cliff falls

On a smaller scale than most landslides are cliff falls (rockfalls), where the cliff face collapses, usually after basal wave erosion has undermined

it. Many cliff falls occur during unusually wet weather (rock formation overloaded by saturation) or unusually cold weather (especially when stresses develop as the result of freezing and thawing of water contained in the rock formation). Cliff falls often occur at night or in winter, perhaps because of stresses caused by shrinkage as the temperature falls. A major cliff fall occurred on Beachy Head, Sussex, in 1999, leaving a white scar on the otherwise grey chalk cliff and producing a fan that ran out almost to the lighthouse (Williams *et al.*, 2004). It was previously possible to walk out to the lighthouse only at very low tides. On volcanic coasts sudden cliff recession may occur, as on the coast of Hawaii in December 2005, when an area of 17.6 ha collapsed into the sea during an eruption, leaving a fresh cliff 18 m high.

At the eastern and western ends of the Isle of Wight cliffs cut in almost vertical chalk strata show paler segments of steeply inclined bedding

Figure 4.32 A loop of boulders below the cliff at Golden Cap, Dorset, marks the limits of a former landslide lobe that has been cut back by marine erosion

planes, where masses of grey weathered chalk have fallen away. Some cliff falls are initiated as breakaways at the cliff crest, but others result from a collapse in the cliff face, sometimes preceded by the formation and widening of vertical cracks (Figure 4.7).

The use of explosives to blast cliff faces, notably where it is deemed necessary to remove dangerous overhanging rock, often increases long term instability by shattering and weakening the rock formation. Limestone cliffs near Llantwit Major, in south Wales, remained unstable after they were blasted in 1969, and it is possible that the collapse of a cave at Gracelands in SW Western Australia was related to previous use of explosives on the cliff. Rockfalls have occurred from time to time on sectors of the Port Campbell siltstone cliffs in Victoria, Australia. Disturbances by seismic testing may have increased the instability of cliffs on this coast.

4.9 Rates of cliff recession

Cliffs recede as the result of basal erosion, slumping, landslides and rockfalls, and the removal of collapsed debris from the cliff base. Cliff recession may accelerate on coasts where sea level is rising, the deepening of nearshore water allowing larger waves to attack the cliffs, but the rate of cliff retreat also depends on rock hardness and the availability of sediment that can be used in abrasion. Cliff recession diminishes on emerging coasts, such as those of the Gulf of Bothnia, where wave erosion has been withdrawing seaward across an emerging sea floor.

Measurements of cliff recession are most readily made at the cliff crest, but it is also possible to measure basal cliff retreat or the recession of the whole profile, the changing gradient depending on the relative rates of retreat of the cliff base and at the cliff crest. The rate at which a cliff retreats depends on the resistance and structure of the outcropping formations, the energy of incident waves, tide range, the presence of a rocky shore or shore platform that affords some protection to the cliff base and the frequency and consistency of basal wave attack. Wave energy is concentrated in a narrow vertical zone on microtidal coasts and dispersed where the tide range increases. On the megatidal shores of the Bay of Fundy there are sloping cliffs in soft glacial drift, cut during the occasional brief episodes when high spring tides bring wave action to the cliff base.

Cliff recession is generally episodic, during phases of strong wave attack. It varies along the coastline with the resistance of coastal rock formations, aspect in relation to wave regimes and the nearshore profile and the incidence of rockfalls and landslides or cliff falls (when the crest of the cliff may retreat, but the cliff base may advance, at least temporarily, by the accumulation of a lobe of slumped material). Some changes are instantaneous; others take place over timescales ranging up to centuries. Most existing cliffs have been cut back within the past 6000 years, when the sea has stood close to its present level. On some coasts it is possible to measure the extent of this recession with reference to remains of the preceding subaerially weathered land surface (oxidised rock) intersecting the outer edge of the shore platform. Gill (1973) used this method to show that the cliffs on the Otway coast in SE Australia had receded 105 m on mudstones and 53 m on sandstones over this 6000 year period. It may be possible to reconstruct the profile before a cliff was cut, as on the flanks of a coastal or island volcano, or on a slope-over-wall coast, and so estimate the extent of recession.

Rates of cliff recession can be measured by repeated profile surveys, by photogrammetric analysis (see, e.g., Jones, Cameron and Fisher, 1993), or by measurements on inserted pegs or micro-erosion meters, which can be linked

Table 4.1 Average rates of cliff recession

- 1 mm/yr on cliffs cut in granite
- 1 mm–1 cm/yr on limestone
- 1 cm/yr on shale
- 10 cm–1 m/yr on Chalk and Tertiary sedimentary rocks
- 1–10 m/yr on glacial drift deposits
- at least 10 m/yr on volcanic ash

Source: Sunamura (1992).

to computers. Micro-erosion meters have been used to measure changes on cliffs and shore platforms in Yorkshire (Robinson, 1976) and on the Otway coast in Australia (Gill and Lang, 1983). Cliff recession is usually expressed as averages in metres per year, but the actual retreat of the cliff crest is episodic and localised as each rockfall occurs. Measurements of cliff recession may be linear, areal or volumetric, and are usually expressed as annual averages (Table 4.1).

Cliff-crest recession in glacial drift on sectors of the Polish coast has averaged a metre a year, with up to 5 m in a stormy year. Recession of cliffs in similar glacial drift of up to 100 m/yr has been measured on Wasque Point, a stormy headland on the south coast of Martha's Vineyard Island in Massachusetts (Kaye, 1973). May and Heeps (1985) found rates of cliff-top recession ranging up to more than 1 m/yr at various sites on Chalk coasts, and an areal loss of 3264 m² over 12 years on Chalk cliffs in Sussex, 97 per cent of which was removed in winter periods, and 42 per cent in two particularly cold, wet and stormy winters. On the Yorkshire coast Robinson (1977) observed quarrying by waves at the base of Liassic shale cliffs unprotected by a beach, and abrasion by waves armed with rock particles formed a basal notch by recession at up to 4.7 cm/yr. Cliff erosion rates were 15–18.5 times higher where there was backshore

Figure 4.33 Slumping clay cliffs at Covehithe in Suffolk. The rate of recession on cliffs of this kind is related to cliff height. During the 1953 North Sea storm surge Williams (1956) recorded 12 m of recession on cliffs 12 m high and 27 m of recession on cliffs 3 m high on this sector of the Suffolk coast

beach material, moving to and fro on a cliff-base abrasion ramp sloping seaward at more than 2.5°. It is necessary to take account of variations in cliff height and calculate the volume of rock removed as a cliff is cut back (Figure 4.33).

Because of the intermittency of such slumping and lateral variations in its effects, measurements of average annual rates of cliff retreat can be misleading. Along the Port Campbell cliffs in Australia the mean rate of cliff recession is only a few centimetres per year, achieved by a series of occasional localised rockfalls, when segments up to 200 m long and 12 m wide have suddenly collapsed into the sea (Baker, 1943). One such collapse in 1939 near Sentinel Rock left a freshly exposed scar that is still discernible as a less weathered cliff sector 60 years later, its base aproned by fallen debris being slowly consumed by marine erosion. Evidence of such slumping

is present on about five per cent of the length of the Port Campbell cliffs, suggesting a recurrence interval of many decades for such major events.

The cliff crest and cliff base at Scarborough Bluffs on the north shore of Lake Ontario have retreated at an average rate of up to 0.5 m/yr, mainly during episodes of storm wave attack in high lake level phases and as the result of recurrent slumping (Carter and Guy, 1988). Variations in slope profile on a retreating cliff are shown in Figure 4.34. Slope-base recession on Calvert Cliffs, Maryland, by wave undercutting is more than 1 m/yr, and by freeze–thaw alternations about 0.5 m/yr (Wilcock, Miller and Shea, 1998). Cliff recession slackens and cliff profiles may become degraded (i.e. slope angle diminished) as shore platforms widen and become protective. On the Dorset coast vertical

BLUFF PROFILE vs TIME

Figure 4.34 Surveyed profiles of North Shore Bluffs, Point Bruce, on the coast of Lake Erie, Ontario, Canada, show variations in the average rate of retreat of the cliff crest and the cliff base (after Quigley and De Nardo, 1980)

cliffs in limestone and shale pass laterally into sloping cliffs where they are fronted by shore platforms.

Sea walls have been built to halt cliff recession, particularly at seaside resorts, some of which now stand forward from the coastline because of continuing cliff recession on the adjacent unprotected coastline. Often the cliff beyond the end of a sea wall is cut back more rapidly as deepening nearshore water allows larger waves to break along its base. Examples of this are seen on glacial drift coasts at Withernsea and Mundesley in eastern England and Ustronie Morskie on the Polish coast. Near Barmouth in North Wales a boulder-armoured caravan park protrudes from an otherwise cut-back coast.

Accumulation of large quantities of fallen rock or deposited beach material on the backshore serves to protect the base of a cliff from wave attack because storm wave energy is expended on the beach. Smaller quantities of beach material that can be mobilised and hurled on the cliff base during storms accentuate abrasion by waves. Herein lies the risk of remov-

ing beach material from the shore. Dredging of shingle from the nearshore area at Hallsands in South Devon for construction work at Plymouth dockyard during the 1890s led to accelerated beach and cliff erosion and storm wave destruction of the fishing village of Hallsands, that used to stand on a coastal terrace near the base of the cliffs. On the Holderness coast in eastern England, Pringle (1981) correlated phases of accelerated cliff erosion and steepening with the passage of low sectors of beach (ords) in the southward-drifting beach system. On the chalk cliffs of Normandy in France a wide flint shingle beach is protective, but where it narrows the gravel is thrown against the cliff base, undermining it and accelerating cliff retreat (Costa, Hénaff and Lageat, 2006).

It has been suggested that some retreating cliffs maintain their general form (with cyclic variations) over timescales of about a century, showing a dynamic equilibrium between the landforms and the processes causing their recession (Cambers, 1976; Brunsden and Jones 1980). Recession of vertical cliffs by intermittent rockfalls and basal rejuvenation provides a similar sequence.

4.10 Summary

Cliffs are influenced by geology (lithology and structure), and their profiles and rates of recession are determined by rock resistance and exposure to weathering and erosion. Cliffs are dissected along zones of weakness, particularly bedding, joint and fault planes. Bluffs are cliffs reduced by subaerial erosion to coastal slopes that are generally well vegetated; they form where basal marine erosion is halted by emergence or accretion of a protective foreground. Slope-over-wall profiles may be structurally controlled, but have also formed where cliffs were degraded by periglacial processes,

forming a rubble mantle that has been undercut by later marine erosion. Such features form on cold coasts, subject to freeze–thaw processes, with ice cliffs locally. By contrast, humid tropical coasts have bluffs in deeply weathered rock, usually with a forest cover.

Cliff dissection results in the formation of caves, blowholes, gorges, natural arches, stacks and truncated valleys. Cliffs may be cut back into plateaus or hilly hinterlands. Coastal landslides and cliff falls occur on soft rock, or where the dip is seaward, and result in sudden recession of the cliff crest. Rates of cliff retreat depend on several factors, including lithology and structure, exposure to strong wave action, presence of a protective beach and sea level changes. These rates can be measured with reference to historical maps and remote sensing, and by repeated surveys.

5

Shore platforms

5.1 Shore processes

Most cliffs are fronted by rocky shore topography, including various kinds of shore platform. The processes at work on rocky shores are generally destructive (erosional), but some serve to protect shore rock outcrops. Some erosional processes lead towards shore planation (the formation of shore platforms) while others dissect shore topography.

5.1.1 Shore abrasion

Waves can achieve abrasion on soft rocks such as clay and shale by hydrodynamic action and sluicing, but they are far more effective when they move rock fragments (sand, shingle, cobbles) to and fro across the shore. This can result in the cutting of a smooth (or slightly grooved) seaward-sloping ramp or shore platform, often truncating the geological formations that outcrop on the shore (Figure 5.1). Waves armed with rock fragments are powerful agents of abrasion that can excavated caves, clefts and crevices. The rock fragments may be of local derivation, eroded from the cliff or quarried from the shore platform, or they may have been brought in by longshore drifting from adjacent parts of

the coast, or shoreward drifting from the sea floor.

The effects of wave abrasion can be seen on shores where the more resistant elements of rock persist as reefs and stacks above a smoothly abraded rock surface. Abrasion ramps, usually a few metres wide, slope gently up to the base of a receding cliff, where there may be an abrasion notch. As irregularities are removed abrasion may produce a smooth shore platform, which can be worn down to a lower level, particularly if the cliff behind it is receding. Lowering of an abrasion ramp or platform on homogeneous rock can be measured by repeated profile surveys, or with the aid of a micro-erosion meter (Spate et al., 1985; Stephenson, 1997).

The potency of wave abrasion is seen where rock fragments that have become trapped in a crevice on the shore are repeatedly moved by wave action in such a way as to excavate smoothly worn circular potholes, the rock fragments becoming well rounded in the process (Figure 5.2). Mutual abrasion where shore boulder accumulations are jostled by wave action leads to the formation of fitting boulders, interlocked in a three-dimensional jigsaw pattern (Hills, 1970). Pebbles rolled to and fro by wave action as the tide rises and falls can also cut grooves across the shore at right angles to the coastline or along

Coastal Geomorphology: An Introduction, Second Edition. Eric Bird.
© 2008 by John Wiley & Sons, Ltd

Figure 5.1 Cliff and shore platform features

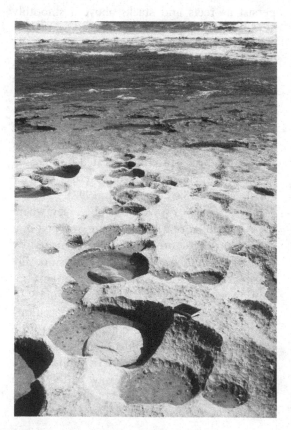

Figure 5.2 Potholes on a dune calcarenite shore near Rye in SE Australia, scoured by boulders circulated by wave action at high tide

the outcrops of bedding planes, joints, faults or weaker formations, particularly where these run at an angle to the coastline. Such abrasion can impede the formation of a shore platform, or dissect an existing platform, as where potholes or grooves are widened into clefts and bays interrupting a shore platform.

Wave quarrying is a form of abrasion that occurs when loosened rock fragments are excavated from a rock surface by wave action (Figure 5.3). It can produce a smooth slope where it lays bare a bedding plane or other flat or gently sloping surface on relatively resistant rock (a structural shore platform), but more often it produces irregularities, with pits, pools, clefts and cavities left where rock material has been removed. When wrack and kelp are torn off by storm waves their holdfasts may retain rock fragments, which are cast up on the shore or float out to sea. Examples of wave quarrying include the excavation of flint nodules from a chalk cliff or shore platform and the disaggregation of blocks of granite, sandstone, or limestone. An example is the dissection of blocks cut out along rectilinear joints in stratified limestone on the shore at Nash Point in South Wales. On arctic coasts disintegration and washing out of rocky material from thawing cliffs of permafrost is known as thermoquarrying.

Figure 5.3 Wave quarrying has dislodged sandstone blocks from the edge of the tessellated pavement, a shore platform on the coast of the Tasman Peninsula, Tasmania. The shore profile is stepped, the almost horizontal sandstone strata showing minor cliffs behind a shore platform that is an exhumed bedding plane

Wave abrasion and wave quarrying are much reduced where marine vegetation has colonised the shore, protecting the rock surface and attenuating wave action, and can be impeded where dense communities of shelly organisms such as barnacles congregate on the shore rocks.

5.1.2 *Weathering on the shore*

Weathering processes are at work on coastal rock outcrops that are not protected by a soil or vegetation cover in the intertidal zone, and at the cliff base up to above mean high tide level. Intricate pitting of rock surfaces results from repeated wetting and drying, which can lead to physical disintegration (decrepitation), releasing rock fragments that are swept away by wave sluicing. A distinction should be made between the wave abrasion of rock outcrops (discussed above) and this wave sluicing of unconsolidated weathered rock material. The wetting and drying process is most effective in semi-arid zones, where frequent desiccation in dry weather alternates with wetting by occasional rain as well as the highest tides, swash and sea spray. Rock formations disintegrate and are sluiced away by wave action down to the level of permanent saturation in coastal rock formations, which is generally just below high tide level, and in this way weathering can result in the formation of a shore platform.

The process of wetting and drying is often accompanied by salt weathering. Salt is deposited on rock surfaces and in cavities in the zone splashed by sea spray, and as it crystallises, absorbing water, its volume increases, creating pressure that widens fissures and detaches rock fragments. The plucked rock surface becomes pitted, and shallow basins may form, with undercut rims. Salt spray weathering can be observed on tropical shores where abrasion by waves is generally weak (because of low wave energy and the rarity of gravelly material) and the strong sun rapidly dries off rock surfaces wetted by saline spray. Tricart (1959) considered that salt spray weathering was a dominant process shaping shore features on the tropical coast of Brazil. It is difficult to separate the physical effects of wetting and drying from the physico-chemical effects of salt crystallisation from drying spray on sea coasts, where wetting and drying is accompanied by salt corrosion.

Sandstone outcrops wetted by sea spray and rainfall, and then dried out, become pitted and honeycombed as sand grains are loosened by the decomposition of the cementing material that formerly bound them. Other fine grained rock formations, such as siltstones, mudstones, shales, schists, phyllites and basalts, show similar superficial decomposition and disintegration. Wetting and drying and salt crystallisation are not effective at lower levels, where the rock formations are permanently saturated by seawater or ground water. The effects of the two weathering processes are seen on basalt boulders on an emerged shore terrace near Lara on the west coast of Port Phillip Bay, Australia, which have been undercut marginally and planed off locally at the level of groundwater saturation. On the South Devon coast, Mottershead (1998) measured changes on a schist outcrop where weathering had lowered a surface that had been covered by an oil spill and left small pinnacles capped by the oil layer. Using relics of the oil layer as a reference level, he deduced that the surface was lowered at the rate of 0.625 mm/yr over a seven year period (Mottershead, 1989). Measurements with micro-erosion meters on mudrock shore platforms on the Kaikoura Peninsula, New Zealand, showed microscale (up to 3.4 mm) raising and lowering of the rock surface, attributed to expansion and contraction of rock subject to alternate wetting and drying (Stephenson et al., 2004).

Weathering can be rapid on the humid tropical shores, as on the disintegrating surfaces of granite outcrops in NE Queensland, notably near Mackay. Weathering features include intricate pitting, fluting, honeycombing (Figure 4.6), fretted boxwork and the excavation of large cavities known as tafoni. The delicacy of weathering processes is shown where slightly harder components are left standing above the platform surface, such as the silicified tree trunks and stumps exposed on the shore platform in Jurassic tuffaceous sandstone in the Petrified Forest at Ciro Bay on the Catlin Coast, New Zealand.

5.1.3 Solution processes on the shore

Rock outcrops on cliffs and shore platforms can be dissolved by rainwater, sea spray and seawater. Solution occurs particularly on limestones, including emerged coral and algal reefs and dune calcarenites (Figure 5.4). Rainwater absorbs carbon dioxide from the atmosphere (carbonic acid) and can dissolve limestone (mainly calcium carbonate).

Seawater off limestone coasts and groundwater flowing from limestone cliffs is normally saturated with carbonates, and thus incapable of dissolving limestone on the shore, but limestone outcrops are certainly attacked by solution in the intertidal zone. Some cliff-base notches cut in limestone have been attributed to solution rather than abrasion, because they are at least as well developed on the side sheltered from strong wave action (as on the lee side of islands and stacks). The problem of how seawater, already saturated with carbonates, can achieve further solution of limestone on the shore was analysed by Revelle and Emery (1957) on Bikini Atoll. They found strong diurnal alternations in the dissolving capacity of seawater, related partly to

Figure 5.4 Karstic weathering has formed lapies on a limestone coast near Sorrento in SE Australia. The shore platform ends in a notch, overhung by a visor

variations in the temperature and carbon dioxide content of water in the shore zone. Nocturnal cooling of seawater increases its capacity to take up carbon dioxide (which is more soluble in water at lower temperatures), permitting it to dissolve more limestone, and the biochemical activities of marine organisms lead to the production of carbon dioxide (from plant and animal respiration), which is used by the plants (chiefly algae) in photosynthesis by day but which accumulates at night when the absence of sunlight halts photosynthesis. This nocturnal increase in the acidity of coastal water permits limestone to be taken into solution, and calcium carbonate dissolved during the night may be precipitated by day, when temperature rises, photosynthesis revives and the capacity of seawater to retain dissolved carbonate diminishes. Much of the precipitated carbonate is quickly carried away from the limestone shore by waves and currents: the opaline sea off the Normandy coast owes its colour to precipitated carbonate particles. As nocturnal solution processes operate mainly in the intertidal zone, they remove calcium carbonate in solution down to a level (close to low tide) at which permanently saturated and submerged limestone is being dissolved more slowly, if at all, by seawater. In this way, solution can produce a shore platform.

Solution of limestone shores may be achieved by surf in breaking waves and by sea spray, thrown into the air as the waves break, particularly during storms and when strong winds blow onshore. Surf and sea spray are aerated and dispersed forms of seawater, and associated atmospheric carbon dioxide may render them corrosive. Limestone surfaces subjected frequently to the impact of sea spray become pitted and irregular in a manner suggesting that solution is in progress, forming karst topography on the shore. This was observed at Diamond Bay, Australia, after a rock pinnacle was broken off just above high tide level during a 1968 storm on a cliff of dark, intricately weathered dune calcarenite. This formed a pale, freshly exposed smooth surface of unweathered sandstone, but within a decade the exposed surface had become colonised by algae and was as dark as the surrounding rock, and by 1993 it was as irregular and intricately pitted (muricate) as the surrounding weathered rock surface.

Limestone outcrops may also show fluting (rillenkarren), with vertical channels 2–4 cm wide separated by sharp ridges, formed by solution in seepage and runoff. Similar vertical grooves and ridges on granite outcrops formed by runoff erosion rather than solution are termed pseudokarren.

Although solution features are usually prominent on limestone shores, they are also subject to abrasion and wave quarrying, particularly near the base of cliffs, where ramps and structural benches are often seen. There are variations related to the lithology of the limestone, notably its hardness and purity, and to coastal climate, with solution more intense on humid than on arid coasts, and contrasts between tropical and cold environments. Other processes active on limestone shores, including bioerosion and induration, are discussed in the following sections.

5.1.4 Bioerosion on the shore

Marine organisms certainly contribute to the etching of limestone on the shore, and Emery (1960) indicated that in favourable conditions biochemical processes can consume rock at least as quickly as physical and purely chemical erosion. The various organisms that inhabit the intertidal zone (especially shelly fauna) contribute to erosion by drilling, scraping, plucking and grazing, and solution by exuded fluids. Algae can excavate notches and perforate limestone, grazing molluscs (limpets, mussels and winkles) scrape the surface, dislodging rock fragments and excavating cavities, and marine worms, sponges and sea urchins all drill or quarry the rock. Where shells occupy hollows on a limestone surface it is tempting to deduce that they have excavated these hollows, but this cannot apply where they are found occupying similar hollows on hard flint surfaces on chalk coasts.

Bioerosion occurs when algal mats that have formed on rock surfaces dry out and peel off with adhering rock particles, especially in dry, windy weather during periods when the zone above high neap tides is not submerged. Some would go as far as to suggest that a notch can be almost literally eaten out by the shore fauna in the high tide zone (Healy, 1968). Shore organisms occur across the intertidal zone and into subtidal environments, but are most active and vigorous between low tide and the splashed zone above high tide. Biogenic effects are most obvious on soft limestones and calcareous sandstones and siltstones on low energy shores, for on stormy coasts the organisms are inhibited, and their effects obliterated by wave abrasion.

Bioerosion processes can contribute to the weathering and removal of rock, and thus to the formation of a shore platform, but as they often extend below the level of shore planation by abrasion, weathering or solution, they also contribute to the dissection of a shore platform.

Rates of bioerosion on limestone, dune calcarenite and emerged coral have been summarised by Trenhaile (1987, Table 4.3). Lowering of a rock surface by grazing organisms can attain several mm/yr, and more localised drilling can be even more rapid. Some marine organisms provide bioprotection where rock surfaces are encrusted by algae or barnacles that impede weathering and erosion, preventing wetting and drying. Cliff-base ledges (trottoirs) and projections (corniches) are features protected and built by marine organisms (Section 13.10).

5.1.5 Frost shattering on the shore

On coasts in cold regions, repeated freezing and thawing result in physical weathering of rock outcrops on cliffs (Section 4.4) and on the shore. This is partly because of thermal contraction and expansion and partly because of pressures created by the formation and melting of interstitial ice (thermoquarrying or thermal abrasion). Where shore ice forms in winter and melts in spring it causes plucking and dislocation of rock fragments. Disintegrated rock is later removed by wave sluicing, or mobilised by waves to achieve abrasion of the rocky shore. Such weathering may result in the formation of a shore platform at the level where freeze–thaw oscillations cease. On the Chalk shore platforms of the Sussex coast, Robinson and Jerwood (1987) found that frost shattering during the cold months of January 1985 and February 1986 contributed to erosion and spalling on chalk boulders, the effects diminishing seaward to the low tide line. Shore ice abrasion has contributed to the cutting of low tide cliffs at the outer edge of shore platforms cut in shale on the stormy Gaspé coast in Quebec (Trenhaile, 1987). Movement of shore ice and icebound boulders by waves as the tide rises and falls (about 13 m at spring tides) has accompanied frost shattering to form the wide shore platform cut in Carboniferous sandstones on the coast at Joggins in Nova Scotia.

5.1.6 Induration on the shore

Induration of shore rocks increases their resistance to weathering processes and wave abrasion, and has contributed to the persistence of rock outcrops in cliffs, on stacks and islets, and on shore platforms and rocky shore topography. Various kinds of rock have become indurated by on the shore, the outcrops becoming much harder than those seen in quarries or encountered in boreholes in the same geological for-

mation a short distance inland. Case-hardening may result from superficial enrichment of shore outcrops by the precipitation of carbonates, or of ferruginous or siliceous cementing compounds. The process is well known to quarrymen: in South Devon the Beer Stone is a fine grained soft homogeneous chalk, moist when it is cut in underground caverns: it is carved and sculptured before being allowed to dry and harden under exposure to the atmosphere.

Induration has contributed to the preservation of shore platforms where rock outcrops have been hardened by the precipitation of cementing materials, notably calcium carbonate or iron oxides.

5.2 Shore platforms

Many cliffs are bordered by shore platforms that extend across the intertidal zone. They have been shaped by the various processes described in the previous section. The term wave-cut platform should be restricted to features shaped entirely by the hydraulic action of waves, usually seen only on soft rock formations such as clay. Erosional processes such as abrasion by waves armed with sand or gravel, tidal scour and weathering have contributed to the shaping of most shore platforms. Generally these processes operate slowly, and the presence of a shore platform indicates that the coast has been relatively stable for a substantial period, perhaps the whole of the time (up to 6000 years) that the sea has stood at its present level in Holocene times. Laboratory simulations of weathering and erosion processes on shore platforms have been conducted by Moses *et al.* (2006).

Some shore platforms are seaward sloping, others subhorizontal, with a sharp drop at the outer edge; some are structural in the sense that they coincide with the upper surface of a resistant formation (which may be horizontal or gently sloping). Seaward-sloping shore platforms

Figure 5.5 Shore platform types. (a) Seaward-sloping shore platforms. (b) Subhorizontal high tide shore platforms. (c) Subhorizontal low tide shore platforms. (d) Plunging cliffs with no platform on the shore

(Figure 5.5(a)) are usually planation surfaces (cut across shore structures) that extend from the high tide line at the base of the cliff to a level below and beyond the low tide line. It has been suggested that a shore platform of this kind may be bordered by a wave-built terrace constructed by deposition at its outer margin, but although this can be found on some lake and reservoir shores such a nearshore terrace is rarely formed off sea coasts (Dietz, 1963).

Structural shore platforms (Figure 5.6) coincide with the upper surfaces of resistant rock formations that may dip seaward, landward or alongshore. There are also more or less horizontal shore platforms, standing close to high tide level (submerged only briefly at high tides), which may have a sharp drop or low tide cliff (often >3 m) at the seaward edge. These are termed high tide shore platforms (Figure 5.5(b)). Similar subhorizontal shore platforms are found close to low tide level (submerged for most of the tidal cycle), particularly on limestones, and generally these also end seaward in a sharp drop: they may be termed low tide shore platforms (Figure 5.5(c)). Shore platforms that are narrow (up to about 10 m wide) are called benches,

and may be structural ledges coinciding with the outcrop of a resistant rock formation. On some coasts the shore topography includes sloping ramps and flat segments, as well as steps or micro-cliffs, uprising ridges and stacks, and dissecting features such as grooves, gulches and clefts. In high latitudes, particularly in Scandinavia, there are strandflats (Section 5.2.5), almost flat shore platforms that grade landward into gently shelving coastal plains. Some shore platforms are of artificial origin, as on the shores of Valetta Harbour in Malta. Finally, there are coasts where cliffs descend into deep water close inshore without any platforms or intertidal rocky areas. These are termed plunging cliffs (Figure 5.5(d)), and will be discussed after the evolution of shore platforms has been considered.

There is the problem of how smooth and continuous a planed rock surface should be for the term shore platform to be applied. Many shore platforms are dissected by grooves and channels, or diversified by small rising outcrops. As these increase in dimensions it may be more appropriate to use the terms incipient shore platform (with higher residual ridges and stacks)

Figure 5.6 A structural shore platform on a horizontal layer of Jurassic limestone at Broad Ledge, near Kimmeridge in Dorset, the outer edge undergoing dissection by waves. The cliff is steeply inclined, whereas on adjacent sectors where there is no protective shore platform the cliff stands vertically

or dissected shore platform (with wide transverse gulches and furrows parallel to the coastline). It is difficult to define these quantitatively, but one suggestion is that the term shore platform should be used only where the platform surface is sufficiently well formed, intact and undissected for a four-wheel drive vehicle to be driven across it. Where less than 10 per cent of the shore is a smooth platform and the rest irregular the term rocky shore could be used.

The various kinds of shore platform and rocky shore may vary along the coast in relation to geology, aspect and morphogenic factors such as tide range and wave regime. Contrasts in shore profile related to geology are evident where there are varied geological outcrops along a coastline and other factors are relatively uniform, and contrasts related to morphogenic factors

may occur on a particular rock outcrop, especially fringing headlands and islands. Such contrasts have been analysed along the southern New South Wales coast (Bird and Dent, 1966) and along the Victorian coast (Hills, 1971) in SE Australia.

Shore platforms may vary in relation to lateral changes in rock resistance. This is well illustrated on the coast at Mount Martha, Port Phillip Bay, Australia, where the coast intersects a granodiorite intrusion, the upper part of which has been strongly weathered and decomposed to clay. Where the clay descends to sea level at the northern and southern ends of the intrusion there are irregular receding cliffs, fronted by seaward-sloping abrasion ramps, but in the central sector, where the granodiorite is more resistant, it forms steep cliffs that descend into deep nearshore water (Jutson, 1940).

5.2.1 Seaward-sloping shore platforms

Seaward-sloping shore platforms (Figure 5.5(a)) – Type A as defined by Sunamura (1992) – extend from the cliff base at about high spring tide level and slope gently, but not always uniformly, to pass beneath low spring tide level, where there may be a sharp drop (low tide cliff) or a gradual decline to the sea floor. Some such platforms may be structural, coinciding with the surface of a seaward-dipping hard rock formation exposed by removal of overlying softer rock, but most have been planed across rock structures. They have been shaped primarily by abrasion (by waves armed with sand or gravel), with only minor contributions from weathering, bioerosion or solution processes. The width of such platforms at low tide is determined by the relative rates of retreat of the cliff base and the seaward edge, and is thus influenced by exposure to wave action, tide range and transverse shore gradient, rock resistance and the length of time that processes have operated at present sea level (Trenhaile, 1987).

Some shore platforms diminish in gradient as the cliff recedes; others retain their slope as the sea floor is worn down at a rate that matches cliff recession. Shore platform width may also be influenced by the dip and resistance of outcropping rocks. Gently sloping abrasion platforms, typically 50–100 m wide, have been cut in front of chalk cliffs where stratified Chalk dips gently seaward on the Kent coast in Thanet (So, 1965), and along the Sussex coast east of Brighton (Figure 1.2). Their width is also related to tide range, here about 4 m. Shore platforms are poorly developed on the Chalk of the Dorset coast, where the chalk outcrop restricted by a steep dip and tide range is small, but a seaward-sloping shore platform has been cut across steeply dipping Chalk at Whitecliff Point at the eastern end of the Isle of Wight: it is not clear why there is no such platform on

similar steeply-dipping Chalk at Ballard Point in Dorset.

On soft rock outcrops the seaward-sloping shore platform has a slightly concave shore profile descending from the cliff base and declining in gradient seaward below the low tide line. This profile is similar to the longitudinal profile of a river valley or the transverse profile of a beach, and may be considered as the theoretical end-product of marine erosion (the eventual profile).

There is often a beach at the base of a cliff, and an intertidal veneer of sand and gravel. Such a profile is seen fronting cliffs of glacial drift on the east coast of England, notably in Holderness and on clay outcrops on the Dorset coast, as on the Weald Clay in Worbarrow Bay. A similar shore profile fronts the soft receding sand cliffs of Alum Bay and Whitecliff Bay, on the Isle of Wight. It is thus probable that this is the profile that would develop in front of cliffed coasts generally, were it not for the complications of structure, resistant rock outcrops, weathering effects and induration.

As a cliff recedes wave abrasion commonly cuts a basal notch, the lower part of which is a seaward-sloping abrasion ramp that declines into the usually gentler gradient of the shore platform and its undersea continuations. The abrasion ramp typically shows freshly scoured rock, without a weathered surface, plant growth or marine organisms. Abrasion ramps sometimes have potholes scoured where sand or gravel have been rotated by wave action to enlarge a crevice (Figure 5.2). Occasionally abrasion ramps become covered with seaweed where there is no longer sand and gravel to be mobilised by waves to maintain abrasion.

On the Liassic rocks of the Yorkshire coast, Robinson (1977) found that the cliff-base ramp was being lowered by abrasion at up to 3 cm/yr. Cliff recession is generally accompanied by downwearing of the shore platform in such a way as to maintain the transverse profile and the seaward slope. Persistence of small mesas (steep-sided flat-topped residuals) capped by flint layers, rising above the general level of the shore platform on the coast, near Birling Gap in Sussex, are an indication of this down-wearing. The adjacent sea floor is also worn down by abrasion, so that the whole coastal profile migrates landward.

There are variations in the upper limits of shore platforms around the coasts of England and Wales. The junction between the cliff base and shore platform is often obscured by beach deposits, but where it is exposed it is generally a sharp break of slope close to high tide level. In detail the level varies from about mean high neap tide and just above mean high spring tide, and can show lateral variation of up to 4 m. It rises with tide range because high tide reaches a higher level above Ordnance Datum. On the Chalk coast of Thanet the junction between cliff base and shore platform is up to 3.6 m higher at the heads of bays than on intervening headlands (Wright, 1970). However, there are sites where the junction is higher on headlands than in adjacent bays, possibly because of storm wave action operating to higher levels (So, 1965), or because the water table is higher on the headland, and weathering processes (such as wetting and drying) are shaping platforms at a higher level.

More often, wave abrasion has cut a gently sloping shore platform across the structure of coastal formations, so that the surface shows a pattern of truncated rock outcrops. Seaward-sloping shore platforms have been cut across steeply dipping Old Red Sandstone striking east to west along the south coast of Wales west of Tenby, notably at Manorbier, where the planed-off strata run parallel to the coastline. On the west-facing Atlantic coast of Cornwall and Devon between Widemouth Bay and Hartland Point seaward-sloping shore platforms have been cut across intricately folded sandstones and shales, truncating numerous anti-clines and synclines that trend east–west, at right angles to the coastline. Similar platforms occur near San Sebastian on the north coast of Spain, and on Aoshima Island, Kyushu, Japan (Sunamura, 1992), while swirling patterns of planed-off strata are seen on the abrasion platform cut in strongly folded Lias limestones on the South Wales coast near Nash Point.

The chalk cliffs and shore platforms on the English and French coasts bordering the English Channel are largely the product of wave abrasion, the waves being armed with flint nodules eroded out of the Chalk (Prêcheur, 1960; So, 1965). Chalk is a relatively homogeneous limestone formation, apart from the flint nodules, that are usually in layers along the bedding planes. As they weather out of the Chalk they are mobilised by breaking waves and used as tools of abrasion as the tide rises and falls. They cut grooves at right angles to the coastline on the chalk shore platforms of the Kent and Sussex coast. Fresh white chalk exposed on the shore platform and on the cliff-base ramp after stormy periods is evidence of wearing and scouring by waves armed with flints, but other processes (such as weathering and solution) are also at work, and their effects can be detected wherever wave abrasion diminishes.

The low tide cliff is present only locally, and does not coincide with any resistant rock outcrop. It may mark the position of the cliff about 6000 years ago, before the existing cliff receded, or it may have been protected by a mantle of periglacial deposits (Coombe Rock) when the Holocene marine transgression initiated cliff recession at a higher level, and exposed subsequently (Hénaff, Lageat and Costa, 2006).

Examination of fallen boulders shows that the chalk surface has been pitted by solution due to rainfall, sea spray and aerated surf, and that the rock has been modified by the physical and biochemical effects of shore flora (chiefly marine algae) and fauna (limpets, mussels and winkles). These processes play an important part in the reduction and eventual disappearance of chalk, releasing further flint nodules that are then used in abrasion.

Seaward-sloping shore platforms are sometimes found where structureless sandstone or shale formations have been eroded by wave action, but their development requires a delicate balance between rock resistance and the intensity of wave abrasion. Relatively resistant formations may be eroded into steep cliffs and seaward-sloping shore platforms on a high wave energy coast but not where wave energy is low.

Seaward-sloping shore platforms are rarely found on massive igneous rock outcrops, but exfoliation (the peeling away of surface layers of rock along fractures that develop parallel to the rock surface) can reduce a granitic reef or shore outcrop to a broadly domed platform just above high tide level, probably because the process of exfoliation comes to an end when it encounters saturated rock. Exfoliation can occur at various scales, ranging from less than 1 cm on boulders (desquamation) to more than 1 m on massive rock outcrops (sheeting). It has been attributed variously to unloading (as overlying rock is removed), insolation (thermal expansion and contraction) and hydration (weathering by groundwater).

There are shore ramps on sectors of the granite coast of the Land's End Peninsula and at Cape Woolamai in SE Australia where the granite is more closely and intricately jointed, and more readily quarried by strong wave action. Similar features are seen where igneous rock formations are deeply weathered, and the underlying solid rock is exposed on the coast by marine erosion as a shore platform. There are also segments of structural shore platform that coincide with exhumed flat or gently sloping joint planes in granite, as at Giant's Castle, St Mary's, Isles of Scilly.

Seaward-sloping shore platforms may be cut in relatively weak formations on a low wave energy coast, but higher wave energy may destroy them. There are vertical cliffs on sandstones with little or no development of fronting shore platforms on the Bridport Sands at West Bay in Dorset and the Hastings Beds on the Sussex coast. In each case the rock outcrops are sufficiently coherent (and slightly indurated) to sustain a vertical cliff, but not sufficiently resistant to form a shore platform. Similar features are seen on cliffs cut into volcanic ash and agglomerate, as on Anak Krakatau in Indonesia (Figure 4.27).

Under such a balance of conditions the morphology of the receding coast can be maintained during recession of cliffs and downwearing of intertidal platforms and the adjacent sea floor. It is possible that the Holderness coast, cut into soft glacial drift deposits, is maintaining its whole profile as the coastline recedes. The cliffs here have been receding at rates of 1 to 2 m/yr, and if this has been maintained since the sea reached its present level the coastline has retreated 6–12 km over 6000 years, and the sea floor at the initial coastline has been lowered at least 10 m. Elsewhere the evolution of a receding coast is usually complicated by the effects of structure, lithology and weathering processes, but the balance of these may sometimes yield a deceptively simple cliff-and-platform profile, as on certain Chalk coasts (Figure 1.2), where solution and bioerosion have accompanied wave abrasion.

More often there are variations in structure and lithology in the shore zone that persist in irregularities of profile with seaward-sloping segments of shore platform developed locally as planation corridors between ridges of harder rock, and channels where less resistant outcrops have been excavated (Figure 5.7). Increasing wave energy exploits structural and lithological contrasts, forming an irregular rocky shore with ribs and reefs of harder rock between deeper clefts and channels, as on the southern coast of the Lizard Peninsula, where the cliffs are fronted by reefs and islets of hard Palaeozoic and Pre-Cambrian rock. There are rocky reefs and skerries instead of shore platforms in the shore and

Figure 5.7 Corridor planation between ridges of more resistant rock on the New South Wales coast

nearshore area. Similar features are seen on the granitic coast of the Bullers of Buchan in eastern Scotland.

On some coasts the present seaward-sloping shore platform is developing as the result of dissection of a similar higher shore platform cut in Late Pleistocene times when the sea stood a little above present high tide level. An example has been documented from Caamaño, Galicia, by Trenhaile *et al.* (1999). As dissection proceeds, the shore becomes very rugged and irregular, with remnants of the emerged platform separated by transverse gullies and corridors, along which a smooth slope, cut by abrasion, ascends landward. The modern shore platform is being cut at a lower level, and where the emerged platform has been completely removed the modern seaward-sloping shore platform runs back to a sharp junction at the base of the cliff, sometimes with an undercut notch. Evolution of a seaward-sloping shore platform by the dissection and re-

duction of an emerged shore platform is seen on parts of the coast of SW England, particularly in Gerrans Bay and Falmouth Bay in South Cornwall, where the emerged shore platform carries a Pleistocene beach deposit (Figure 3.3). At Whitsand Bay segments of the emerged platform are preserved on ribs of Dartmouth Slate, which run out from the cliff base between sandy coves and segments of the modern seaward-sloping shore platform.

Emerged shore platforms of Pleistocene and Holocene age occur on many coasts, some related to a Late Pleistocene interglacial high sea level, others the outcome of land uplift. The Pleistocene emerged shore platform at about 3 m above present sea level in southern Britain is found at various other levels, ranging up to 6 m in northern Britain, where isostatic rebound has followed deglaciation (McCann, 1964).

On the eastern side of Cape Otway, Victoria, Australia, a Pleistocene cliff and shore platform cut in Cretaceous sandstones were buried by dunes that spilled over the headland during the Last Glacial low sea level phase and became lithified as dune calcarenite (Section 9.10). After the sea rose to its present level about 6000 years ago the dune calcarenite was cut back to form a cliff and shore platform within which are segments of the exhumed Pleistocene cliff and shore platform in Cretaceous sandstones.

5.2.2 Structural shore platforms

Structural shore platforms are found where waves have exposed the surface of a flat or gently dipping resistant rock formation, usually a bedding plane, as on Broad Ledge near Kimmeridge (Figure 5.6). On the coast of Malta limestone shore platforms have been exposed by the removal of overlying clay. The structure and lithology of coastal rock formations influence the development of shore platform profiles, notably where sandstones or limestones are interbedded with softer shales, as in the Lias of the Lyme Regis coast. Wave action has excavated bedding planes to produce structural ledges on the upper surface of each resistant layer, terminating in small scarps. Some are horizontal, some slope seaward or landward, and others alongshore. Segments with intricate scarps and dip slopes on Liassic limestones and shales are interspersed with seaward-sloping shore platforms at Robin Hood's Bay in Yorkshire. Limestone strata also form a stepped shore topography at Kimmeridge Ledges on the Dorset coast, each ledge disintegrating into large angular blocks as the underlying shales are etched out by wave scour.

Where soft formations contain minor resistant layers, these may persist as segments of structural shore platform. This is well illustrated on the SW coast of the Isle of Wight, where the soft Wealden shales contain thin sandstone horizons that rise and fall gently across the Brighstone Anticline and form structural shore plat-

forms exposed at low tide, as at the Pine Raft at Hanover Point (Bird, 1997). Structural shore platforms are also found locally on the upper surfaces of horizontal or gently dipping Triassic sandstones of the Devon coast between Torquay and Sidmouth and on the Carboniferous limestone of Northumbria, where undulating ledges follow shore outcrops of thick layered strata.

In Australia, stepped cliff profiles have developed on outcrops of Triassic sandstones in the Sydney district, and there are structural shore platforms that coincide with the upper surfaces of resistant strata: they are horizontal where the bedding is flat, and inclined where the rocks are dipping. It is possible to follow a particular ledge along the shore, down the dip of a resistant sandstone layer, until it passes below sea level, when the ledge on the next higher resistant rock outcrop begins to dominate the shore profile as it declines into the sea. This kind of coastal topography results from storm wave abrasion along bedding planes and joints and the removal of dislocated rock masses to lay bare a structural shore platform. Where the rock formation is horizontal the shore platform is similar to the Type B shore platforms discussed below, but where the rock formation dips gently seaward the profile may resemble a seaward-sloping (Type A) shore platform, passing below low tide level, differing in that abrasion has not truncated shore outcrops. Where ledges or platforms have developed on flat or gently dipping formations above high tide level they could be emerged features formed during an earlier phase of higher relative sea level, but on high wave energy coasts storm waves can cut structural ledges at various levels (Figure 4.11).

Structural platforms are often seen on coasts in volcanic rocks, where the platforms coincide with the upper surfaces of lava flows, which may be dissected along joint planes and eventually reduced to blocks of basalt on the shore. There are good examples on the northern and western coasts of Skye in Scotland and on Phillip Island in Australia. Abrasion, weathering and bioerosion have contributed to the shaping of structural shore platforms, and wave sluicing (and occasionally wind action) is necessary to sweep away weak, weathered or unconsolidated material to lay bare the upper surface of a resistant rock formation.

5.2.3 Subhorizontal high tide shore platforms

The term high tide shore platform (Figure 5.5(b)) is a brief way of describing a platform that is horizontal, or nearly horizontal (with a very gradual seaward slope, usually less than 1°), developed at, or slightly above, mean high tide level (Type B of Sunamura, 1992). Such subhorizontal platforms are exposed to subaerial processes for much of the tidal cycle, but overwashed by storm waves and submerged by high spring tides (Figure 5.8). They typically end abruptly seaward in a steep drop (low tide cliff), below which the sea floor declines, usually with a broad

and gentle concave profile. A sandy sea floor is indicated by the clouds of sand that rise in suspension with the up-current in incoming wave fronts (Figure 2.3): the sand is generally a veneer on a rocky substrate.

These high tide shore platforms are well developed in microtidal, low wave energy environments, particularly on sandstones, mudrocks and other permeable fine grained rock formations including basalt and consolidated volcanic ash. Originally observed and studied in New Zealand, Hawaii and Australia, they occur widely on the islands and shores of the Pacific and Indian Oceans, and have also been noted locally on the Atlantic coast. They are quite distinct from seaward-sloping shore platforms, and although some are structural (coinciding with the upper surface of an outcropping rock formation) most truncate local geological structures and cannot be explained in terms of lithological control.

High tide shore platforms are produced by water layer levelling processes, especially wetting and drying, rather than wave abrasion, although

Figure 5.8 Subhorizontal high tide shore platform on the Otways coast in SE Australia, cut across fine-grained Cretaceous sandstones and mudstones that dip about 10° landward. There is a minor cliff at the seaward edge

wave sluicing removes the weathered material down to a particular level (the upper level of rock saturation, no longer subject to recurrent drying) to lay bare the horizontal platform. Wetting and drying has contributed to the lowering and flattening of shore platforms on basalt and sandstone in SE Australia (Hills, 1971) and on mudrocks on the Kaikoura Peninsula in New Zealand (Stephenson and Kirk, 2000).

Wetting and drying is usually accompanied by salt crystallisation, and the two processes cause pitting and honeycomb weathering on the cliff face and on upstanding rock outcrops on high tide shore platforms. The pits are usually small (diameter 2–5 mm), but can be several centimetres wide, grading upward to tafoni (Section 5.1.2), and the ridges between the pits are preserved by induration. Such weathering produces rock fragments, mainly of silt and sand size, which are sluiced away by waves at high tide. Pools and channels form on the platform surface and become enlarged and integrated as their overhanging rims recede, so that gradually the rock surface is reduced to a high tide shore platform. The process is also known as water layer weathering, and as the level of rock saturation need not coincide with bedding planes it offers a mechanism by which shore planation may form platforms that transgress local geological structures. It also accounts for the fact that high tide shore platforms are almost horizontal, often with a raised rim or rampart at the outer edge where the rock is permanently saturated to a higher level by breaking waves. This is often more pronounced where the rocks are dipping seaward (landward scarp) or landward (seaward scarp) than where they are horizontal. In places the seaward rampart has been removed by erosion, but where it persists there is an implication that the outer edge of the shore platform has been cut back little since the sea attained its present level in Holocene times, and that cliff recession has been accompanied by the lowering and flattening of the shore platform behind

the rampart as the result of weathering processes and the sluicing away of weathered material by wave action.

Shore platforms of this kind are poorly represented around the British Isles, probably because of the relatively high tide ranges, the stormy seas and the rarity of suitable lithologies. Another factor is the widespread availability of rocky debris, particularly shingle, which aids abrasion and results in the cutting of seaward-sloping intertidal shore platforms. Type B shore platforms occur in the subantarctic South Shetland Islands on sheltered parts of the coast where there is intertidal shore ice plucking and bulldozing of rock outcrops, abrasive wave action being inhibited by the freezing of the sea for several months each year. On more exposed parts of the coast stronger wave action forms sloping intertidal Type A platforms (Hansom, 1983).

In places high tide (Type B) shore platforms pass laterally into seaward-sloping (Type A) shore platforms, notably where there is an increase in rock fragments that can be moved to and fro by waves to achieve abrasion, where there is exposure to stronger wave action, or where the shore is dominated by a structural feature such as a seaward-sloping hard rock formation. On the New South Wales coast high tide shore platforms give place to seaward-sloping shore platforms on the more exposed sectors, and where wave abrasion is facilitated by the presence of locally derived shingle.

Rates of lowering of subhorizontal shore platforms have been measured on the Yorkshire coast, where Robinson (1977) found that a subhorizontal platform up to 200 m wide had been downwasting at 1–2 mm/yr. Stephenson and Kirk (2000) measured rates of erosion on shore platforms cut in mudstone at Kaikoura Peninsula, New Zealand, and found that subhorizontal platforms were being lowered at 0.733 mm/yr, whereas seaward-sloping shore platforms had a lowering rate of 1.983 mm/yr.

Shore platform width is easily measured on subhorizontal shore platforms that end in a sharp drop at the outer edge, but most seaward sloping shore platforms pass evenly below low tide level, and the outer boundary is arbitrary. In general shore platform width is related to (a) relative rates of recession of the backing cliff and the seaward margin, (b) variations in wave energy with coastal aspect, configuration or protection by islands or reefs and (c) variations in rock structure (notably the dip), lithology and hardness alongshore.

Changes in the width of a subhorizontal shore platform can be measured with micro-erosion meters. On the mudrock coast of Kaikoura in New Zealand Stephenson (2001) found that the cliff base retreated by 0.05–0.91 mm/yr between 1942 and 1994, without any recession of the seaward edge in these 52 years, so that the subhorizontal shore platforms had thus widened.

Measurements of rates of shore platform lowering are more difficult where the rock formations are closely jointed and thinly bedded, as on many limestone and sandstone coasts, so that there is spasmodic detachment and removal of blocks of rock (Figure 5.3). Micro-erosion meters can be used on platforms of massive or homogeneous rock, but there is a problem where organisms such as barnacles or algae, which can cause bioerosion or bioprotection, are extensive on the shore platform. They can be removed (at the risk of causing some erosion) to determine vertical changes on the rock surface at the site they had colonised, but these measurements do not indicate the relative contributions of physical, chemical and biological processes to such changes.

It has been suggested that high tide shore platforms could be essentially storm wave abrasion platforms, produced by waves driven across them during storms when the cliff at the rear is cut back; in calmer weather, wave action is limited to the outer edge, that gradually recedes. It is difficult to accept this as an explanation for high tide shore platforms, except in the special case where a structural platform coincides with the upper surface of a horizontal rock formation at high tide level. Storm waves are unlikely to concentrate energy at any particular level because they are occasional, of varying duration, and come in a variety of dimensions, operating over a height range related to the rise and fall of tides. Horizontal or gently inclined platforms that truncate local geological structures (Figure 5.9) cannot be explained in terms of storm wave attack, which has a destructive influence on these features.

Evolution of high tide shore platforms as the result of weathering processes is indicated where they are as well, or better, developed on sectors of the coast that are sheltered from strong storm wave activity. On the New South Wales coast the strongest storm waves arrive from the SE, but the high tide shore platforms are at least as broad and often better developed on the northern sides of headlands and offshore islands, as on Broulee Island. On the more exposed southern sides the platforms show evidence of dissection and destruction by wave abrasion at their outer margin and quarrying by waves along joints and bedding planes. There is evidence that this recession results in the extension of the concave abrasion platform at a lower level, beyond the seaward drop, which is extending landward as the result of this cutting back of the high tide shore platform (Bird and Dent, 1966). Wave abrasion, operating alone, tends to develop the simple profile of a steep cliff bordered by a seaward-sloping inter-tidal shore platform, within constraints imposed by the structure and lithology of coastal outcrops.

It is possible that subhorizontal shore platforms at or slightly above mean high tide level were originally formed as seaward-sloping shore platforms by wave abrasion at an earlier phase when sea level was higher, and have subsequently been flattened by weathering with the sea at its present level and diminished wave energy. This

Figure 5.9 Shore platform cut across steeply dipping hard Devonian mudstones on the coast of Cape Liptrap, Victoria, Australia, backed by steep vegetated bluffs that were probably active cliffs during an earlier phase when sea level was higher

could have occurred where the Holocene marine transgression attained a maximum slightly above present sea level before dropping back during a phase of Holocene emergence, and where high tide shore platforms are backed by degraded cliffs that have not been kept fresh by marine attack. Dissection of the outer edge of the platforms can then be interpreted as the result of the cutting of a new abrasion platform at a lower level following emergence.

Some high tide shore platforms may have been formed as the result of dissection and downwearing of similar subhorizontal shore platforms that developed when the sea stood at a higher level in late Pleistocene times (Section 3.9), in which case fragments of older and higher shore platforms may persist locally as emerged terraces. However, many high tide shore platforms are backed by actively receding cliffs, of-

ten with a small basal abrasion ramp, and it is not necessary to invoke an episode of higher sea level in Holocene times to account for their development.

High tide shore platforms are not found on soft rock outcrops such as clay or sand, where wave abrasion proceeds too rapidly for weathering effects to persist, or on massive hard rock outcrops such as granite or quartzite, on which weathering is very slow.

The width of a high tide shore platform (see above) is determined by the relative rates of recession of the cliff at the rear (by removal of weathered material and occasional storm wave abrasion) and along the seaward margin (by more continuous wave action on permanently saturated and unweathered rock). It is also related to the height of the cliff (i.e. the volume of rock to be removed as the platform is cut:

Figure 5.10 Old Hat island at Paihia, Bay of Islands, New Zealand

small where the shore platform has developed as the result of dissection and downwasting of an earlier shore platform cut at a higher relative sea level) and the length of time since the sea reached its present level (Trenhaile *et al.*, 1999).

Variations in high tide shore platforms are related to the lithology and structure of coastal rock outcrops, the effectiveness of weathering processes, nearshore topography, wave regime and tide range, as well as the availability of abrasive debris. These factors vary intricately on a coast of irregular configuration, with local variations in aspect resulting in changes in the width and transverse gradient of the shore platform, the persistence of upstanding rock outcrops and the degree of dissection by wave abrasion and quarrying, particularly along the seaward margin. Broad, flat high tide shore platforms cut in basalt on the ocean coast near Flinders in Australia, become more irregular, with stronger

structural features, on the adjacent more sheltered shore of Westernport Bay.

Development of shore platforms by weathering and the washing away of disintegrated material down to the level of rock saturation is seen on Old Hat islands, where residual central hills are encircled by flat rock ledges. Old Hat islands have formed on impermeable, homogeneous rock formations in the low wave energy landlocked Bay of Islands in New Zealand (Cotton, 1974), where Paihia is a well known example (Figure 5.10). Other examples are seen on the Pacific coast of the United States, notably at Sekiu on the coast of the Strait of Juan de Fuca in Washington State. They have formed where the land had previously been deeply weathered: wave sluicing has removed fine grained weathered material to expose a shore platform on the upper limit of unweathered rock. This implies that the depth of prior weathering coincided with present day high tide level, but there is a possibility that some shore platforms at high tide level have formed where weathered rock has been re-indurated by secondary cementation. An example of this is seen at Corinella in Westernport Bay, Australia.

Little attention has been given to shore platforms on emerging coasts. On Cape Flattery, on the coast of Washington, USA, there are seaward-sloping shore platforms and weathering subhorizontal shore platforms in the present intertidal zone on a coast that has been rising tectonically at about 1.6 mm/yr. Bird and Schwartz (2000) concluded that these platforms had been lowered by abrasion and weathering at about the same rate as the falling relative sea level.

Some shore platforms cut in ferruginous sandstone owe their persistence to induration, resulting from intertidal precipitation of iron compounds. Thornton and Stephenson (2006) found that higher parts of the shore platforms on Cretaceous sandstone and mudstone on the Otway coast in Victoria, Australia, correlated with increased rock strength, as measured with a rebounding Schmidt hammer. Such variations in rock strength may be inherent in the lithology of shore outcrops, or they may arise from local induration.

5.2.4 Subhorizontal low tide shore platforms

Low tide shore platforms are horizontal or almost horizontal platforms (with a very gradual seaward slope), exposed only for a relatively brief period when the sea falls to low tide level (Figure 5.5(c)). They are also Type B platforms, and are called low tide shore platforms because they are subaerially exposed only when the tide is low. They are best developed where the tide range is small (microtidal coasts) on limestone, emerged coral and dune calcarenite coasts. Low tide shore platforms may be broad and almost flat to the cliff base, where there is

frequently a notch overhung by a visor (Figure 13.5). As on high tide shore platforms there is sometimes a slightly higher rampart at the outer edge, in this case formed by an encrustation of algae in a zone that is kept wet by wave splashing even at low tide. There is usually a sharp drop at the seaward edge, down to a sea floor with a broad gently concave profile, possibly a developing seaward-sloping subtidal platform. Undermining of the cliff along the seaward edge results in the collapse of the outer margin of the shore platform, often along arcuate fractures.

Subhorizontal shore platforms of this kind have developed by the removal of limestone in solution. Corrosion of shore limestones can be achieved by rainwater (at low tide), by aerated surf and sea spray, and by seawater when its carbon dioxide content increases, notably during nocturnal cooling. Percolating groundwater and daytime seawater are usually saturated with dissolved carbonates, and unable to corrode limestone outcrops; they may instead precipitate carbonates. Limestone shore platforms are developed by marine solution processes on the coasts of arid regions, so solution by rainwater is not essential, but it has undoubtedly contributed to shore limestone corrosion in humid environments.

Limestone cliffs and rocky shore outcrops show surface pitting and irregular dissection into networks of sharp edged ridges and pinnacles (lapies) as the result of corrosion, especially by sea spray (Figure 5.4). Contrasts in the features of limestone coasts in tropical and temperate environments were identified by Guilcher (1958). On limestone coasts in cool temperate regions (where seawater is colder and less saturated with carbonates), as on the Burren coast in western Ireland, the shore rocks are intricately corroded, with numerous pinnacles and small pools. On warm temperate coasts, as in Portugal and Morocco, solution of shore limestone produces pitted rock outcrops

Figure 5.11 Notch and visor around a mushroom rock at Sorrento, SE Australia

and flat-floored pools (at various levels) with overhanging rims that may be residual limestone with or without algal or vermetid encrustations. As the pitted rock outcrops are consumed the flat-floored pools grow larger, and coalesce, eventually forming a broad shore platform at about mid-tide level. On tropical limestone coasts (including emerged coral reefs), notches produced by solution are prominent, with overhanging visors bearing pinnacles formed by solution in sea spray and rain. They are conspicuous on the limestone cliffs of Phang-nga Bay in Thailand (Figure 3.2), on the great limestone cliffs of northern Palawan in the Philippines and on emerged coral reef islands in the Pacific Ocean.

A flat floored notch with an overhanging visor at the cliff base and around rock outcrops rising above the shore platform can be produced by solution processes, which are most rapid (removing up to a millimetre of rock annu-

ally) just above mid-tide level (Hodgkin, 1964). Notches of this kind are well developed on tropical coasts, particularly behind shore platforms cut into emerged coral reefs. They also occur on limestone coasts around the Mediterranean Sea. The importance of solution processes is indicated where such notches extend around stacks and islets (forming mushroom rocks), and are at least as well developed on the side sheltered from incoming waves (Figure 5.11). As the notch is cut back, the shore platform is extended. The level of the shore platform is determined by the downward limit to which solution processes are effective, and at this level the precipitation of carbonates begins. It is noteworthy that coral reefs are built up, and emerged coral reefs planed off, to this same level, as seen on the shores of Mbudya Island in Tanzania.

The level of planation (mainly by solution) on limestone coasts is thus slightly lower than planation (mainly by weathering) on fine

Figure 5.12   Cliff and shore platform cut in Pleistocene dune calcarenite at Jubilee Point, near Sorrento, SE Australia. The cliff shows a sequence of dune sandstones (with excavated caves), intervening calcrete layers and a capping of unconsolidated Holocene dunes

grained sedimentary and volcanic rocks. At Cape Schanck in Australia the rising tide submerges subhorizontal low tide platforms cut in Pleistocene dune calcarenite before it reaches similar high tide platforms cut in Tertiary basalt.

Shore platforms on limestone coasts often coincide with an indurated horizon that develops where the precipitation of carbonates where downward-percolating groundwater meets carbonate-saturated seawater and solidifies and hardens the rock outcrop. This is most obvious on Pleistocene dune calcarenites (Section 9.10), as on the Nepean ocean coast in Victoria, Australia (Figure 5.12), where the shore platform truncates dune sandstone bedding inclined at various angles and directions, but coincides with the surface of a horizon where the calcareous sandstone has been indurated by carbonate precipitation. The calcarenite is less

resistant both above and below this horizon, which thus becomes a structural influence on shore platform evolution. Induration also occurs on blocks and boulders of dune calcarenite that fall to the shore. Locally, an unusual form of shore pothole develops where soil pipes that had been excavated into old land surfaces by solution below plant root systems (the surrounding calcarenite becoming indurated by associated carbonate precipitation) are exhumed and washed out as circular rimmed vase-like depressions.

Apart from solution, other weathering processes are at work on limestone coasts. Wetting and drying contributes to the disintegration of limestone surfaces, particularly on dune calcarenites, but ceases at high tide level, above the limit of solution processes. Salt crystallisation also contributes to the erosion of sea-splashed surfaces that dry out sufficiently for crystals to form. Bioerosion is certainly active, with numerous shelly organisms attacking the rock surface in the intertidal zone, where many species are present. Shelly organisms certainly contribute to pitting and the excavation of notches (Spencer, 1988), but they do not play a major role in shore planation because they are also present, though in reduced numbers, below mid-tide level. Instead, they may form bioconstructional features such as the algal (mainly *Lithophyllum*) and vermetid ledges projecting from the base on notches on some Mediterranean limestone coasts, notably in Crete (Kelletat and Zimmerman, 1991).

Wave sluicing at high tide washes away weathered, eroded and precipitated fine grained sediment from limestone shore platforms. Where sand and gravel are present wave abrasion becomes effective, and waves armed with rock debris cut an inclined abrasion ramp towards the rear of the platform, rising to the base of the receding cliff. As cliff recession proceeds the inclined abrasion ramp is lowered and flattened by solution and the shore platform is extended landward. While cliff-base and stack-base notches on limestone coasts are largely the outcome of solution processes, wave abrasion can modify them, producing abrasion notches with scoured sloping ramps, particularly on sectors exposed to strong wave attack. Quarrying by waves can dissect shore platforms, especially along joints or faults, and by undermining the cliff at the seaward edge, where the platform margin eventually collapses. As on high tide shore platforms, strong wave action armed with abrasive debris can disrupt limestone shore platforms and replace them by seaward-sloping shore platforms of the kind seen on chalk coasts, where waves achieve abrasion by mobilising flint nodules. Such sloping platforms are more likely to develop on stormy high wave energy coasts and on mesotidal and macrotidal coasts than where the tide range is small.

5.2.5 Strandflats

Strandflats are extensive coastal platforms of problematical origin found in front of mountainous slopes on the fiord coasts of Norway, Spitzbergen, Iceland and Greenland. They are up to 50 km wide, and have a relative relief of ±200 m. They are not strictly shore platforms, although they may grade into seaward-sloping shore platforms that extend beneath shallow coastal waters. They may be partly emerged shore platforms, for numerous rocky hills (hutberge, interpreted as former islands) rise above them. They occur only on cold coasts, and must be related to past glacial or periglacial processes. It is possible that they developed in Pleistocene times as the result of plucking and disintegration of coastal rock outcrops by sea ice that rose and fell with the tide, followed by the sweeping away of debris by wave action when the ice melted (Tietze, 1962). Alternatively, they may be the outcome of prolonged coastal periglaciation, repeated freezing and thawing producing a frost-shattered rock mantle that was sluiced away by the sea, exposing the underlying unweathered rock. It is not clear why strandflats have formed on some glaciated coasts but not on others: they are not seen in Alaska or British Columbia, in Patagonia, or in South Island, New Zealand.

5.2.6 Rocky and boulder-strewn shores

Some coasts have irregular rocky shores, either where shore platforms have been greatly dissected, or where they have failed to form because of an intricate geological structure and much variation in rock type and hardness. The surface is rugged, with many ridges, promontories and stacks, and complex patterns of grooves and channels. There are often intervening small bays, some of which may contain beaches of sand or gravel, and usually there are many dislodged blocks and boulders, so that some shores can be described as boulder strewn. The boulders may have been produced by disintegration of shore rock outcrops or imported as glacial erratics, relinquished from a former ice sheet, as on parts of the Baltic coast.

5.3 Plunging cliffs

Plunging cliffs (Figure 5.5(d)) pass steeply beneath low tide level without any shore platforms or rocky shore outcrops. It is necessary to explain why the processes that have produced the various kinds of shore platform have not succeeded in forming one where there are plunging cliffs. There are several possible explanations.

(a) Plunging cliffs can be produced by recent faulting, the cliff face being the exposed plane of the fault on the up-throw side, the down-thrown block having subsided beneath the sea. There has been insufficient time for marine erosion to cut a shore platform at present sea level. Plunging cliffs along the line of the Wellington fault, on the western shore of Port Nicholson in New Zealand, formed in this way. They slope at about 55°, show little evidence of marine modification at the intertidal level and descend to the 12 fathom (about 20 m) submarine contour, close inshore and parallel to the coastline.

(b) It is possible that powerful waves breaking against a soft rock coast (such as the cliffed drumlins in western Ireland) produce plunging cliffs by cutting a platform below low tide level (Guilcher, 1966). This may explain why the Nullarbor cliffs in southern Australia, cut in soft Miocene limestone, have no shore platforms, but descend to a submarine platform a few metres below low tide level, that declines gently sea-

ward. The Portland Stone cliffs of the south coast of Purbeck in Dorset plunge into deep nearshore water where the underlying soft formations (Portland Sand and part of the Kimmeridge Clay) have been swept away by strong waves and currents to leave submerged limestone ledges. Where stacks rise from the nearshore sea floor, rather than from a shore platform exposed at low tide (Figure 4.24), the implication is that cliff recession has been accompanied by downwearing of the shore to produce a platform below low tide level.

(c) Plunging cliffs may be formed by tectonic subsidence of coastal regions, partially submerging escarpments or hilly country, so that coastal slopes descend to a submerged undulating foreground. The plunging cliffs of Lyttleton Harbour and Banks Peninsula in New Zealand originated as the result of subsidence in this area during and since the Holocene marine transgression.

(d) The explosive eruption of Krakatau, a volcano in Sunda Strait, Indonesia, in 1883 left steep plunging cliffs on residual islands facing into the caldera, and similar features are seen on Santorini, north of Crete, which exploded about 3500 years ago.

(e) Coasts built by recent volcanic activity, as on the island of Hawaii, show plunging cliffs on sectors where there has not yet been time for shore platforms to have formed.

(f) The absence of shore platforms on very sheltered sectors of plunging cliffs bordering rias and fiords is explained by low wave energy across short fetches following marine submergence of glaciated topography.

(g) Plunging cliffs on coasts where rock outcrops are extremely resistant is probably due to the fact that the period of up to 6000 years since the sea attained its present level has

Figure 5.13 The 600 m high plunging cliff on quartzite at Slieve League, on the coast of Donegal, NW Ireland, is a glaciated coastal slope: largely a corrie wall shaped by glacial and periglacial action. It was invaded by the sea in Holocene times and then only slightly modified by basal marine erosion following the Holocene marine transgression

been too brief for marine processes to have cut platforms. Examples of this are seen on the granites on the Land's End Peninsula and the Bullers of Buchan on the east coast of Scotland, where steep rocky coastal slopes descend beneath the sea. Similar features are seen on the various igneous and metamorphic rocks that form the high and rugged cliffs on the Lizard Peninsula in Cornwall, exposed to strong wave action from the Atlantic Ocean, and the high cliffs of Moher and Slieve League in NW Ireland also plunge into the sea (Figure 5.13). There are several sectors of steep, plunging cliff on the flanks of resistant limestone massifs bordering Mediterranean coasts: for example in Sardinia. In Australia the plunging cliffs on massive granite on Wilsons Promontory in Victoria are the partly submerged slopes of granite hills and mountains.

(h) Some hard rock cliffs on the west coast of Scotland and in the Hebrides plunge well below present sea level to submerged shore platforms that were formed when the cliffs were cut back under glacial or periglacial conditions and sea level was lower. Such planation was probably rapid, and similar to the processes that formed the strandflats seen on emerging coasts, as in Norway.

(i) Alternatively, the cutting of a shore platform at present sea level could have been prevented by continuing land subsidence or frequent sea level changes. In formerly glaciated regions shore planation has not occurred because of continuing isostatic uplift: the sea has stood only briefly at its present level.

5.4 Summary

Processes at work on rocky shores are largely erosional, dominated by abrasion by waves armed with sand and gravel. These can quarry rock fragments and cut grooves, clefts and potholes, as well as sloping ramps at the base of a receding cliff. In addition there are various weathering processes, including repeated wetting and drying, salt crystallisation and solution, resulting in the pitting, fluting and disintegration of rock outcrops, and bioerosion by marine organisms. On cold coasts there are also freeze–thaw processes and the effects of shore ice. Although most shore processes are destructive, some can result in the induration of shore rocks.

These various processes can shape shore platforms, some of which are seaward sloping (generally formed by abrasion) while others are subhorizontal (generally formed by weathering). Where they coincide with a flat or gently sloping rock surface, usually a bedding plane, they are termed structural. Subhorizontal shore platforms may stand at or near high tide level (formed mainly by wetting and drying), others just above low tide level (formed mainly by solution). On some coasts emerged shore platforms (cut when sea level was relatively higher) are undergoing dissection as newer platforms develop in relation to present sea level.

Strandflats are broad relatively flat surfaces fronting mountain slopes in formerly glaciated areas, shaped by frost shattering or sea ice. Irregular shores where platforms have not yet developed are rocky and boulder strewn. Cliffs that descend directly into deep nearshore water without any platforms or rocky shores are termed plunging cliffs, and are generally the result of faulting, subsidence, volcanic activity or prior glaciation on hard rock coasts.

6
Beaches

6.1 Introduction

A beach is an accumulation on the shore of generally loose, unconsolidated sediment, ranging in size from very fine sand up to pebbles, cobbles and occasionally boulders, often with shelly material. Beaches fringe about 40 per cent of the world's coastline, and generally consist of unconsolidated deposits of sand and gravel on the shore. Some are long and almost straight or gently curved; others are shorter, and include sharply curved pocket beaches in bays or coves between rocky headlands. Many are exposed to the open ocean or stormy seas, but others are sheltered in bays or behind islands or reefs. Beach systems deal with the interactions between beaches and the processes (waves, currents, tides and winds) that work on them.

Beaches form the seaward fringes of barriers (Chapter 8), which are banks of beach material deposited across inlets and embayments to enclose lagoons and swamps. Some beaches are bordered by deep water close inshore; others face shallow or shoaly water, often with bars, which are intertidal or subtidal ridges of sand or gravel running parallel, or at an angle to, the high tide shoreline (Section 8.8). There are beaches that have been fairly stable over periods of years or decades, but most show rapid changes, especially in stormy weather. Some beaches are obviously gaining or losing sediment; others consist of sediment in transit (migrating along the coast), and others remain in position and may be relict, without any present-day sediment supply. Many beaches change in plan (i.e. the shape seen on a map or vertical air photograph) and profile (transverse to the shore), either rapidly over a few hours or days, or slowly over several decades or centuries. Some changes are cyclic (the beach returning to the same plan and profile over varying periods); others are prograding (advancing seaward by deposition) or receding as the result of continuing erosion. There are also artificial beaches, formed by the dumping of sand or gravel brought from the land, or alongshore or offshore sources.

There is a problem in defining the seaward limit of a beach because beach sediment extends for a varying distance offshore. Short (1999) defined a beach as 'a wave-deposited accumulation of sediment lying between modal wave base and the upper swash limit, where wave base is the maximum depth at which waves can transport beach material shoreward'. However, since this wave base is conventionally where water depth is half the wavelength (Section 2.2), this definition sets a seaward limit on many ocean coasts in water at least 50 m deep, perhaps several kilometres

Coastal Geomorphology: An Introduction, Second Edition. Eric Bird.
© 2008 by John Wiley & Sons, Ltd

offshore. Most accounts of beaches admit their extension below lowest tide level and for some distance (usually not stated) beyond the breaker zone.

A beach profile is shaped by swash and backwash as waves break upon the shore, and varies in response to wave conditions. Constructive waves shape convex profiles, often with one or more swash-built berms (ridges or beach terraces) parallel to the coastline, while erosive waves cut scarps and shape concave profiles.

Beach profiles can be surveyed using graduated poles and a level, a theodolite, an electronic distance measurer, a global positioning system (GPS) or a wheeled vehicle designed to register rise and fall along a beach transect. Profiles can be surveyed alongside a pier or groyne, but there is a risk that they are anomalous in the vicinity of such structures. Beach morphology can be mapped by conventional survey methods, or by using GPS in traverses that can be translated into morphological maps by computer. Alternatively, remote sensing techniques such as x-band radar, airborne laser terrain mapping (ALTM) and light detection and ranging (LIDAR) have been used by the United States Geological Survey and other agencies. The profiles should ideally extend down the beach and out beneath the breaker zone, but nearshore surveys can present difficulties when wave action is strong. Repeated profile surveys can be made to determine patterns and rates of change on a beach. There is a problem of achieving sufficient resolution to be able to measure vertical changes (aggradation or degradation) or horizontal changes (progradation or retrogradation) of a few centimetres, particularly in the nearshore zone, but vertical changes of ± 10 cm have been measured (Leatherman, Whitman and Zhang, 2005; Davidson-Arnott, 2005).

6.2 Beach sediments

Beach sediments consist of sand or gravel particles of various sizes (Table 6.1), the proportions

Table 6.1 Beach grain size categories. The Wentworth scale of particle diameters. The ø scale is based on the negative logarithm (to base 2) of the particle diameter in millimetres (ø = log 2d), so that coarser particles have negative values

Wentworth scale category	Particle diameter (mm)	ø scale
Boulders	>256	below −8ø
Cobbles	64–256	−6ø to −8ø
Pebbles	4–64	−2ø to −6ø
Granules	2–4	−1ø to −2ø
Very coarse sand	1–2	0ø to −1ø
Coarse sand	1/2–1	1ø to 0ø
Medium sand	1/4–1/2	2ø to 1ø
Fine sand	1/8–1/4	3ø to 2ø
Very fine sand	1/16–1/8	4ø to 3ø
Silt	1/256–1/16	8ø to 4ø
Clay	<1/256	above 8ø

An alternative is a decimal scale centred on 2 mm, in which sand ranges from 0.2 to 2.0 mm, but this does not correspond well to generally perceived grain size categories. The sand range excludes sediment that would be generally classified as fine to very fine sand, and the coarser (>2.0 mm) and finer (<0.2 mm) divisions do not match widely accepted categories of pebbles and cobbles or of silt and clay.

Panel 6.1   Granulometric analysis of beach sediment

A sample of beach sediment, washed and dried, is passed through a set of sieves of diminishing mesh diameter to divide it into size grades, which are weighed separately. The results are presented as a graph of grain size distribution (Figure 6.1), the steepness of which increases with the degree of sorting. Beach sediments are generally well sorted, so the bulk of a sample falls within a particular size grade (fine sand in Figure 6.1) and the central part of the graph rises steeply. The grain size distribution of a beach sediment is usually asymmetrical and negatively skewed (the mean grain size being coarser than the median), as the result of the removal of fine particles by wave (and on sandy beaches wind) action.

The shape of the curve can be characterised numerically by using the median diameter (P_{50}, the 50th percentile) and selected higher and lower values, such as the 16th and 84th percentiles, as an indication of the relative proportions of coarser and finer particles. In Figure 6.1 the median (P_{50}) is 1.4, P_{16} is 0.7 and P_{84} 1.9. Values for the mean, sorting (standard deviation) and skewness can be calculated from the following formulae:

$$\text{mean} = \tfrac{1}{2}(P_{16} + P_{84}) - \text{in this case } \tfrac{1}{2}(0.7 + 1.9) = 1.3$$

$$\text{sorting} = \tfrac{1}{2}(P_{84} - P_{16}) - \text{in this case } \tfrac{1}{2}(1.9 - 0.7) = 0.3$$

$$\text{skewness} = \frac{\text{mean} - \text{median}}{\text{sorting}} - \text{in this case} \frac{(1.3 - 1.4)}{0.3} = -0.33$$

There are descriptive categories for sorting and skewness. A sediment with sorting less than 0.35 is very well sorted, 0.35–0.50 well sorted, 0.50–1.00 moderately well sorted, 1.00–2.00 poorly sorted, 2.00–4.00 very poorly sorted and >4.00 extremely poorly sorted. A sediment with skewness −1.00 to -0.30 is strongly negatively skewed, −0.30 to −0.10 negatively skewed, −0.10 to 0.10 nearly symmetrical, 0.10 to 0.30 positively skewed and 0.30 to 1.00 strongly positively skewed.

The sample in Figure 6.1 is therefore very well sorted (0.30) and strongly negatively skewed (−0.33). These parameters can be calculated with the aid of a computer program, but this example indicates the principles involved.

of which can be determined by grain size (granulometric) analysis, as shown and explained in Panel 6.1 and Figure 6.1. Some are coarse, dominated by cobbles and pebbles, others finer, with various grades of sandy sediment, granules being relatively rare. Some are uniform (i.e. well sorted), with granulometric analysis showing a high proportion of a particular size grade; others are more varied in texture, sometimes with contrasting zones of coarser and finer material along the beach face. Beaches are better sorted on high wave energy coasts, particularly if they are swash dominated (Section 6.9).

Many of the world's beaches are sandy, but coarser particles (gravels, which comprise granules, pebbles or cobbles, generally of stone but sometimes shelly) are often present, and may be scattered across a sandy beach or arranged in patterns such as cusps or ridges running parallel to the shore. These mixed sand and shingle beaches are highly variable, and their patterns may change within a few hours in response to varying waves and tides. Sand and gravel particles may be angular or subangular in shape, but usually become rounded by abrasion. Collisions may result in pitted surfaces (percussion marks), but as sand and gravel are agitated by wave action the particles are worn smooth and gradually diminished by attrition. As attrition and rounding proceed, sand grains become smooth and highly polished, while pebbles and cobbles tend to become slightly flattened, and thinner at

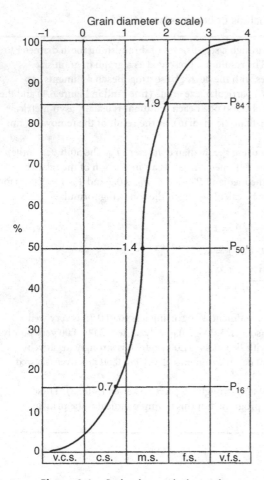

Figure 6.1 Grain size analysis graph

the beach face washed regularly by wave action, up to just above high tide level, angular or suban-gular material persisting at the top and bottom of the beach. This has been observed at Delec in Brittany and below the quarry at Grassy on King Island, Tasmania. Rates of abrasion of flint pebbles have been measured on beaches in East Sussex (Dornbusch *et al.*, 2002, 2004). In Brit-tany Guilcher (1958) observed that beach grav-els remained poorly rounded and poorly sorted even on beaches that had attained smooth, gen-tly curved outlines.

Beach gravels remain angular on arctic and antarctic coasts, where rounding by wave ac-tion is halted during prolonged winter sea freez-ing, and frost shattering occurs: the addition of frost-shattered and solifluction gravel outweighs the rounding and smoothing effects of summer wave action. These beaches are also less well sorted and often include muddy zones. Freeze–thaw processes form cracks and mounds and produce stone polygons on the poorly sorted beaches.

Attrition of pebbles may produce fine pow-dery material, as on Chalk coasts, where the sea is often green or milky with chalk in suspension (hence the Opal Coast in northern France). On beaches no longer receiving sediment, attrition gradually reduces the beach volume, lowering and flattening the beach profile.

Beaches composed entirely of well rounded pebbles and cobbles are known as shingle beaches, especially in Britain: they have also been described as coarse clastic beaches (Orford *et al.*, 2002). There are also boulder beaches, with heaps of stones ranging in diameter up to more than a metre, as on parts of the Dorset coast, where they are derived from disintegrating lime-stone and sandstone outcrops.

Descriptions of beaches sometimes consider the shape of beach sediment particles (usually sampled from the modal size range). An index of roundness can be obtained by dividing the

right angles to their longest axis. The rate of such wearing depends on the hardness and structure of the grains, quartz sand and flint gravel be-ing relatively durable, whereas sandstone frag-ments may disintegrate and disperse. Pebbles of dense, heavy rock, such as flint or basalt, are more slowly transported and reduced by attri-tion than those of softer sandstone. There are also pebbles and cobbles of soft material such as clay or peat, eroded from cliffs cut in fractured outcrops of these materials.

Angular gravel dumped on the shore from coastal quarries becomes rounded pebbles on minimum diameter by the maximum length (for a spherical pebble 1.0) and an index of flatness by dividing the sum of the length and width by twice the thickness (minimum flatness, a spherical pebble, 1.0).

Most pebbles on shingle beaches are ovoid, with length and breadth exceeding width: spherical pebbles are very rare. Some beaches consisting of well sorted and well rounded, highly polished sand grains emit a squeaking noise when walked on, as at Whistling Sands on the Lleyn Peninsula in North Wales, and Squeaky Beach on the west coast of Wilson's Promontory, Australia.

Beaches dominated by granules (grain diameter 2–4 mm) are comparatively rare, but they do occur locally where there is a source of material of this calibre (e.g. quartz and feldspar grains in decomposed granite) or where a vein of quartz or basalt has disintegrated into gravel that has been reduced to this size range by attrition. Vault Beach, near Dodman Point in Cornwall, is dominated by quartz granules, derived from narrow quartz veins in the Dodman Phyllite. Conventionally, deposits of sediment finer than sand (i.e. silt and clay) are regarded as muddy shores rather than beaches.

Lithification of sandy beach sediment by interstitial precipitation (usually of carbonates) leads to the formation of layers of beach rock (beach sandstone), while cementation of pebbles forms beach conglomerate and cementation of angular gravel yields beach breccia (Section 6.6).

6.3 Evolution of beaches

In Chapter 3 it was shown that the sea has risen and fallen several times around the world's coastline during the past few million years, so that former coastlines, some with beaches, are found at various levels above and below present mean sea level. Existing beaches are geologically of recent origin, having formed as the Holocene marine transgression slackened, or gave place to a sea level still-stand: on most coasts about 6000 years ago. Sand and gravel from various sources have been delivered to beaches in coves and bays along steep coasts, except where there are cliffs plunging into deep water, or where the shore is too rocky and rugged to retain a beach. Waves reflected by steep coasts prevent beach deposition, although there is often a sandy sea floor seaward of a steep rocky coast, as on the Land's End Peninsula, where the nearshore profile is too steep for waves to carry sand shoreward to form a beach.

Roy *et al.* (1994) suggested that sandy beaches occur where the transverse gradient of the coast is between 0.1 and 0.8°. Where the gradient is gentler waves cause shoaling and the formation of sand bars, while on steeper gradients waves move sediment offshore and erosion may result in exposure of bedrock. Shingle beaches can occur on much steeper transverse gradients.

Perched beaches form where swash deposits sand or shingle between mid-tide and just above high tide level, fronted by a rocky or bouldery shore. Examples are seen in the granitic Isles of Scilly off SW England, notably at Wingletang Beach on St Agnes. At Bunbury on the west coast of Western Australia basalt ledges at mid-tide level cause wave reflection that inhibits the formation and persistence of beaches in front of the rock outcrop, but allows swash to shape a perched upper beach behind it.

Beaches have also formed along the fringes of coastal lowlands, except where wave energy is too low, and the shore becomes muddy and marshy. Beaches may be coarser on high wave energy coasts, but there are many sandy beaches on high wave energy coasts where there is (or has been) a sandy supply but coarser sediment is unavailable, while gravelly beaches occur on low wave energy coasts where there is a source of coarse material. Beaches are absent altogether where there is no niche for deposition (as on plunging cliffs), where no sand or shingle has been supplied, or where they have been removed by erosion. A change in coastline orientation may be accompanied by the disappearance of beaches because sand drifting alongshore passes on out to the sea floor instead of moving round to the next coastline sector.

Some beaches remain narrow, fringing cliffs and steep coastal slopes or bordering alluvial plains and wetlands, while others have widened with the addition of successively formed backshore beach ridges (Section 6.18), that may bear dunes built of sand winnowed from the shore.

There are relationships between patterns of refracted waves approaching the shore and the sediment characteristics of beaches. Where convergence of wave orthogonals (Figure 2.2) indicates augmented wave energy (i.e. larger waves breaking on the beach) beaches become generally steeper, higher and better sorted; erosion is more severe and divergence of longshore currents causes sediment dispersal. Divergence of orthogonals indicates a low wave energy coast, with the reverse of these conditions: lower beaches with gentler gradients, generally finer and less well sorted beach sediment, reduced erosion or perhaps deposition, and convergent longshore currents bringing in beach sediment. These relationships are complicated, however, by other factors, such as the nature of available sediment.

6.4 Provenance of beach sediments

The origin of the various kinds of beach sediment can be determined with reference to their petrological and mineralogical characteristics, and to patterns of sediment flow produced by waves and currents on the coast and in nearshore areas. Beaches have received their sediments from various sources (Figure 6.2).

Some have been supplied with sand and gravel washed down to the coast by rivers (Section 6.4.1): either large rivers draining a catchment that yields an abundance of such sediment (e.g. the De Grey River in NW Australia) or where rivers drain steep hinterlands (e.g. on the west coast of South Island, New Zealand). Others consist of material derived from the erosion of nearby cliff and foreshore outcrops (Section 6.4.2), particularly where these include weathering sandstones and conglomerates. Sand and gravel have been washed in from the sea floor by waves and currents to form beaches, particularly on oceanic coasts; many of these are shelly or calcareous, derived from marine organisms (Section 6.4.3). In a few places beaches have been supplied with sand delivered by winds blowing from the hinterland (Section 6.4.4). In recent decades many beaches have been augmented by the arrival of sediment produced as the result of human activities, such as agriculture or mining on the coast or in the hinterland (Section 6.4.5). Some beaches have been artificially nourished or replenished, especially at seaside resorts. While many beaches are still receiving sediment from one or more of these sources (Section 6.4.6), some have become relict, and now consist of deposits that accumulated in the past, but are no longer arriving (Section 6.4.7). The following sections exemplify these several categories.

6.4.1 Beaches supplied with fluvial sediment

Fluvial supply of sediment to beaches occurs where sand and/or gravel washed down to the mouth of a river is carried along the coast by waves that arrive at an angle to the shore (Figure 6.3).

Alternatively, rivers may sweep sediment out to the sea floor, where it is reworked by waves and currents, and some of it (mainly the coarser sand and gravel) delivered to the nearby coast. There is

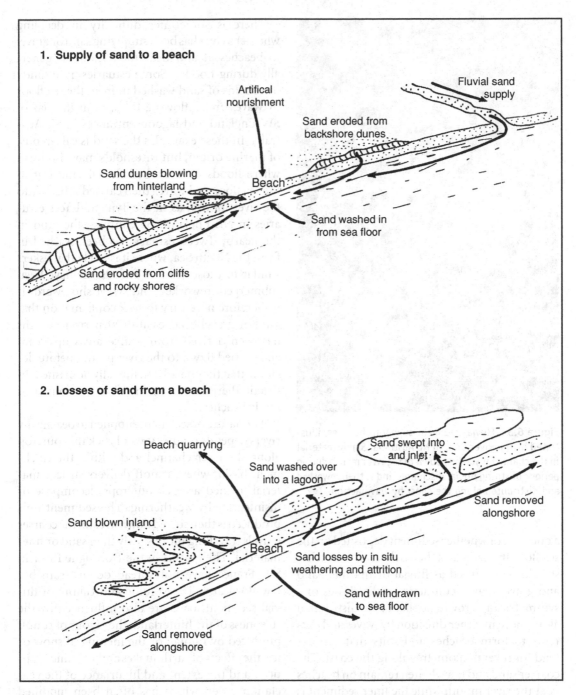

Figure 6.2 Gains and losses of sand from a beach. Shingle beaches have similar gains and losses, except for those produced by wind action

Figure 6.3 Fluvial sediment supply to beaches, illustrated by the River Don, in NE Australia. Sandy material in the channel is washed down to the river mouth during periods of strong fluvial discharge, and is then reworked and distributed along the coast by wave action

There is sometimes difficulty in deciding whether a river has been supplying sand or gravel to beaches at and near their mouths (generally during floods). Some estuaries have sandy thresholds of sand washed in from the sea floor by waves and inflowing tides, as in the rias of SW England and lagoon entrances in SE Australia. In these examples the sand is calcareous, of marine origin, but thresholds may also form where floods have washed fluvial sand out to the sea floor and waves then carried it back into the river mouth, as in the Tyne and Tees estuaries in NE England. By contrast, the sand in the nearby Tweed estuary has been washed in from the North Sea, where it is likely that eskers similar to those seen in Tweedsdale have been submerged, reworked and swept shoreward. It is therefore necessary to seek confirmation that sand or gravel beaches at or near river mouths has been derived from source areas upstream and carried down to the river mouth before deciding that they have been fluvially nourished. In Britain there are few (if any) fluvially nourished sandy beaches.

The nature of sediment supplied to beaches by rivers depends on the types of rock that outcrop along the river channel and within the catchment basin, where runoff delivers surface material formed as rock outcrops decompose or disintegrate by weathering. The sediment thus produced is then sorted downstream, the coarser gravels travelling more slowly than sand or finer material. The gravels in the Dordogne River in SW France have not spread downstream below Bordeaux. The volume and calibre of fluvial sediment loads are also influenced by the steepness of the hinterland, the vigour of runoff produced by rainfall or the melting of snow or ice, the effects of earthquakes and volcanic eruptions and the extent and luxuriance of the vegetation cover, which has often been modified by agriculture, forestry, mining or urban development. Examples of each of these are given below.

a question of whether sediment deposited on the sea floor by rivers and then delivered to beaches should be regarded as fluvial or marine. Sand and gravel may accumulate on the shores of a symmetrically growing delta, or be distributed alongshore in either direction by waves and currents to form beaches and spits that can extend for several kilometres along the coast. The coarser sand and gravel often remain on beaches near the river mouth while the finer sediment is carried further along the shore. Rivers that flow into inlets, estuaries or lagoons may deposit their loads of sand or gravel before reaching the sea.

Catchments including weathered granite or sandstone produce sandy sediment, often dominated by quartz, the most durable of common rock minerals and so the most widespread, dominating most sandy beaches. Quartz sand grains are often red, brown or yellow in colour because they are coated with iron oxides, but when these are removed by abrasion or weathering the sands become white or glassy. Feldspars and micas are less durable, but may persist in some beach sediments, especially those receiving abundant material from steep hinterlands and those where feldspathic sandstones (arkoses) outcrop in river catchments, as on the Otway and Gippsland coasts in Victoria, Australia. There are varying proportions of heavy (ferromagnesian) minerals in material derived from such rocks as granite, and these may become concentrated in beach sediments as layers of mineral sand, such as the rutile and ilmenite found on parts of the Australian coast, magnetite in southern Japan and gold in Alaska. Sand brought down from volcanic hinterlands also reflects the composition of the source rocks: weathered basalt, for example, yields grey or black sand dominated by the dark mineral olivine, as on the beaches of Hawaii, Indonesia and Kamchatka.

Sand from the Loire and the Gironde rivers has been delivered to beaches on the west coast of France, the Tagus brings sand down to the Portuguese coast and there are fluvially fed sandy beaches around the Mediterranean and on the Pacific coast of the United States, notably those that have been supplied by the Columbia River, extending southward into Oregon and northward into Washington State. Sandy beaches supplied with fluvial sand derived from sandstone catchments are extensive in Southern California, and in eastern Australia, particularly in Queensland, where beaches include substantial quantities of sand and some gravel brought down by rivers from steep catchments east of the Great Dividing Range. Beaches near the mouths of the Shoalhaven, Moruya, Bega and Towamba Rivers

in New South Wales consist largely of fluvially supplied quartz and feldspar sand originating from weathered granite. Beaches composed primarily of fluvially supplied sand are also seen on coasts adjacent to the mouths of the Gascoyne, Ashburton and De Grey Rivers in NW Australia, each of which has a large catchment of sandy desert from which substantial quantities of sand are carried downstream occasionally, when rapid runoff is generated by the heavy rainfall that accompanies tropical cyclones (Bird, 1978b). Occasional sudden downpours flood wadis in eastern Egypt and Sinai, washing terrigenous sand down to beaches on the shores of the Red Sea. In Kenya sand supplied by the Galana River has drifted southward to prograde the beach at the resort of Malindi (Figure 6.4).

Shingle beaches are found where rivers drain catchments with fissile rock outcrops or extensive gravel deposits and flow strongly enough to deliver this stony material to the shore. In the North Island of New Zealand, Quaternary volcanic deposits include much coarse gravelly material, some of which has been carried downstream by such rivers as the Mohaka, and deposited to form pebbly beaches along the north coast of Hawke Bay (Bird, 1996b). Where the hinterland is steep and mountainous, as on the west coast of South Island, New Zealand, or the Caucasian Black Sea coast, strongly flowing rivers carry gravel as well as sand down to the coast, to be incorporated in beaches.

Shingle beaches are rare in the humid tropics because coastal and hinterland rock outcrops are generally deeply weathered and contain little gravel. There are exceptions on coasts bordering steep hinterlands, as along the north coast of New Guinea below the Torricelli Mountains, and at Lae, which has a cobble beach supplied with gravels from the Markham River.

Beach gravels are common in areas where past or present glaciation or periglaciation has produced large quantities of stony morainic debris that can be washed down to the coast, as

Figure 6.4 Prograded beach at Malindi, Kenya. The vegetation has advanced on to the backshore as the beach has been widened by sand accretion

in southern Alaska where streams swollen by spring melt water carry large quantities of sand and gravel derived from moraines and glacial outwash down to the coast to form beaches on the shores of Icy Bay (Molnia, 1985). Glacifluvial gravels are the source of shingle beaches on the Canterbury coast in South Island, New Zealand, and on the coast of Patagonia. Shingle beaches on the north coast of Scotland include gravelly material derived from Pleistocene glacifluvial deposits and brought down by the Rivers Spey and Findhorn. In Peru, pebble beaches have been derived from Pleistocene conglomerates and supplied to the coast by Andean rivers during phases of wetter climate and stronger runoff. In northern Tasmania, pebble and cobble beaches have been derived from glacifluvial gravels (notably from closely jointed dolerites) carried down to the coast by rivers and distributed alongshore by wave action.

Beaches receive accessions of fluvial sediment during episodes of flooding after heavy rainfall or the sudden melting of snow or ice in the river catchment. Floodwaters carry sand and gravel out to form shoals off river mouths, and when these shoals are reworked by waves and currents after the floods abate some of the sand or gravel is washed onshore to be added to the beach. On steep coasts, such as those in southern Brazil, beaches are periodically renourished with sand brought down by river floods from a steep hinterland where soil erosion has generated an abundance of fluvial sediment. In 1967 torrential rainfall in the Serra do Mar mountains south west of Rio de Janeiro caused extensive erosion, catastrophic river flooding, and the delivery of large quantities of sand, silt and clay to the sea, notably in Caraguatatuba Bay. Wave action subsequently sorted this material and washed the sandy fraction onshore, to be added to the beaches (Cruz *et al.*, 1985). Soil erosion within catchments has increased and coarsened the sediment yield of many rivers. In Gippsland, Australia, rivers that formerly delivered only silt and clay to their mouths have trains of sand and gravel produced by soil erosion moving downstream.

Earthquakes in the hinterland can increase fluvial sediment yields. The severe tremors that shook the Torricelli Mountains in northern New Guinea in 1907 and 1935 caused massive landslides that delivered large quantities of sand and gravel to rivers draining northward to the coast, where beaches were widened, and beach material subsequently drifted alongshore from river mouths, eastward to Cape Wom.

Volcanic eruptions within a river basin can also generate downstream movement of large quantities of sand and gravel. In the North Island of New Zealand river loads have been occasionally increased as the result of episodes of volcanic activity. The Tarawera volcanic eruption in 1886 produced sands and gravels that moved down rivers such as the Rangitaiki, and beaches on the shores of the Bay of Plenty (notably on Matakana Island) prograded when this material reached the coast (McLean, 1978). In a similar way successive eruptions of Merapi volcano, north of Jogjakarta in Java, have resulted in downstream flow of sand and gravel in the Opak River, the arrival of this sediment at the river mouth being followed by the progradation of beaches of grey volcanic sand at and west of Parangtritis (Bird and Ongkosongo, 1980). In SE Iceland glacifluvial streams have had their sand and gravel loads increased by the eruptions of hinterland volcanoes, which also cause catastrophic ice melting and river flooding, leading to the formation of wide outwash plains, known as sandurs, the seaward fringes of which are subsequently reworked by wave action to form beaches. However, although the Mount St Helens eruptions in 1980 delivered vast quantities of sediment to the Columbia and other rivers flowing down to the

Pacific coast, much of it was silt and clay, and there has been little ensuing progradation of the sandy beaches in Oregon and Washington.

Runoff is much more rapid from unvegetated slopes than from slopes that carry a plant cover that intercepts, retards and recycles rainfall and thus diminishes fluvial sediment yields. Rivers draining catchments where the vegetation cover has been reduced or removed deliver larger and coarser sediment loads to the coast. Around the Mediterranean, for example, deforestation, overgrazing and excessive cultivation of hinterlands over the past 2000 years has led to increased runoff, soil erosion and river flooding, so larger quantities of fluvial sediment have reached the mouths of rivers, and beaches have prograded. In Greece and Turkey rapid deposition at and around river mouths led, for example, to the historical infilling of the Maliakos Gulf (Bird, 1985a). Such augmentation of fluvial sediment supply depends on the continued presence of weathered material on the hinterland slopes, and where this is removed entirely the sediment yield diminishes. The Argentina River, on the Ligurian coast of Italy, had a phase of rapid delta growth and beach progradation when sediment yield from the steep, eroding catchment was high, but in recent decades the supply has declined as bedrock became widely exposed upstream.

Some beaches have been enlarged by the accumulation of sand or gravel brought down by rivers from mining areas in the hinterland. On the island of Bougainville, north of New Guinea, tailings from copper mines increased the load of the Kawerong River, and were carried downstream to be deposited on beaches at and around the river mouth at Jaba, on the shores of Empress Augusta Bay (Brown, 1974). Beaches have been augmented on the Pahang delta on the east coast of Peninsular Malaysia by the arrival of large quantities of sand generated by alluvial tin mining upstream. In New Caledonia extensive hilltop quarrying of nickel and chromium ores has resulted in massive spillage of rocky waste down steep slopes into river valleys and the delivery of increased and coarsened fluvial loads to prograde beaches at and near river mouths, as at Karembé on the western coast and Thio and Houailou on the eastern coast (Bird, Dubois and Iltis, 1984). In Chile the arrival of sandy tailings washed downstream from copper mines resulted in progradation of the beaches in Chañaral Bay (Paskoff and Petiot, 1990), and in SW Ireland waste from a copper mine was carried down the Ballydonegan River to form a beach on the coast of Kerry. In Cornwall the quarrying of china clay from Hensbarrow Downs, a granite upland north of St Austell, produced large quantities of sand and gravel waste with much quartz and feldspar, some of which was carried down rivers to prograde beaches at Par and Pentewan, while tin mining waste was piped through a coastal ridge to form an artificial beach in Carlyon Bay (Everard, 1962). Similar effects are seen in the George and Ringarooma Rivers draining a tin mining region in NE Tasmania. Beaches supplied with sand or gravel from mining waste have convex profiles as long as the supply continues, but when it ceases erosion begins and they become concave.

The contribution of river sediment to beaches has thus been substantial in many parts of the world, but few beaches are entirely of fluvial origin and many have been nourished primarily from other sources now to be considered.

6.4.2 Beaches supplied from eroding cliffs and foreshores

Beach sediment derived from the erosion of cliffs and foreshores has characteristics related partly to the lithology of the outcropping formations and partly to the energy of the waves and currents that erode these outcrops and carry sediment along the shore. Sand eroded from cliffs cut in soft sandstone has nourished beaches at Point Reyes on the coast of Southern California, and where cliffs of Tertiary sand line the shores of Bournemouth Bay in Dorset.

In general, gravel beaches are found where the coastal rock formations have yielded material of suitable size, such as fragments broken from thin resistant layers in sedimentary rocks, or intricately fissured formations, or pebbly conglomerates. Weathered granite yields quartz and feldspar sand, forming cobbles and pebbles only where it disintegrates along intricate closely spaced joint planes into blocks and angular fragments that become rounded as cobbles and pebbles. At Kimmeridge in Dorset limestone ledges on the shore have broken up along joint planes to form a blocky beach. Closely jointed basalt also yields blocks that become rounded by abrasion to cobbles and pebbles, but otherwise weathers by superficial flaking to black sand. The rounded boulders on the beach near Giants Causeway are derived from corestones in weathered basalt rather than directly from the columnar structures. Shingle beaches are not found where the coastal rock outcrops are homogeneous, as on massive granites, or where they are soft and fine grained.

Extensive shingle beaches have been derived from the nodules of hard flint eroded from Chalk cliffs on the south coast of England and the north coast of France (Dornbusch et al., 2006). As the cliffs are cut back the flint nodules released by weathering and erosion form gravelly beaches, which are agitated by wave action and soon become well rounded shingle. Flint nodules exposed in chalk cliffs are black, often with a white rind, and recently derived flint cobbles and pebbles on adjacent beaches are black or blue. The shingle that has drifted eastward along the south coast of England to form accumulations such as the cuspate foreland at Dungeness contains large quantities of brown flints that have not come directly from the Chalk. They have had a much longer and more complicated history, having been weathered during residence in various

Tertiary or Quaternary gravel deposits on land, or on the emerged sea floor during low sea level phases: the change from black to brown results from oxidation of iron compounds, and there has often also been some leaching of silica. The rate of change is not known, but it seems unlikely that many brown flints originated from Holocene erosion of Chalk outcrops.

As well as flint, the shingle beaches of SE England include chert from the Upper and Lower Greensands and quartzites, sandstones and limestones from various Jurassic, Cretaceous and Tertiary rock outcrops along the coast. Each of these rock types is present in beach material in bays along the cliffy coasts of Dorset. Shingle has been supplied from conglomerates or breccias in coastal rock outcrops, as at Budleigh Salterton in Devon, where a cobble beach has been derived from coarse Triassic (desert wadi) gravels exposed in the adjacent cliffs.

Cliffs cut in glacial deposits (rubble drift, gravelly moraines or drumlins moulded by ice movement) have produced sand and gravel delivered to beaches around Puget Sound and other similar coasts, as in New England, the Danish archipelago and the southern shores of the Baltic, notably in Poland. In the British Isles sand and gravel from cliffs of glacial drift have supplied beaches on the North Sea coast (Clayton, 1989) and on the west coast, particularly around the Irish Sea.

Beaches occur where glacial moraines intersect the coastline and cliffs cut into these deposits have provided a source of sand and gravel that has been spread along the coast. Examples have been noted in southern Norway along the Ra moraine, and in Poland and Estonia (Orviku et al., 1995), as well as on the New England coast. The redistribution of material from a morainic zone has nourished beaches such as in Dingle Bay and Tralee Bay on the west coast of Ireland and spits such as Blakeney Point in Norfolk and Whiteness Head in Scotland. Sand, gravel and boulders eroded from cliffs cut in Pleis-

tocene glacial drift on the east coast of England at Holderness coast has been supplied to local beaches and drifted south to Spurn Head. In North Wales the bouldery beaches of Barmouth Bay and the Harlech coast have been eroded from cliffs of glacial boulder clay, and the beach of sand backed by pebbles and cobbles in Hell's Mouth (Porth Neigwl) in North Wales has been derived from erosion of cliffs of glacial drift. Erratic boulders (i.e. not of local derivation) are found on beaches and in the nearshore zone on the formerly glaciated shores of the Baltic Sea and Puget Sound.

Beaches on a cliffed coast cut in massive rock formations, homogeneous limestone or fine grained sediment may be derived from an outcrop of a rock formation alongshore yielding sand or gravel. On the Port Campbell coast in Australia soft Miocene calcareous sediments form vertical cliffs, and there are few beaches except where cliff recession has intersected a dune calcarenite ridge, from which sandy sediment has been derived.

Erratic gravel and boulders of unusual origin are seen on the shores of Petrifaction Bay, on Flinders Island, Australia, where granite blocks that had been transported in a Tertiary lava flow are being eroded out of clay cliffs in the now deeply weathered volcanic rock. On the north coast of France near Port d'Ailly, west of Dieppe, the beach below Chalk cliffs contains scattered boulders of quartz sandstone (known as sarsens) that have fallen from a capping of Tertiary deposits.

In SW England beaches include sand and gravel derived from cliffs cut in the periglacial deposits (frost-shattered earthy rubble known as head) that mantle coastal slopes, and quartzite pebbles have come from disintegrating outcrops of vein quartz from cliffs and shore outcrops in the Devonian rocks. Many beaches in Devon and Cornwall incorporate sand and gravel from Pleistocene emerged beach deposits that stand a few metres above high tide level. These were buried by Late Pleistocene periglacial head deposits and subsequently exposed and cut back as cliffs by marine erosion, as on the shores of Falmouth Bay in Cornwall. On the similar coasts of Brittany, sand and gravel beaches formed during the last interglacial stage (Eemian) were on a larger scale than their modern counterparts, which are partly derived from them (Guilcher, 1958).

On granite coasts the contrast between the upper, leached pale layers of weathered sandy material (including slope wash) and the underlying darker layers, containing illuvial organic matter and iron oxides, can yield contrasts in the colour of derived beaches, as at White Beach and Yellow Beach near Lady Barron, on Flinders Island, Tasmania.

Gravel beaches are rare on humid tropical coasts, but they occur where there is a suitable source, such as the lateritic ironstone crusts that outcrop in coastal cliffs and rocky shores in the Darwin district in Australia and at Port Dickson in Malaysia. Sand and gravel beaches on the shores of the young volcano at Anak Krakatau, Indonesia, have been derived from the erosion of cliffs cut in unconsolidated volcanic ash and agglomerate (Bird and Rosengren, 1984). There are similar beaches around the modern volcanic island of Surtsey, that began to form off the south coast of Iceland in 1963 (Norrman, 1980).

Some beaches have incorporated sand or shingle derived from Pleistocene barrier formations (Section 8.6) truncated by marine erosion, as in the sandy cliffs near Seal Rocks in northern New South Wales. Beaches have been supplied with sand, gravel and boulders by landslides that form protruding lobes, usually with much silt and clay, as on the shores of Lyme Bay and on the North Norfolk coast in England. As these are consumed by wave action the finer sediment is dispersed, and the sand, gravel and boulders are released to nearby beaches.

6.4.3 Beaches supplied with sediment from the sea floor

Shoreward drifting of sand and shingle from the sea floor has contributed to beach deposits, and is most obvious on coasts where there is no supply from rivers or melting glaciers, or from cliff and shore erosion. It has long been known that gravel ballast dumped offshore from ships can move shoreward to be deposited on the beach (Johnson, 1919, quoting Murray 1883).

Sediment washed into beaches from the sea floor includes sand or gravel eroded from submerged geological outcrops or collected from unconsolidated bottom deposits. The latter include sediment that originated from the hinterland, and was spread across the continental shelf by river outflow, glaciers or wind action during phases of lower sea level in Pleistocene times. Shoreward drifting has supplied sediment derived from organisms that live on the sea floor, including shells and biogenic calcareous material, to many beaches (see below).

Just as there are river catchments on land from which coastal sediment has been derived and delivered to the coast at river mouths, so are there sea floor catchments on the continental shelf from which beach material has been, and may still be, carried shoreward, and delivered to parts of the coast. The sandy beaches of Picardy in NE France have received sand from Palaeocene and Eocene formations on the floor of the English Channel to the west. The contrast between the wide calcareous beaches and dunes on the west coast of the Penthièvre isthmus (on the Presqu'ile de Quibéron in Brittany) and the narrow quartzose beaches on the east coast is related to distinct sea floor catchments. On King Island, Tasmania, the contrast between calcareous beaches and dunes on the west coast and the quartzose beaches and dunes on the east coast is related to contrasting sea floor catchments, the continental shelf to the west being an area of calcareous biogenic sediment production (cold upwelling water from the Southern Ocean and a meagre terrigenous input from a low lying, arid continent) and the floor of Bass Strait to the east (strewn with quartzose sediments derived from weathered granites and fluvial deposits at low sea level stages). On the west coast of Sardinia the beach sands are generally calcareous, but at Mari Ermi there is quartzose sand derived from offshore granite around Isola di Mal di Ventre. In NW Iceland beaches of brown calcareous sand from the sea floor contrast with beaches of black or grey volcanic sand and gravel derived from rivers or cliff erosion.

There is the problem (Section 6.4.1) of whether sand and gravel delivered to the sea floor by rivers and then swept up on to beaches should be regarded as of fluvial or marine origin. A similar problem arises where sand has been washed into the sea by streams flowing across beaches during rainy periods (as on the beaches of NW Australia, where transverse gullies are cut in the wet season) or by melting snow or ice, forming nearshore fans that are subsequently reworked by waves and swept back on to the beach.

The depth from which material can drift shoreward varies in relation to sea floor topography, wave and current movements on the sea floor and the size, shape and specific gravity of the available sediment. It is also influenced by the presence of sea floor vegetation, notably sea grasses (Section 10.6) that can inhibit sediment flow. These grow luxuriantly in the Mediterranean and off the South Australian coast. In the Mediterranean, Van Straaten (1959) found that waves move sand on to beaches on the French coast from a depth of 9 m, and on coasts receiving long ocean swell the process may be effective from greater depths than this. Off southern California there is only limited and occasional movement of sea-floor sand by ocean swell in depths exceeding 18 m, indicating the probable maximum limit. Sand dumped in water 12 m

deep off the New Jersey coast was not delivered to the beach (Harris, 1955), and was therefore beyond the limit for shoreward drifting there.

The importance of shoreward drifting of sediment to beaches was deduced in the Isles of Scilly by Barrow (1906) because of the lack of eroding cliffs and sediment-yielding rivers, sand and gravel (including shelly debris) having been swept in from the surrounding sea floor by wave action. Shoreward drifting has occurred where the nearshore waters are shallow, or are becoming shallower because of land uplift or a falling sea level. This can be seen in Scandinavia, where isostatic uplift is causing emergence, so that sand and gravel from submerged but shallowing glacial drift deposits are being carried shoreward on to beaches, as at Storsand in Sweden, Brusand in Norway and Kalajoki in Finland. Similar emergence has led to the formation of a wide sandy beach on the SW shore of the island of Laesø in the Kattegat (Møller, 1985). During the phase of lowering of the Caspian Sea between 1930 and 1977 shoreward movement of sea floor sand took place, and beaches were widened by deposition as the shores emerged, and backshore dunes formed. Around the Great Lakes similar beach progradation has occurred as the result of shoreward drifting during each of several phases of falling lake level (Olson, 1958; Dubois, 1977).

On the coasts of Scotland, where postglacial isostatic recovery is in progress, emerged Holocene shingle beaches and beach ridges are often bordered seaward by younger sand deposits in beaches and dunes, indicating a diminution in the calibre of sediment carried shoreward from the shallowing sea floor.

Widespread shoreward drifting took place on many coasts during the Holocene marine transgression, when the sea advanced across shoaly topography and waves washed sediment on to the shore, forming beaches and dunes (Figure 6.5(a)–(c)). As the sea rose some of the beach material was washed or blown landward, and it

Figure 6.5   On many coasts sand drifted shoreward from sea floor shoals during the world-wide Late Pleistocene–Holocene marine transgression (a), and for a time after this transgression came to an end (b), so that beaches prograded. With the attainment of a smooth concave sea floor profile progradation ceased (c) and the landward migration of this profile has since been accompanied by beach erosion and coastline retreat (d). Where a sea wall has been built along the eroded shore to halt coastline recession there has been further lowering of the beach (e)

Figure 6.6 The Loe Bar, a shingle barrier enclosing Loe Pool on the south coast of Cornwall. Lobes of shingle have been washed over the barrier by storm surges and tsunamis

is probable that minor oscillations of sea level occurred, facilitating shoreward sweeping of sand and gravel on to beaches. Chesil Beach in Dorset is an example of a shingle barrier (Section 8.6) that formed and was driven shoreward as the sea rose, with a lagoon (The Fleet) on its landward side backed by a submerging, indented hinterland coast that was never exposed to wave action from the English Channel to the SW. The Loe Bar in Cornwall (Figure 6.6) and the Slapton barrier beach in South Devon consist of flint shingle swept up from the sea floor on to a coast that has no cliff or shore sources of flint (Figure 6.7). These beaches are probably relict, no longer receiving shingle from the sea floor.

Glacial drift deposits submerged on the floor of the North Sea by the Holocene marine transgression have been reworked by wave action, with sand and gravel carried shoreward to beaches on the east coasts of Scotland and England, and on the Dutch, German and Danish North Sea coasts. Along the Atlantic coasts of Britain and Europe aprons of periglacial head deposits that extended out on to the sea floor during the Last Glacial low sea level phase (Figure 4.15) have been submerged and reworked by wave action to yield sandy material that in some places (Corrubedo and Laxe in Galicia, for example, and Figuera do Foz in Portugal) is still moving onshore to beaches.

Shoreward drifting of sand and gravel has produced beaches in suitable niches, such as shallow bays, coves and inlets on steep and rocky coasts. On oceanic coasts swell has shaped long gently curving beaches, often backed by wide sandy plains with multiple beach ridges and dunes that formed after the Holocene marine transgression came to an end. The bulk of shoreward sand drifting on the Australian coast took place during this transgression, when the nearshore zone migrated shoreward over unconsolidated deposits stranded during the

Figure 6.7 Slapton Ley, a coastal lagoon in South Devon, separated from the sea by a narrow barrier of fine shingle

previous sea regression. Beach progradation occurred in the ensuing still-stand as ocean waves swept sand in from nearshore shoals. Relics of such nearshore shoals persist in sheltered areas to the east of Wilson's Promontory and in the shallows between Robbins Island and Hunter Island off NW Tasmania, and in both cases sand is still moving onshore to beaches.

On the Gippsland coast in SE Australia, where most of the rivers deposit their loads in estuaries or lagoons such as the Gippsland Lakes (Section 11.8), behind wide dune-capped Holocene coastal barriers, and where eroding cliffs are very limited, the Ninety Mile Beach has been formed almost entirely of sand swept in from the sea floor during the Holocene marine transgression and the ensuing still-stand. Successively formed beach or dune ridges backing the Ninety Mile Beach include evidence of Holocene progradation, which in places is still continuing.

Other examples of coasts where sand from nearshore shoals is being carried onshore to be added to beaches include SW Denmark, where the sandy islands of Fanø, Mandø and Romø have beaches prograding in this way, parts of the southern coast of Florida and near Montevideo in Uruguay. Sand is also being washed in from a shallow sea floor to prograde beaches on the shores of Carmarthen Bay in South Wales, the wide sandy beaches on either side of Holy Island in Northumberland and in Studland Bay, Dorset. Progradation has taken place along several kilometres of beach at Tentsmuir in Scotland, and as the beach does not widen towards the mouth of the River Tay it is deduced that shoreward drifting from the sea floor has been more important than fluvial sand supply.

Sand supplied from the sea floor may earlier have been of terrigenous origin, laid down during phases of low sea level and subsequently reworked by wave action during marine transgressions. On the New South Wales coast, sand delivered from the sea floor has smaller proportions of the less resistant feldspar and mica grains than freshly supplied fluvial sand, the beaches becoming predominantly quartzose away from river mouths.

Shoreward drifting is impeded by fringing coral reefs or shore platforms, sediment accumulating against the outer margin of such structures being dispersed alongshore until some of it drifts up through gaps to form pocket beaches on the backing coastline. Short (1999) has described a beach deposited at the base of a cliff behind a shore platform as a platform beach. At Singatoka in Fiji a beach of black volcanic sand has been washed in from the sea floor through a gap in the fringing coral reef. Here, as on many oceanic islands of volcanic origin, there are alternations of black sand beaches with pale calcareous sand derived from corals or shells.

Sandy beaches in southern Australia have been supplied largely by shoreward drifting from the sea floor, with only minor contributions from cliff and rocky shore erosion or from fluvial sources. On the Otways coast in Victoria, for example, Davis (1989) found up to 15 per cent feldspar in beaches derived partly from eroding cliffs and shore platforms in Cretaceous feldspathic sandstones and mudstones, but beaches on this part of the coast consist largely (usually more than 90 per cent) of light brown quartz and carbonate sand washed from the sea floor into small coves and inlets along the rocky shore. Weathering of the feldspathic rocks on the Otways coast yields mainly fine grained sediment (silt and clay), and there are no beaches of cliff-derived sand along the cliff base behind the shore platforms. In general, beaches on the coast of SE Australia contain small amounts of quartz and feldspar sand derived from cliff and rocky shore erosion, and varying proportions of biogenic carbonate sand. Their similarity along cliffed coasts of granite, basalt, sandstone and limestone suggests that they are largely of sea floor origin.

Shelly and calcareous beaches

Beach deposits originating from the sea floor include sand or gravel derived from marine organisms, notably shells, which may be intact or broken, or comminuted to calcareous sand, by the time they arrive on the beach. On tropical coasts there are beaches of coralline and algal gravel derived from the disintegration of fringing and nearshore reefs. The sand-sized discs formed by the filamentous alga *Halimeda* are common on many coralline beaches, and abundant on beaches in Brazil. Shell Beach in Shark Bay, Western Australia, consists entirely of whole and broken shells washed in from the adjacent sea floor.

Shelly and calcareous beaches are best developed where the supply of inorganic (mineral) sand is meagre, as on the subarid coasts of SW Australia where there are few rivers and rocky sectors do not yield sandy sediment. This is one of several coastal sectors where calcareous beaches were a source of sand for dunes and dune calcarenites in Pleistocene times. When a terrigenous sediment supply is reduced or halted (e.g. by the damming of a river or the walling of an eroding cliff) the beach becomes progressively more calcareous.

Some shelly beaches are composed of rock-dwelling species, and occur where there are shore or nearshore rock outcrops. Others have come from shallow sandy and muddy environments in sheltered bays and estuaries (Gell, 1978). Cockles live in nearshore muddy habitats, maintaining their position until they die, when the shells are extracted by wave action and washed ashore, as on the shelly beach at Sembilang, north of the Klang delta in Malaysia. On the Texas coast there are shelly beaches left by deflation of a sandy matrix by onshore winds. Other examples

Figure 6.8 A shelly beach and intertidal zone (Traigh Mhôr) on the Island of Barra in the Outer Hebrides, Scotland. The intertidal sand flat is firm enough to be used as an airfield at low tide

of shelly beaches are seen on Herm in the Channel Islands, at Barricane Beach near Woolacombe in North Devon, and on the shores of the Firth of Forth near Edinburgh. There are mussel shell beaches on the rocky coast of Maine and bordering the Sea of Azov, sandy beaches composed of oolites formed in the clear warm seas of the Bahama Banks in the Atlantic, and shell grit beaches on the shores of the Hebrides in western Scotland. Light foraminiferal sands have formed beaches on the west coast of Ireland, as at Dogs Bay in Galway. Cockle shells are abundant on beaches and intertidal sands on Traigh Mhor, on the Hebridean island of Barra (Figure 6.8). Shelly beaches are also common on the shores of estuaries and lagoons, sometimes bordering salt marshes or emplaced on them as cheniers (Section 6.19), as on the Essex coast.

Coarse calcareous sand, known locally as calcified seaweed, is extensive on the sea floor off south Cornwall, particularly in Falmouth Bay. It is biogenic sand, formed by the alga *Lithothamnium calcareum*, but little of it drifts on to the beaches along the present coast, which are dominated by quartzose sand and gravel and material eroded from coastal rock outcrops. It is dredged for agricultural use as a fertiliser off Cornwall and Brittany (where it is known as maerl), where extraction has led to erosion of beaches in the Bay of the Seine (Cressard and Augris, 1982).

Beach and sea floor accumulations of *Lithothamnium* (with other algae and shell grit) form the maerls of Brittany, and shelly *Crepidula* species introduced from the United States (possibly during the Normandy landings in 1944) have become so common that they now form wide shelly beaches on the shores of the Bay of Mont Saint Michel in NW France.

Shelly and calcareous sand and gravel are gradually consumed by disintegration, abrasion, attrition and solution, particularly in rainwater, so beaches composed of this material persist only if there is a continuing supply from the sea floor.

In southern and western Australia there are calcareous sandy beaches consisting partly of shelly debris, and partly of biogenic material from sand-sized organisms such as foraminifera and bryozoa carried in from the continental shelf by wave action. In Pleistocene times similar calcareous beaches supplied dune sands that were lithified to form the dune calcarenites mentioned previously, a rock sufficiently coherent and resistant to have been eroded into cliffs, shore platforms and nearshore reefs (Figure 5.12). Disintegrating calcarenite provides another source of beach sand, and is a prominent constituent of bay and cove beaches interspersed with cliffs along calcarenite coasts, particularly in Western Australia.

Shelly beaches often contain varying proportions of inorganic sand or gravel, but beaches bordering low lying (i.e. uncliffed) desert coasts may be strongly calcareous because the meagre runoff has yielded so little fluvial terrigenous sediment supply to the shore, beaches being dominated by material washed in from the sea floor. The Eighty Mile Beach, in NW Australia, exemplifies this, and there are shelly beaches bordering the dry northern shores of Spencer Gulf near Whyalla in South Australia. As sources of natural beach nourishment diminish it is likely that the proportion of biogenic sediment on beaches will increase on coasts where nearshore waters provide a rich environment for shelly organisms.

Sea-grass hay on beaches

Where sea grasses such as *Zostera* or *Posidonia* grow abundantly offshore, their stems, leaves and rhizomes may be torn away by storm waves and washed on to the beach. Heaps or ridges of seagrass hay are frequently piled up on beaches,

notably around the Mediterranean Sea and on the southern shores of Australia, particularly on dune calcarenite coasts. The hay absorbs wave action and temporarily protects the beach. Rotting sea grass produces acidic water that dissolves shelly sediment and removes iron oxide coatings from sand grains. Intertwined plant debris may become aggregated and rounded into balls that look like brown cobbles. Sea-grass hay is regarded as a nuisance on resort beaches, but can be collected for use as fertiliser or thermal insulation. The growth of nearshore sea grasses can be suppressed by dumping sand on them.

6.4.4 Beaches supplied with wind-blown sand

Beaches have been supplied with sand blown from the hinterland where there is a suitable source of unconsolidated sand with little or no retaining vegetation, and winds blow from the land to the sea. This happens on arid coasts, as in Angola, where barchans spilling on to the shores of Tiger Bay have added sand to local beaches, on the desert coasts of Namibia between Sandwich Harbour and Conception Bay, in Qatar in SE Arabia and in Mauritania. Desert dust from the Sahara is carried far out to sea, but it is unlikely that beaches and dunes on the Canary Islands have been supplied with wind-blown sand more than 100 km from West Africa.

Where dunes have been built by onshore winds, sand may occasionally be swept back to the beach and into the sea by winds that blow from the land. An example of this is seen on the north-facing shores of the Slowinski National Park (Figure 6.9), on the Polish Baltic coast, where a wide beach has been partly nourished with sand blown by southerly or SW winds from poorly vegetated backshore dunes (Borowca and Rotnicki, 1994). There is similar eastward movement of dunes from the backshore to beaches on Prince Edward Island in Canada, and

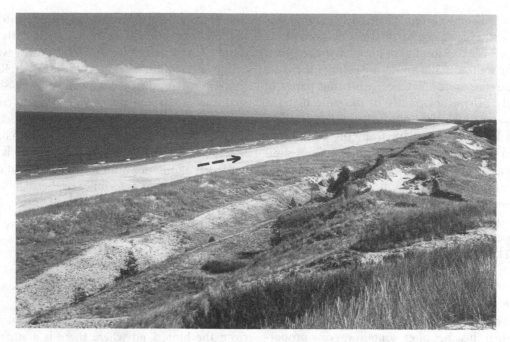

Figure 6.9 Westerly winds blow sand along the beach in the Slowinski National Park on the Baltic coast

westerly winds occasionally sweep sand from backshore dunes to beaches north and south of Holy Island on the NE coast of England, and on the Sands of Forvie in eastern Scotland. Sand blown from backshore dunes tends to flatten beach and nearshore profiles, thereby diminishing incoming waves and reducing beach erosion.

Where the prevailing winds blow more or less alongshore, dunes may drift across promontories and headlands to nourish beaches on the lee coast. There are examples of this on the south-facing Cape Coast of South Africa, notably at Port Elizabeth, where sand driven by westerly winds across Cape Recife has been spilling on to the eastern beaches. There are similar situations in Uruguay, where dunes spilling on to the shore are nourishing the beach near Castillos, in Paracas Bay, Peru, where wind-blown sand has spilled across the Paracas Peninsula and on the Victorian coast in SE Australia, notably at Cape Woolamai and on the Yanakie Isthmus, Wilson's Promontory (Figure 6.10). On Cape Ot-

way, Pleistocene dune calcarenites extend across the headland, indicating that this process was active at earlier stages.

6.4.5 Beaches made or modified by human activities

Many beaches contain small proportions of sand or gravel formed from fragments of glass, concrete, brick and earthenware produced by human activities. These are prominent on coasts with a long history of human settlement, as around the Mediterranean, or close to large urban or industrial centres. At Workington, on the coast of Cumbria, NW England, the beach is dominated by basic slag waste from a former steelworks. While the steelworks was active, dumping of this material prograded the coastline, comparison of 1884 maps with 1981 air photographs showing an advance of up to 200 m. The slag tip has since been cliffed by marine

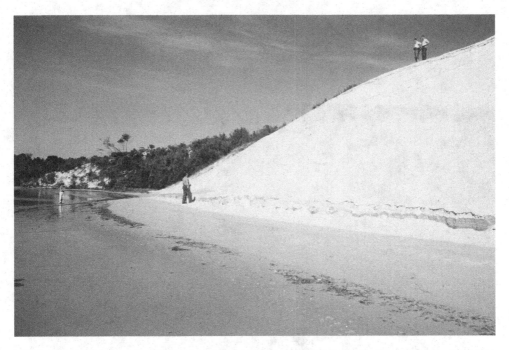

Figure 6.10 Wind-blown sand spilling from the Yanakie dunes, Wilson's Promontory, Victoria, Australia, on to the shore of Corner Inlet, where it nourishes sandy beaches

erosion, and as the cliffs retreat sand and gravel derived from the waste drifts northward along-shore, augmenting the natural beach as far as the Pell Mell breakwater at the mouth of the Derwent River (Empsall, 1989). Beaches near ports have sometimes been augmented by sand or gravel from ballast brought in by ships and dumped before taking on a cargo, as at the for-mer coal port of Saundersfoot in SW Wales and the china clay port at Charlestown in Cornwall. The beach in Oriental Bay near Wellington, New Zealand, has been received sediment from ships' ballast dumped offshore.

Some beaches have been supplied with flu-vial sediment increased by mining activities in the hinterland (Section 6.4.1), and other beaches have been directly supplied with material from quarry waste spilling on to the shore. At Hoed in Denmark wave action has distributed gravelly flint waste dumped from a coastal quarry cut in chalky glacial drift, and built it into a series of low parallel beach ridges fronted by a shingle beach extending north and south from the quarry (Fig-ure 6.11) (Bird and Christiansen, 1982). Gravelly waste from coastal quarries tipped over cliffs into the sea on the east coast of the Lizard Peninsula in Cornwall has accumulated in adjacent coves at Porthallow and Porthoustock as widened grav-elly beaches (Figure 6.12). The beach at Rapid Bay in South Australia widened by up to 230 m as the result of spillage of quarry waste (Bourman, 1990). On the west coast of Cor-sica the arrival of waste debris from an asbestos mine produced a beach in a cove where none existed previously (Paskoff, 1994). Other exam-ples include the Nganga Negara tin mine site on the west coast of peninsular Malaysia, where the dredging of tin from a depositional apron and coastal plain fronting the granitic Segari Hills generated large quantities of quartzose sand,

Figure 6.11 Shingle beach ridges on the coast at Hoed in Denmark, formed by accretion of gravel waste dumped from a coastal quarry in chalky glacial drift. The house (arrowed) stands on the 18th century coastline behind the prograded beach ridge plain

gravel and boulders, that were heaped as a high tailings bank along the coastal fringe. Waves have reworked this material to form a gently shelving sandy beach, finer and better sorted than the tailings sediment. Waste material from a large granite and diorite quarry at Ronez, on the north coast of Jersey, has spilled down to the shore and formed a beach downdrift to the east.

Waste dumped on the shore from coal mines has developed into a beach of black pebbles and sand at Lynemouth in Northumberland, and similar material from coastal collieries has been deposited on several parts of the Durham coast, notably between Seaham and Easington and at Horden. On Brownsea Island, in Poole Harbour, Dorset, wave action has distributed debris from a former pipeworks to build a beach of broken subangular earthenware. Antique brown pottery fragments are mixed with white pebbles in shingle beaches south of the Corinth isthmus in

Figure 6.12 Progradation of the beach in the cove at Porthoustock, Cornwall, was due to the arrival of gravelly material derived from waste tipped into the sea from nearby coastal quarries

Greece, and rounded glass and porcelain pebbles on Glass Beach at Fort Bragg in California were derived from garbage formerly dumped over a cliff on to shore. Many beaches contain material derived from garbage of various kinds, including fragments of broken glass, metal, brick and plastic, derived from bottles, cans, containers and other litter dumped on the shore, carried down by rivers or washed in from the sea.

Driftwood, including sawn timber, is extensive on beaches bordering high, forested hinterlands, as in British Columbia and the Washington and Oregon coasts. On the beaches of the Westland coast, New Zealand, driftwood consists mainly of trees washed down to river mouths, but coasts adjacent to lumbering areas in western Canada and northern Russia are heaped with sawn timber washed up on the shore. It is erosive when it lies at an angle to the coastline, and can be agitated by waves as an abrasion tool, but protective where it lies parallel to the coastline.

Artificially nourished (or renourished) beaches, formed partly or wholly by the dumping of sand or gravel, brought to the shore from inland quarries, alongshore sources or dredged from the sea floor, on the shore are now extensive, particularly in the United States, Europe and Australia. Artificial beaches can be formed from sediment obtained from the sea floor (notably the dredged approaches to ports) and dumped on or near the coast, as at Odessa on the Ukrainian Black Sea coast (Shuisky, 1994).

6.4.6 Beaches of mixed origin

Most beaches contain sand or gravel from more than one of these sources, the proportions received from rivers, eroding cliffs and shores, the sea floor or aeolian inputs varying considerably along coastlines. In Southern California beaches have been fed with fluvial sand from such rivers as the Santa Clara and Ventura, but they also include much material eroded from coastal cliff outcrops, as well as sediment, including shelly debris, swept in from the sea floor. On the Cape Coast of South Africa beaches include sand of aeolian origin where dunes are spilling along the shore, as well as fluvial, marine and cliff-derived sediment. Fluvially supplied beaches downdrift from a river mouth may gradually become mixed with sediment from other sources, such as eroding cliffs or the sea floor, as on the shores of Hawke Bay in New Zealand (Figure 6.25). The fluvial contribution may remain identifiable by mineralogical evidence, as on the Ninety Mile Beach in SE Australia, which consists largely of well rounded quartz sand and shell fragments washed in from the sea floor, but in the vicinity of the mouth of the Snowy River also contains fluvial sand, supplied by occasional river floods, with distinctive minerals, including augite. North Queensland beach sediments are varying mixtures of sand from rivers, eroding cliffs and nearshore reefs. Samples from Garners Beach, near Bingil Bay, contain quartz and feldspar sand supplied by nearby rivers, fragments from local rock outcrops, particles of disintegrated beach rock, ferruginous sand-rock from an eroded Pleistocene sandy beach formation and coralline and shelly material from an adjacent reef. On the western shores of the Arabian Gulf the beaches consist largely of sand blown from the desert, but also include shelly and inorganic sediment washed in from the sea floor. Near Broome in NW Australia there are beaches derived from the red pindan sands (quartzose sand with an iron oxide stain) in elongated desert dune ridges that have been submerged and reworked by wave action to produce clean white quartz where the iron oxide stain has been removed by abrasion. In Roebuck Bay and along the Eighty Mile Beach this quartzose sand is mixed with pale grey calcareous biogenic sand washed up from the sea floor. Similar mingling of red desert sand and pale shelly marine gravel is seen at the northern ends of the South Australian Gulfs (Spencer Gulf and Gulf St Vincent). Pumice derived from submarine volcanic eruptions floats on seawater, and is a common, and often far travelled, constituent of many oceanic beaches, particularly around the Pacific.

6.4.7 Relict beaches

Some beaches have become relict, the sediment sources that originally supplied them being no longer available. This may be due to the natural or artificial diversion of a river, the construction of a dam or the implementation of successful anti-erosion works in a river catchment, so that the former fluvial supply of sediment to the beach has ceased. Alternatively, it can be the result of the halting of cliff erosion by sea wall construction, or emergence due to land uplift or sea level lowering, resulting in withdrawal of wave attack from the cliff base. On many oceanic beaches there is evidence that sediment is no longer being washed in from the sea floor because the supply from unconsolidated shoaly deposits has run out (Figure 6.5(c)). Most British shingle beaches are relict in the sense of being coarse material that was collected and delivered to the coast during the later stages of the Holocene marine transgression, and soon after it came to an end, but in the vicinity of eroding Chalk cliffs or Tertiary or Quaternary gravel deposits beaches are still receiving shingle.

Relict beaches, continually reworked by wave action, can become very well sorted, or develop lateral grading in grain size in relation to incident wave regimes. Chesil Beach, on the shores of Lyme Bay, is essentially a relict beach that it is well sorted, with lateral grading from small pebbles in the west to large cobbles in the SE (Figure 6.13). Such adjustments are impeded where there is a continuing supply of fresh sediment. There is also a tendency for the grain size of beach sediment to diminish as the result of gradual attrition by wave action (more rapidly in the intertidal to high tide swash zone). Pebbles on the beach south of Bridgwater Bay in Somerset diminish in size as the result of attrition as they drift alongshore. Slapton Beach in Devon is an example of a well sorted relict beach of coarse sand and fine shingle, the calibre of which has been reduced by prolonged attrition as the result

Figure 6.13 Lateral grading on Chesil Beach, Dorset. The scale in both photographs is a one foot (approximately 30 cm) ruler. On the left is the beach near the western end (*A* in Figure 6.24), where the mean pebble diameter is between 1 and 2 cm, and on the right the beach towards the SE end (*B* in Figure 6.24), where the mean pebble diameter is about 5 cm

of agitation by wave action. Flint cobbles have
been gradually reduced by attrition to pebbles
and eventually flint sand on the beach on the SW
coast of the Isle of Wight, while the diminution
of basalt cobbles to pebbles, granules and sand
can be traced along beaches at Flinders on the
coast of the Mornington Peninsula in Australia.

6.5 Weathering of beach material

Beach sand or shingle may be weathered phys-
ically by disintegration or breakage resulting
from collisions or impacts with solid rock outcrops or shattering by freeze–thaw processes.
Sand and shingle may also be weathered chemically by decomposition or solution. Pebbles and
cobbles may be drilled or scoured by marine organisms. On quartz sand grains it is possible to
use electron microscopy to distinguish between
percussion marks caused by abrasion due to impacts and etching produced by corrosion.

Physical weathering of sand or shingle diminishes mean grain size and reduces the volume of
a beach in much the same way as attrition. Beach
volume is also reduced by chemical weathering,
notably the dissolving and removal of carbonates (shells, coral, or limestone fragments) by
percolating rainwater or corrosive groundwater.
As has been noted (Section 5.1.3), seawater is
usually saturated with dissolved carbonates, but
at lower temperatures (at night, during winter
or in cold climates), and in aerated sea spray,
the carbon dioxide content rises, causing acidi-
fication, which leads to further solution of carbonates. Cobbles and pebbles of chalk and limestone diminish in size on beaches as the result
of solution (which may produce surface etching) and attrition (which usually causes rounding and smoothing), but it can be difficult to
measure the relative rates of these two processes
on a beach.

6.5.1 Beaches on cold coasts

Beaches are essentially similar on coasts in various climatic environments, except in cold regions where wave action ceases (at least in winter) because of the freezing of the sea. When
the ice fringe melts, wave action shapes characteristic beach forms, including beach ridges,
spits and tombolos, as on the northern coasts of
Alaska and in Antarctica. Around Antarctica the
sea ice margin is up to 1 000 km wide, and separates from the shore during the summer, when
waves generated across a limited fetch may break
on beaches. However, coastal waters are generally calm, or have winds blowing offshore from
the anticyclonic continental uplands, which do
not generate shoreward waves, and diminish
any incoming waves: low wave energy coasts
predominate, and the sea is generally stormier
offshore.

Arctic and antarctic beaches are dominated
by physical weathering (notably frost shattering) and may receive ice-rafted erratic rock fragments and the debris yield of melting coastal
glaciers in summer. Shaping of beaches is subject to shore ice effects, including the piling up
of sea ice driven onshore by winds and waves,
which can form a beach or boulder ridge several metres high. Ice accumulation can also form
irregular mounds on and within beaches, and
when it melts these subside to form depressions
or potholes (Short, 1999). On the shores of the
Gulf of St Lawrence boulder barricades are elongated nearshore rows of boulders deposited by
the grounding of boulder-laden ice rafts when
the ice breaks up in spring (Dionne, 2003; Rosen,
2005). There are also boulder pavements, assemblages of rocks in the intertidal zone on cold climate coasts formed by impaction beneath shore
ice, usually smoothed and often striated (Hansom, 2005). Ice-rafting also delivers material
such as driftwood and salt marsh fragments to
shores in high latitudes.

Figure 6.14 Beach rock exposed by erosion on the shore of a sandy cay on Gili Bidara, Lombok, Indonesia

6.6 Beach rock

Precipitation of carbonates in the zone of fluctuating water table within a beach (related to the rise and fall of tides and alternations of wet and dry weather) can cement beach sand into hard sandstone layers known as beach rock, which may be exposed by subsequent erosion (Figure 6.14). Beach rock (a form of beach calcarenite) has been found on the shores of the Caribbean Sea, around the Mediterranean, the Red Sea and the Arabian Gulf and on the coasts of Brazil, South Africa and Australia. It has been quarried for use as building stone from the beaches of Kuwait.

Formation of beach rock is assisted by high evaporation, which causes upward movement of water and dissolved carbonates in the beach sand (Stoddart and Cann, 1965). Beach rock forms as a layer of beach sand becomes consolidated by secondary deposition of calcium carbonate (as calcite or aragonite) precipitated from groundwater in the zone between high and low tide level. Precipitation of calcium carbonate may be aided or caused by the activity of micro-organisms, such as bacteria, that inhabit the beach close to the water table.

Beach rock is frequently found on tropical beaches, especially on coral cays (Section 6.6). Cementation of beach sediment occurs in warm environments where the interstitial water has a temperature exceeding 20 °C for at least half the year, but occasionally beach rock is found on temperate coasts where the cementing calcite has been supplied by seepage of carbonate-rich water from the hinterland. An example is seen on the beach in Harlyn Bay, on the north coast of Cornwall, where slabs of beach rock cemented by carbonate-rich groundwater seeping from backshore dunes have been exposed by beach erosion near high tide. On the dune calcarenite coasts of southern Australia beach rock occurs in association with carbonates deposited by seepage from the cliffs, but it is necessary to distinguish beach rock from calcrete layers that outcrop in the dune calcarenite. In the Baie d'Audierne, Brittany, shelly conglomerate formed by carbonate precipitation has been eroded into jagged outcrops.

Cementation can proceed rapidly, for artefacts such as bottles have been found incorporated in beach rock. On Mediterranean coasts beach rock outcrops often include fragments of pottery from ancient civilisations. Where the cemented beach material includes angular gravel (often coralline) it is termed a beach breccia, and where rounded pebbles are enclosed, a beach conglomerate. Exposed by beach erosion as seaward-sloping layers of calcareous sandstone (possibly further hardened on subaerial exposure), beach rock can be undercut by wave action and broken into flagstones, which may be thrown up beach by tsunamis or storm surges, as at Port Hedland (Figure 2.7). Patterns of eroded beach rock can be used to trace changes in beach outline, particularly on cays. Beach rock found above the present intertidal level is an indication that coastal emergence has taken place, while sea floor outcrops of beach rock are an indication of submergence. A horizontal variant of beach rock, known as cay sandstone, occurs where carbonates leached by rainwater from the upper layers of a beach are precipitated below, as at Belize in the Caribbean and Diego Garcia in the Indian Ocean.

On coasts of ferruginous sandstone, sand and gravel may be cemented by precipitated iron oxides to form a beach rock (ferricrete) or beach conglomerate. Another form of induration is seen where the beach surface develops a coherent (biscuit-like) crust as the result of interstitial deposition of fine grained sediment or precipitated salt that binds the sand or gravel. Where such a crust has developed, wave action on the beach face may cut a small cliff, capped by the indurated layer: a process known as beach scarping. The profile of beaches formed by deposition of coal-mining waste on the Durham coast in NE England includes a small scarp cut into a surficial layer of sand bound by clayey downwash. Beach surface crusts formed by the precipitation of salt in dry weather may inhibit movement of sand by wind action, but these crusts are usually too soft and friable to impede wave action.

6.7 Nearshore processes

Sediment that has been supplied to a beach is subject to various processes that change its calibre and composition, and result in the shaping of beach morphology. Waves that break parallel to the coastline move sand or gravel either shoreward, when the swash is stronger than the backwash, or seaward, when the backwash is stronger, thereby producing alternations of onshore and offshore drifting.

Waves that arrive at an angle to the coastline produce a transverse swash, running diagonally up the beach, followed by a backwash that retreats directly down into the sea. This results in the zigzag movement of beach material alongshore (beach drifting), that is accompanied by sediment flow along the nearshore zone, generated by the longshore currents that accompany obliquely arriving waves. The combined effect of these processes is longshore drifting (also known as littoral drift, but the term longshore drifting is more accurate because littoral drift could be in any direction across the shore) of sediment to beaches and spits downdrift. Longshore drifting is most rapid when wave crests approach the shore at an angle of between 40 and 50°, where the coastline is straight or gently curved and unbroken by headlands, inlets or estuaries, and where the nearshore sea floor profile is smooth. It increases with wave energy and is aided by a small tide range, which results in more continuous and concentrated wave action than where the zone of breaking waves rises and falls over a substantial tide zone.

Beach sediments may move first one way, then the other, according to the direction from which the waves approach. If waves arrive as frequently from one direction as the other the resultant (or net) drifting over a period will be negligible, but usually one direction predominates. Such long term drifting is indicated by the longshore growth of spits, the deflection of river mouths or the accumulation of beach sediment alongside headlands or breakwaters. The predominance of southward longshore drifting on the east coast of England is shown by the southward growth of spits at Spurn Head and Orfordness, the southward deflection of the River Yare at Yarmouth and the accumulation of beach material on the northern sides of harbour breakwaters at Lowestoft and Southwold. It is interrupted by predominant westward longshore drifting on the north Norfolk coast, shown by the westward growth of spits at Blakeney Point and Scolt Head Island. This divergence of longshore drifting could be related to the impact of NE waves from the North Sea arriving on the large salient of East Anglia and generating westward and southward longshore drifting from the vicinity of Happisburgh. Divergent longshore drifting is also indicated on the Fylde coast in Lancashire by the growth of spits at Rossall Point (Fleetwood) to the north and Lytham to the south.

The beaches of southern England are trains of shingle that drift mainly eastward. Drifting shingle accumulates alongside breakwaters, headlands and landslide lobes (which act as temporary breakwaters), and is interrupted by rocky or bouldery shores, at least until sufficient sand or gravel have arrived to fill in crevices and form a smooth enough surface for drifting to proceed. At Seven Rock Point, near Lyme Regis in Dorset, shingle drifting eastward from Pinhay Bay was delayed for several years by an irregular bouldery shore below a landslide, but resumed after

pebbles had filled the gaps between the boulders. At Sea Palling in Norfolk longshore drifting has been interrupted by a tombolo that has formed on the coast in the lee of an offshore breakwater. Deposition has occurred on the updrift side of the tombolo and beach erosion downdrift.

On the north Queensland coast in Australia the dominant waves generated by the prevailing SE trade winds have drifted sand northward from river mouths to form a succession of beaches and spits that deflect the river mouths northward. The Burdekin delta, for example, has sandy beaches extending northward from the mouths of each of its distributary channels, culminating in the long recurved Bowling Green spit. On the Adelaide coast in South Australia SW waves move sand northward to accumulate on the prograding beach at Largs Bay, where longshore drifting has been intercepted on the southern side of the Outer Harbour breakwater.

Beaches showing alternations of longshore drifting as waves arrive from each direction occur on the Pacific coast of North America because of seasonal changes in the direction of incident swell. In Half Moon Bay and on Boomer Beach in California waves from the SW move sand northward in summer and waves from the NW drive it back south in winter. Further north on the Pacific coast the Columbia River has delivered large quantities of sand that have drifted northward in summer when the waves arrive mainly from the SW, and southward in winter when NW waves prevail. The stronger northward drifting has carried sand up the Washington coast to supply beaches as far as Cape Shoalwater, and southward drifting has built the beach that extends down to Tillamook Head.

On the east coast of Port Phillip Bay, Australia, beaches are diminished by erosion at their southern ends and increased by deposition at their northern ends during summer. This pattern is reversed in winter in response to seasonal variations in the direction of the onshore winds that generate wave action (Figure 6.15).

Figure 6.15 Seasonal variations on Black Rock Beach, Port Phillip Bay, Australia, where northward drifting is dominant in the summer (November–April) and southward drifting in winter (May–October). The wave roses are based on the percentage frequency of onshore winds above Beaufort Scale 3 (Figure 6.17), which determine wave action on this part of the coast

6.8 Beach morphology

The shape of a beach changes as waves and currents move beach sediment from one sector to another. In addition to the longshore drifting of sand and gravel when waves arrive at an angle to the shore, sediments are moved to and fro between the beach and the nearshore zone (onshore–offshore drifting) when waves break parallel to the coastline. Currents generated by winds, waves or tides move sediment in the nearshore and intertidal zones as the tide rises and falls. As they move sand or gravel offshore, onshore or alongshore, they promote erosion or accretion, influencing the form of the waves that subsequently break upon the beach, so contributing indirectly to the shaping of beach morphology.

Beach morphology also changes frequently in response to rising and falling tide and to wind action. As the tide rises, stronger waves arrive through deepening water, winnowing fine sediment from the beach face, and as it falls a veneer of fine sediment may be deposited, sometimes forming a seaward-prograding low tide step. Wind action can deliver hinterland sand to the shore (Section 6.4.4), sand to backshore dunes, longshore winds carry it along the beach or offshore winds blow it into the sea. The beach face is often diversified by ephemeral zones of finer and coarser material arranged in various patterns and by minor features such as ridges, terraces and cusps. In general, the coarser the beach material the steeper the beach face, but after a stormy episode the beach face often shows a steep or vertical scarp, particularly when it is cut into damp, coherent sand (scarping, Figure 6.21).

Nearshore processes shape beach morphology (Hardisty, 1990), but there is feedback, whereby beach morphology influences the processes at work in the nearshore zone (Komar, 1976). This adjustment between nearshore processes and beach morphology as these interactions proceed was illustrated by Short (1992) in his analysis of beach systems on the central Netherlands coast.

Beach morphology is three dimensional, but is usually analysed in terms of beach outlines in plan and beach profiles.

6.9 Beach outlines in plan

The orientation of a beach is related partly to wave patterns and partly to the general trend of the coastline, notably the position of prominent headlands. Beaches tend to become oriented at right angles to the strongest incident waves (Lewis, 1938), and often have smoothly curved outlines in plan, concave seaward (Figure 2.3), shaped by the dominant pattern of approaching waves that have been refracted so that they anticipate, and on breaking fit, the plan of the beach. Such swash-aligned beaches are well developed on oceanic coasts, receiving refracted ocean swell. Examples include the Loe Bar in Cornwall, the beach on the west coast of South Uist in the Hebrides and the Ninety Mile Beach and Encounter Bay in Australia. Some curved beaches are shaped by a single dominant swell, such as the SE swell entering Jervis Bay in eastern Australia, which is refracted into patterns that fit the curved outlines of bordering beaches that face in various directions.

Other beaches may owe their curvature to the resultant of ocean swells approaching from more than one direction, as in Half Moon Bay, California, which receives a SW swell in summer and a NW swell in winter. As has been noted (Section 2.2.4), there are smoothly curved beaches on coasts not receiving ocean swell but shaped by locally generated waves, as on the east coast of Port Phillip Bay (Figure 6.16) and the east coast of Lake Wellington in the Gippsland Lakes, Victoria, Australia. Once established, these gently curving beaches maintain their outlines even when beach material is withdrawn seaward from them by storm waves.

Changes in configuration occur on beaches that have not yet become adjusted to the prevailing wave regime. Where the beach outline is more sharply curved than the approaching waves the breakers produce a convergence of longshore drifting of beach sediment towards the centre of the bay, where the beach prograded until it fits the outline of the arriving swell. Convergent longshore drifting in Byobugaura Bay, near Tokyo, has prograded the central part at Katakai in this way, with sediment derived from the erosion of the northern and southern parts, and a similar evolution has taken place at Guilianova on the east coast of Italy. Where the beach outline is less sharply curved than the approaching waves a divergent longshore drifting from the centre is set up until the beach outline fits the

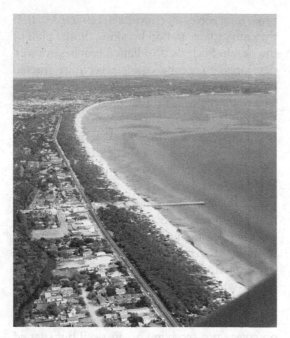

Figure 6.16 The long gently curving sandy beach on the east coast of Port Phillip Bay, extending from Mentone to Frankston, is exposed to wave action arriving from between SSW and NNW over a fetch of up to 60 km. The beach shows seasonal alternations of longshore drifting, southward in winter (NW winds and waves dominant) and northward in summer (SW/SSW winds and waves dominant), but its shaping is related to the resultant effect of onshore winds and waves rather than to any one refracted swell. Parallel nearshore sand bars are present

wave pattern. This has occurred on the shores of the Andalusian Bight, in southern Spain, where erosion of the central part has been balanced by progradation at Matalascanas to the south and Mazagon to the north. Beaches have been reorientated where sand or shingle eroded from one coastal sector has drifted alongshore to accrete on another sector downdrift, particularly where it has been intercepted by headlands and breakwaters.

Beaches shaped by waves generated by local winds (i.e. not receiving ocean swell) have outlines related partly to the direction, strength and frequency of onshore winds, partly to variations in the length of fetch (i.e. open water across which waves are generated by those winds) and partly to wave refraction. The orientation of such beaches as those shown in Figure 6.16 is the outcome of the long term effects of waves arriving from various directions.

These can be expressed as a resultant of wind-generated waves (Figure 6.17), calculated from records of the frequency and strength of onshore winds (Beaufort Scale > 3, that is more than 20 km/hr), that produce only small waves that have little effect on beaches. Typically such beaches are modified by erosion and accretion (deposition) until they become orientated at right angles to this onshore wind resultant. Where there are marked contrasts in fetch these must be taken into account, for strong winds blowing over a short fetch may be less effective in generating wave action than gentler winds blowing over a long fetch. If the direction of longest fetch coincides with the onshore wind resultant, the beach becomes orientated at right angles to this coincident line, but where the two differ the orientation becomes perpendicular to a line that lies between them. On coasts of intricate configuration (as in the Danish archipelago or Puget Sound), wave action is determined more by the fetch than the onshore wind resultant, and many beaches have become aligned at right angles to the maximum fetch (Schou, 1952). The relationship is well illustrated by the shingle beaches (known as ayres) of the Orkney and Shetland Islands. Similar patterns in relation to wind regime and fetch are seen on beaches on the shores of landlocked embayments or coastal lagoons.

Where waves arriving at an angle to the coastline are refracted round headlands they shape asymmetrical (crenulate, half-heart, log-spiral or zeta-form) beaches in the adjacent embayment, as in Venus Bay and Waratah Bay, Victoria, Australia (Figure 6.18). On the New South Wales coast a succession of such beaches shows sharply curved southern ends, with finer sand, a

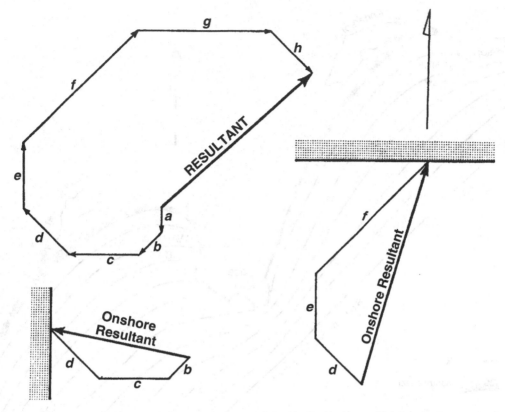

Figure 6.17 Wind resultants formed by drawing vectors (*a* to *h*) obtained by multiplying the frequency of winds in each Beaufort Scale (>3) category by the cube of the mean velocity of that category in a directional diagram. Onshore resultants are obtained by taking only the vectors of winds to which a coastline is exposed

gentler beach face and diminished berm height, and sand coarseness increasing on the straighter northern shores. Although this pattern is related to refracted SE ocean swell there is usually also longshore drifting of sand northward round headlands from one zeta-form beach to the next. A series of asymmetrical beaches on the south coast of the Lleyn Peninsula in Wales occupies embayments cut out by refracted SW waves in glacial drift between bedrock headlands, the most prominent of which is Pen-y-chan, east of Pwllelhi, which has intercepted sand and shingle drifting eastward. On the east coast of Sri Lanka similar successions of asymmetrical beaches have been shaped by refracted SE ocean swell between deltaic salients formed at river outlets.

Where waves are refracted by nearshore islands, offshore reefs or shoals on the sea floor they move into the shore and build cuspate beach outlines in their lee. If these offshore features remain stable the cusps and bays persist, but changes in sea-floor configuration, when currents scour out hollows or build up shoals, or when banks or bars migrate shoreward, seaward or along the coast, modify patterns of wave refraction and so change the outlines of the beach. As a shoal moves along the coast the cuspate beach in its lee is eroded on one side and accreted on the other, so that the cusp also migrates.

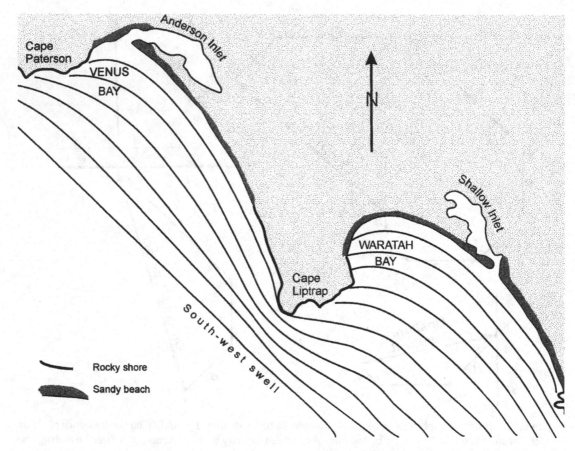

Figure 6.18   The asymmetrical curvature of beaches in Venus Bay and Waratah Bay, Victoria, Australia, is produced by the refraction of ocean swell arriving from the SW

Winterton Ness, on the NE coast of Norfolk, is a sandy cuspate foreland that has been migrating southward. Evidence from successive historical maps shows that its northern coast has been cut back by erosion by waves arriving from the NE, while the southern coast has prograded with the addition of new dune ridges. This migration was accompanied by southward movement of the Ness, which may have been influenced to changing wave patterns produced by variations in offshore shoal topography. By contrast, Benacre Ness, in a similar situation on the Suffolk coast, has been migrating northward as the result of accretion on its northern side of sediment supplied by the predominant southward longshore drifting, and erosion on its southern side. Again, migration of offshore shoals may also have influenced this evolution (Robinson, 1966).

A distinction is made between swash-dominated beaches built parallel to incoming wave crests (particularly refracted ocean swell) with little or no longshore drifting but alternations of onshore–offshore drifting, and drift-dominated beaches with alignments parallel to the line of maximum longshore sediment flow, generated by obliquely incident (typically 40–50°) waves. Beaches exposed to refracted ocean

Figure 6.19 Sand has been washed in from the sea floor to beaches in the SW of Westernport Bay, Australia. Drift-dominated beaches have formed as longshore drifting carried sand along the coasts from Point Leo past Balnarring to Somers and Sandy Point, and from Cat Bay to Observation Point on Phillip Island. Ebb currents have swept sand out SW from Sandy Point to form a long sand bar trailing SW (Middle Bank). The sea floor west of the sand bar has extensive sea-grass vegetation, which now impedes sand drifting, but it is possible that from time to time SE waves sweep sand back to the western coast between Point Leo and Balnarring, thereby completing a long term circulation. Beaches that were eroding on this western coast in the 1940s now show accretion

swell have swash alignments, as do bay-head beaches, but waves moving into embayments build beaches on drift alignments, often terminating in spits. An example of both swash-dominated and drift-dominated beaches in the SW of Westernport Bay, Australia, is given in Figure 6.19.

In general, swash-dominated beaches are smoother in outline than drift-dominated beaches, which are typically sinuous, with intermittent migrating beach lobes and slightly divergent longshore spits. On the coast of Wales the beach in Porth Neigwl on the Lleyn Peninsula is swash-dominated, whereas the beaches on either side of the southern entrance to Menai Strait are drift dominated, ending in spits, and similar features are seen in Portmadoc Bay, to the south. With downdrift progradation such convergent spits can grow to form a curving swash-dominated beach.

Few beaches are entirely swash or drift dominated. Longshore drifting is alternating and balanced on largely swash-dominated beaches such as Chesil Beach in Dorset and Encounter Bay in South Australia, and onshore–offshore (swash–backwash) sequences occur on largely drift-dominated beaches such as Sandy Hook in New Jersey or the beaches north of the Columbia

River in Washington State. The curving beach behind Disaster Bay, New South Wales, a narrow embayment between high parallel sandstone cliffs, is almost entirely swash-dominated, with an outline shaped by incoming refracted ocean swell but too limited in aspect for oblique waves to generate longshore drifting. Numerous parallel dune ridges indicate that progradation has maintained this curved plan.

Most British beaches are drift-dominated with frequent alternations of longshore drifting, swash-dominated beaches being largely confined to the Atlantic coasts receiving ocean swell, such as Rhossili Bay in South Wales. Orfordness is an elongated drift-dominated beach on the East Anglian coast, but it includes multiple parallel beach ridges that indicate occasional swash domination.

Shaping of beach outlines also depends on patterns of sediment supply, with a tendency for beaches to become narrower downdrift from a sediment source such as a river mouth or eroding cliff. Accretion of beach sediment is influenced by the pattern of incoming waves in relation to the source of the material. On deltaic coasts where sediment is being delivered to the mouth of a river, waves arriving parallel to the coastline diverge along both sides of the delta and distribute the fluvially supplied sediment to produce symmetrical beaches, as on the Tagliamento delta on the northern shore of the Adriatic Sea, or trailing lateral spits, as on the shores of the Ebro delta in Spain (Figure 12.5).

6.9.1 Equilibrium beach plans

Where beach outlines have become adjusted to the prevailing pattern of refracted waves they are more stable than those that have not yet achieved such an adjustment, and may attain an equilibrium with processes at work on them. The asymmetrical swash-aligned beaches produced where incoming waves are refracted round a headland,

as on the New South Wales coast in Australia, have been cited as an example (Silvester, 1974), and the inference is that artificial beaches shaped in this bay will be stable. However, where such zeta-form beaches have been established, on the SE coast of Singapore Island and on the NE coast of Brunei, beach erosion has simply continued on the zeta-form alignments.

Equilibrium is defined in the Oxford English Dictionary as a condition of balance between opposing forces, the forces being so arranged that their resultant is zero. Because of the variability of beach features in response to changing processes it is necessary to state a timescale over which such an equilibrium can exist. A cyclic equilibrium is one that returns to its original condition after being disturbed, and a dynamic or shifting equilibrium is one that changes while remaining in balance with driving forces. A beach may show cyclic equilibrium when losses during phases of erosion are balanced by gains during phases of accretion so that it is neither gaining nor losing sediment, or as long as losses are compensated by gains. Dynamic equilibrium may be achieved when a beach is either prograding (advancing seaward by sediment accretion) or eroding (receding landward) but maintaining its outline in plan, but the coastline is not stable under such conditions. Stability can only be attained where there is a sufficient input of sediment to balance episodic losses, and so maintain the outline of a beach in plan and in position.

Beaches that show alternations of longshore drifting, one end being prograded as it receives sediment from the other, in response to changing directions of wave approach, may achieve cyclic equilibrium if they return regularly to earlier configurations, there being no net gain or loss of beach material. On drift-dominated beaches a cyclic equilibrium may be attained where the beach attains a curvature adjusted in such a way that waves impinging on the shore provide sufficient energy to transport the sediment arriving at one end of the beach through to the other, the configuration being maintained. There is an analogy with the graded stream concept, adapted for beaches with long-shore sediment flow, but there are complications where sediment is coming in from the sea floor or where there are losses of sediment offshore.

Equilibrium concepts have been applied to beach outlines in profile (Section 6.10.11) as well as in plan. Beach outlines in plan may persist even when gains or losses steepen or flatten beach profiles. Alternatively, beach profiles may be maintained even though there are gains or losses in beach volume resulting in the advance or retreat of the coastline.

6.10 Beach outlines in profile

Beach profiles can be surveyed by conventional field methods or remote sensing. They are related to the nature of beach sediment and to wave conditions during the preceding few days or weeks: the effects of a severe storm may still be visible several months later. The profiles are shaped largely by wave action, notably the swash generated as waves break upon the shore and the ensuing backwash, that generate onshore and offshore movements of sediment. There is often an upper beach above normal high tide level, only occasionally submerged by large breaking waves or exceptionally high tides, a middle section, often more steeply sloping, and a lower gently sloping or concave section that extends down to and below low tide.

Swash and backwash velocities can be measured with a dynamometer in waves close to the shore, and correlated with onshore and offshore movements of beach sediment, as indicated by profile changes or accumulations in trays placed to intercept beach material drifting shoreward or seaward. It has been noted (Section 2.2.6) that waves with steepness ($H_o : L_o$) in deep water of less than 0.025 produce constructive (spilling) breakers that move sediment shoreward on to the beach, while those with higher ratios (plunging) are withdrawing sediment to the sea floor. Spilling breakers occur where the ratio of breaker height to deep water length $H_b : L_o$ is greater than $2.4(\tan \beta)^{1.8}$, β being the nearshore slope gradient (Sunamura, 1992). However, grain size and beach permeability also influence swash–backwash conditions, and breaking waves are more likely to build up the profile on a shingle beach, where backwash energy is diminished by percolation, than on a sandy beach, where the backwash is more effective in moving sediment back into the sea.

Although storm waves are destructive, withdrawing sand from the beach, they are often accompanied by strong onshore winds that may blow some sand in across the beach. After a gale a sandy beach may be narrowed by wave scour, but its surface may be raised by wind-blown sand accretion on the upper beach, especially where sand has accumulated as a backshore apron against a backing cliff or sea wall.

Beach profiles also vary with tide range. On microtidal coasts, as around the Baltic and Mediterranean, wave action is concentrated within a relatively narrow vertical zone and the beach profile is often steeper than on the same size beach material where tide range is larger. On sheltered sandy microtidal beaches in Western Australia (within bays or behind reefs), profiles vary from gentle concave through stepped to moderately steep (Hegge *et al.*, 1996).

Where waves break over a nearshore reef there is a relatively sheltered lagoon bordering the beach, and typically the broken waves begin to re-form, but are of relatively small dimensions, with constructive swash. There are examples on the south coast of Western Australia near Hopetoun. Some artificially constructed nearshore breakwaters are intended to protect and produce swash-built beaches.

Figure 6.20 Concave profile on a shingle beach at Ringstead in Dorset, cut back by preceding storm wave action. Beach cusps have been formed on the concave slope. View eastward to White Nothe (W)

6.10.1 Cut and fill

Beach profiles are lowered and cut back by strong wave action, especially during stormy periods, when plunging and surging waves, with limited swash and stronger backwash, withdraw sand and shingle to the nearshore zone, leaving a concave profile (Figure 6.20). There is often a microcliff, cut by storm waves (Figure 6.21).

In calmer weather spilling waves (Section 2.2.6), with a strong swash and lesser backwash, move sand and shingle up the beach and rebuild a convex profile. Alternations of beach erosion by plunging or surging waves and beach accretion by spilling waves are known as cut and fill, and are associated with alternations of seaward and shoreward drifting of sediment. Generally the erosion takes place rapidly during brief storm events and the shoreward drifting is a slower process that extends over longer periods.

Cut and fill occurs over timescales varying from a few days to several years, but as storms are usually more frequent in winter many beaches show a seasonal sequence of winter cut and summer fill. On the Pacific coast of the United States sandy beaches are scoured by storm waves in winter to expose rocky and bouldery shores, but in spring gentler wave action in calmer weather restores the sand cover. This was demonstrated by profile monitoring along the beach and nearshore zone alongside the Scripps Institution Pier at La Jolla (Shepard, 1973). On monsoonal coasts beach erosion occurs during storms in the wet season, with recovery in calmer dry season weather. The Darwin beaches in northern Australia are cut back and lowered during the wet summer and rebuilt in the drier winter, and similar seasonal cut and fill is seen on the east coast beaches of Peninsular Malaysia, where erosion prevails during the NE monsoon (Teh Tiong Sa,

Figure 6.21 Microcliff (about 1.5 m high) cut during a storm on the Ninety Mile Beach, Victoria, Australia

1985). In Marsden Bay, NE England, King (1972) found that sand moved shoreward on to beaches when winds blowing offshore flattened incoming waves (constructive), whereas seaward movement occurred when steeper waves (destructive) were generated by winds blowing onshore. It is possible that winds blowing offshore cause surface water outflow that is compensated by inflow on the sea floor. Whether waves of a particular form will push sediment up on to a beach or withdraw it from the beach face depends also on the pre-existing transverse profile of the shore and the grain size of shore sediments. Shoreward movement is more likely where the transverse profile is gentle, and the sediment predominantly fine grained.

Alternations of cut and fill can take place over longer periods. The fortnightly sequence of increasing high tides from high neap to high spring tides is in effect a short term marine transgression, and is often accompanied by the erosion

of beaches. It is followed by a sequence of diminishing high tides from springs to neaps, effectively a regression of the sea, which is often accompanied by beach accretion. Seasonal alternations of beach erosion and accretion occur where there is a sharp contrast between a wet and a dry season, between stronger winter and gentler summer wind and wave action or between the directions of approach of dominant winds in different seasons.

6.10.2 Sweep zone

As a result of cut and fill alternations over varying periods of time the beach profile shows changes in form, being higher and wider after phases of accretion and lower and flatter (with a backing rise or cliff) after phases of erosion. The vertical section of a beach within these alternations (subject to removal and replacement) is known as the sweep zone. It can be measured by making successive surveys across the beach to determine the cross-sectional area between the accreted and eroded profiles, and when this is multiplied by the length of the beach the volume of sand gained or lost can be calculated. Measurements on the Ninety Mile Beach have shown that the seasonal sweep zone has a mean cross-sectional area of 42 m³, indicating a loss of 6 million m² of sand from this beach during stormy phases, which are most frequent in winter, and equivalent replacements during fine weather. On such a beach profile variations are cyclically restored by natural processes over periods of time.

6.10.3 Beach gradient and sediment size

The transverse slope of a beach profile is also related to the grain size of beach sediments. Gravels tend to assume steeper gradients than those on sand (usually less than 5°), chiefly because they are more permeable, and beach face slope increases with pebble size. On Chesil Beach the beach face slope increases from about 5° on fine shingle at the western end to more than 20° on the large cobbles near the SE end.

As waves break the swash sweeps sediment up on to the beach. Backwash tends to carry it back, but the greater permeability of gravel and coarse sand beaches absorbs and diminishes its effectiveness, so that swash-piled sediment is left at relatively steep gradients. Fine sand beaches are more affected by backwash, and develop gentler slopes as more sediment is withdrawn seaward.

Storm waves thus steepen the profiles of shingle (gravel) beaches and flatten those of sandy beaches. On the North Norfolk coast the 1953 storm surge lowered and cut back sandy beaches on Scolt Head Island, but piled up the crests of shingle beaches, parts of which were rolled landward, as at Blakeney Point (Steers, 1964). Both

sectors of coastline retreated, but the shingle beaches were left with steeper profiles than the sandy beaches, although the latter were generally backed by cliffed dunes. On the sandy beaches of the Atlantic coast of Britain, as at Braunton Burrows in North Devon, storm waves are almost entirely destructive, lowering and cutting back sandy beaches, and similar effects have been observed on east coast sandy beaches, as at Goswick Sands and Gibraltar Point.

Nevertheless, there is great variation in sand and shingle beach gradients, depending on preceding wave conditions as well as grain size. Storm waves can cut vertical cliffs in sandy beaches and comb shingle down to a gentle slope.

6.10.4 Upper and lower beach

A beach with a mixture of sand and shingle (or fine and coarse sand) may be sorted by swash and backwash until the profile consists of a coarser, steeper upper beach and a finer and flatter lower beach, exposed as the tide ebbs. The upper beach is often of shingle or coarse sand, clearly demarcated from the lower beach of finer sand, a contrast that results from the differing responses of shingle and sand to storm waves, which pile up swash-built shingle that persists as backwash withdraws the sand (Figure 6.22).

The shingle is permeable, and descends to a seepage zone and a gentler slope of wetter, firmly packed sand, across which it may be possible to drive a car. Backwashed sand is sometimes deposited in the form of a step of relatively coarse sand at the foot of the beach, which may prograde seaward as the tide falls. This is often bordered on the lower beach by shore-parallel ripples shaped by interacting swash and backwash. On some coasts the lower beach passes seaward to broad intertidal bars and troughs (Section 8.8). Where the upper and lower beaches are separated by a shore platform the upper beach

Figure 6.22   An upper beach of shingle and a lower beach of sand on the east coast of Dungeness in SE England, separated by swash–backwash sorting. The arrow indicates the high spring tide line

is generally swash dominated by waves reaching it at high tide. The lower beach, exposed in front of the outer edge of the shore platform at low tide, is subject to lowering by reflection scour, as at Bunbury in Western Australia.

6.10.5 Beach stratification

During stormy phases the finer and lighter components of a beach sediment may be withdrawn by strong backwash, leaving the coarser and heavier material as a lag deposit on the beach face. In calmer weather gentle wave action moves fine sediment back on to the beach face. Where overall beach accretion has taken place such alternations form stratified (or laminated) deposits, with coarser seams between layers of finer sediment, seen when a section is cut across a beach. Within each stratified layer the distinctive sediment is well sorted, being related to a

particular phase of deposition or reworking by waves. If there is a phase of erosion these may outcrop as contrasted zones along the face of the beach.

The internal structure of a beach can be examined by using ground-penetrating radar (GPR), which identifies and charts discontinuities.

6.10.6 Beach berms

Swash deposition can build up and prograde a flat berm (beach terrace) at about high tide level or form a swash berm, a subdued ridge parallel to the coastline along the length of the beach, sometimes backed by a shallow lagoon that drains out as the tide falls. On sandy beaches the force of breaking waves may compact the sand so firmly that it is possible to ride a bicycle, drive a car or land an aircraft on a berm. On shingle beaches breaking storm waves throw some pebbles forward and up the beach, to build a berm at the swash limit.

Coarse sand can be built into berms by storm waves that erode beaches of finer sand. Storm waves wash coarse sand from the beach to the backshore and accompanying onshore gales blow sand landward. The gradient and morphology of the nearshore sea floor are also relevant, for where it is gentle and shallow, storm waves lose some of their energy and break as constructive surf with a strong swash, instead of the plunging breakers seen on steeper shores. Under these conditions even medium to fine sand can be built up as berms along the shore.

On microtidal coasts there is typically a single swash-built berm, whereas on mesotidal coasts (as in East Anglia) there are often two berms, one just above high tide level, the other just above low tide level. This is because swash action is more prolonged at these two levels than when the tide is rising or falling. On macrotidal coasts, especially where the shore is sandy, there are multiple berms, grading into foreshore ridge and runnel topography (Section 8.8). Grain size also influences berm development, so that a shingle beach passing laterally into a sandy beach may have one shingle berm that divides alongshore into two sandy berms.

6.10.7 Beach cusps

The beach face is often divided into zones of shingle, shelly gravel and coarse and fine sand that run parallel to the coastline, sorted and shaped by swash and backwash processes. On some sand and shingle beaches the pattern takes the form of beach cusps, consisting of regular successions of crescentic depressions containing finer sediment, up to a metre deep and opening seaward between cuspate points of slightly coarser material. They are generally found where waves arrive parallel to the coastline, rather than obliquely, and are ephemeral features (persisting

for a few hours to several days), but once formed they influence the patterns of swash as the waves break. They are often found on beaches exposed to ocean swell.

Beach cusps are shaped by breaking waves that generate swash salients with currents that divide at points where coarser sediment is deposited, and coalesce in intervening areas to form backwash that scours out depressions. Their regular spacing results from the interaction of the swash salients with orthogonal edge waves (Guza and Inman, 1975). The spacing of the cuspate points (two to five times breaker height) increases as wave height is augmented, and is also related to wavelength. Miniature cusps about 30 cm apart have been seen on a low energy lagoon shore beach, but generally they are spaced at intervals of 10–20 m, attaining 50 m on beaches receiving ocean swell (as on the Loe Bar in Cornwall). In some beach cusps the erosional depressions are backed by miniature cliffs cut into the beach face (Allen et al., 1996).

Beach cusps are most clearly developed where the beach sediment has a bimodal grain size distribution. They are frequently formed on shingle beaches in Britain where there is finer material occupying the depressions (Figure 6.20). They are less clearly developed where there is a lateral transition to more uniformly sorted beach material: if the proportion of pebbles to sand diminishes the cusps are broken into less regular shingle patches and the depressions become fainter on a well sorted sandy beach. Cusps of coarser sand and depressions of finer sand occur on the sandy beaches on the Gulf and Atlantic coasts of the United States. In Australia they frequently form on steep coarse sandy beaches in embayments along the Sydney coast, and on the sandy shores of landlocked embayments such as Jervis Bay and Botany Bay. Beach cusps on the Perth coast fade when wave height increases during sea breeze episodes and revive in calmer weather when longer ocean swell dominates (Masselink and Pattiaratchi, 1998).

Less regular cuspate re-entrants are excavated in sandy beaches behind the heads of rip currents, where larger waves move in through water deepened by outflow scour. Such local scour has been described on sandy beaches along the Atlantic coast of the United States, as on Siletz Spit in Oregon (Komar, 1983).

6.10.8 Wind action on beaches

Beach profiles are also modified by wind action, when sand is blown along or across the beach, lowering some parts and building up others. Anemometers and intercepting trays can be used to measure quantities of sand moved across or along a beach in relation to the direction, strength and duration of wind action. Sand blown to the backshore can be banked up against cliffs or bluffs, or built into coastal dunes. Dunes may later by cliffed by marine erosion, the beach regaining some of the sand previously lost to the backshore, so that over time there are beach–dune interactions (Section 9.1).

Offshore winds sweep sand back across the beach and into the sea, to be reworked and transported by waves and currents in the nearshore zone. Winds blowing along the beach (Figure 6.9) can carry substantial amounts of sand, especially when the beach face is dry (thereby augmenting the effects of longshore drifting by waves and nearshore currents). This was demonstrated by So (1982) on the sandy beach at Portsea, in Victoria, Australia, and is a common occurrence on windy North Sea sandy beaches at low tide, as at Bray in NE France.

Fine sand is more readily mobilised and transported by wind action than coarse (or heavy mineral) sand, and pebbles (lag gravel) are left behind when beaches of sand and gravel are winnowed, in much the same way as when wave backwash withdraws finer sediment from the beach face. A strong wind can produce a sheet of drifting sand, with sand grains airborne (in suspension), bouncing (saltation), and rolling along the beach. Under these conditions beach barchans (Section 9.6) may form on the shore, and migrate downwind. Movements of wind-blown sand to backshore dunes are discussed in Chapter 9.

6.10.9 Downwashing

Runoff and seepage during heavy rain or as the result of melting snow or ice can lower the beach profile by washing sand or gravel into the sea. On coasts where rainfall is high, seeping groundwater may take the form of springs that wash out irregular pits and form stream-cut furrows on the sandy foreshore at low tide, as on Three Mile Beach in NW Tasmania. In NW Australia near Cape Leveque outflowing streams during torrential summer downpours cut gullies across sandy beaches, forming arcuate lobes of nearshore sand that are reworked by waves and swept back to the shore by wave action in the dry winter. Antarctic beaches are covered by snow in winter, but receive silty solifluction downwash during the summer thaw.

6.10.10 Response to tidal oscillations

Beach profiles change as the tide rises or falls. In general the sea level rise as the tide comes in leads to recession of the beach profile and withdrawal of sediment seaward, while emergence as the tide falls is usually accompanied by shoreward movement of sediment and beach progradation. Similar trends can be detected during the sequence of rising high tides between neaps and springs and of falling high tides between springs and neaps. There are, however, many complicating factors, including the effects of longshore drifting, variability of wave action and rates of sediment input and output. A beach receiving abundant sediment (e.g. from a nearby river mouth or rapidly eroding cliffs) may prograde even when sea level is rising, while a beach that is losing sediment offshore or alongshore may be cut back even when sea level is falling.

6.10.11 Equilibrium beach profiles

The question of beaches attaining equilibrium in plan has been discussed (Section 6.9.1), and similar considerations apply to the evolution of equilibrium in beach profiles. Many coastal geomorphologists and engineers have accepted the idea that beaches can attain a profile of equilibrium, but there is some confusion over what this actually means (Pilkey *et al.*, 1993).

After a phase of sustained and steady wave activity a beach profile becomes a smooth, concave upward curve, the gradient of which depends partly on the grain size of the beach sediment, gravel beaches generally having steeper profiles than sandy beaches. This can be considered an equilibrium beach profile, achieved by adjustment between the beach profile and the waves and currents at work on it. There is a tendency for a beach profile to fluctuate about an equilibrium that may change slowly over time (Dean, 2005). Such a profile can be described as a concave upward curve:

$$h = Ax^{0.67}$$

where h is the still water depth, x is the horizontal distance from the coastline and A is a dimensional shape parameter, the shape of which depends on the grain size of the beach material (Dean, 1991). Bodge (1992) preferred an exponential expression:

$$h = B(1 - e^{-kx})$$

in which B and k are depth and inverse distance. These equations fit many surveyed beach profiles, especially those without nearshore bars and

shoals, which are complications that are difficult to match with mathematical curves. It should be noted that many beach profiles are surveyed when calm conditions have become established following a phase of stronger wave action.

The fact that mathematical curves fit surveyed profiles is not an indication of stability, for as beach profiles steepen or flatten in response to changes in wave regime or granulometric composition they continue to show profiles similar to those generated by the formulae of Dean (1991) or Bodge (1992). The smoothly curved profiles may represent a condition that beaches attain when their granulometric composition has become adjusted to specific wave conditions. However, granulometric composition is modified by gradual attrition as the result of wave agitation, and by sorting, and this will lead to profile changes unless mean grain size is maintained by a continual inflow of coarser sediment as the finer fraction is removed. There is a tendency for larger particles to be concentrated on the beach surface because of agitation by the rising and falling beach water table.

The sequence of cut and fill observed on many beaches, with sediment removed during stormy phases and replaced during subsequent calmer weather, can be considered a cyclic equilibrium to the extent that the profiles are restored by natural processes. This can occur over periods ranging from a few days or weeks (a storm and its aftermath) to a year or more (seasonal alternations).

Geomorphologists have used the term dynamic equilibrium to describe the condition where landforms maintain their shape, even though changes (uplift, erosion, deposition or subsidence) have occurred. A dynamic equilibrium could exist on a beach when the profile remains the same (perhaps with cyclic alternations) while the beach as a whole is either advancing (prograding) or receding (retrograding). Many would argue that this advance (with land gained by accretion) or retreat (with land lost by erosion) of the coastline really indicates disequilibrium. As will be shown later, analyses of global patterns of beach change have indicated that over the past century most of the world's beaches have been retreating; some have been advancing, and a few have remained static or shown no net change. The supposition that a particular beach is (or has recently been) in some kind of equilibrium should therefore be treated with caution.

An equilibrium between the beach profile and nearshore processes may not have been attained before there is a change in processes as the sea becomes stormier or calmer. A new adjustment then begins. As nearshore processes are highly variable, most beach profiles are changing most of the time. This variability is compounded when the whole three-dimensional beach morphology is considered, for equilibrium in both plan and profile must be an unusual condition. When geomorphologists or engineers refer to beach equilibrium concepts they should indicate which kind of equilibrium they envisage, the timescale they are using, and whether they are dealing with the beach in plan or in profile, or the whole three-dimensional beach system.

6.10.12 Eroding and prograding beach profiles

Beaches that are eroding typically show a gently inclined or concave slope backed by a scarp cut into beach sediments or backshore dunes. Sand and gravel removed from a beach by storm waves are spread across the nearshore sea floor as a terrace or deposited as bars and troughs running more or less parallel to the coastline. The intertidal shore profile may thus become wider and flatter after a phase of erosion.

Prograding beaches (Section 6.17) are typically convex, often with berms or beach terraces that are being built seaward and new foredunes developing above high tide level. Sustained accretion raises the profile and increases beach volume as the coastline progrades.

6.11 Beach morphodynamics and beach states

Beach morphodynamics (the study of the mutual adjustment of beach morphology and shore processes, involving sediment transport) can be studied in terms of field measurements and analysed with the aid of computer simulations (Cowell and Thom, 1994). Analysis of incident wave regimes on beaches has shown that wave energy is partially reflected by steep ($>3°$) beaches (especially shingle beaches), whereas gently sloping (generally sandy) shores dissipate wave energy, the waves breaking and spilling across a wide surf zone. The outcome is beach states that correspond to surf scale categories based on the dimensions of breaking waves (Wright and Short, 1984; Short, 1999). Dean (1991) proposed a surf scaling parameter (Ω), based on breaker height (H_b), and obtained the formula

$$\Omega = H_b/w_s T$$

in which w_s is mean sediment fall velocity and T the wave period. This enables beaches to be described as reflective ($\Omega < 1$) where they receive surging breakers and have a high proportion of wave energy reflected from the beach face, or dissipative ($\Omega > 5$) where wave energy from spilling breakers is lost across a wide gently sloping beach. An intermediate category ($\Omega = 1$–5) is recognised (Masselink and Short, 1993).

On the New South Wales coast, where fine–medium grained sandy beaches occupy asymmetrical embayments with microtidal moderate–high wave energy conditions, a distinction can be made between steep sandy beaches facing relatively deep water (without bars) close inshore, which are reflective because part of the incident wave energy (plunging waves) is reflected seaward, and flatter sandy beaches fronted by nearshore sand bars and wide surf zones, which are dissipative because much of the wave energy (spilling waves) is lost as waves arrive through the shallow water. The surf zone, where the waves have broken, is often less than 10 m wide on reflective beaches, but may be at least 100 m wide on dissipative beaches. Reflective beach profiles have upper slopes of between 6 and 12°, often with a distinct, steeper step at the base, then a gentler nearshore slope of 0.5–1°, and with well formed ripples parallel to the coastline. Dissipative beach profiles are broader and flatter, with slopes typically less than 1°, and a nearshore zone diversified by multiple parallel sand bars.

On the Sydney coast many beaches are predominantly steep and reflective in late winter and spring, and predominantly gentle and dissipative, with wide surf zones and sand bars, in summer. The seasonal oscillations vary with degree of exposure (beaches in narrow bays between headlands being more often reflective), sand texture (coarser sand on reflective beaches) and wave height (larger waves develop wider and gentler dissipative shore profiles). Asymmetrical bay beaches receiving SE wave action (particularly in winter) develop steeper reflective profiles in the southern part and wider dissipative profiles, with sand bars, on the more exposed central and northern parts.

Sandy beaches are more likely to be dissipative than shingle beaches, especially at low tide. They have gentler shore profiles than shingle beaches, usually with one or more sand bars, and they produce longer edge waves and more widely spaced rip currents and beach cusps. Shingle beaches are steeper and reflective, with little if any surf zone and no bars.

Studies of beaches in other parts of Australia and around the world have shown an association of high, steep waves and fine sand on dissipative beaches, while long, low waves and coarser sediment (including shingle) characterise reflec-

tive beaches. The intermediate category is found where the combination of moderate–high wave energy and fine–medium sand results in a transitional beach type. In Britain shingle beaches, such as Chesil Beach, are reflective, and the wide sandy shores of Atlantic coast beaches usually dissipative. Some shingle-backed sandy shores are reflective at high tide and dissipative at low tide, as at Porth Neigwl in North Wales.

On beaches with sediment of a particular size, wave conditions largely determine the beach state, which can change from reflective through intermediate to dissipative and back again with variations in weather. Reflective beaches can be cut back by moderate wave action, whereas dissipative beaches are scoured only when swash levels are augmented by storms, a heavy swell or exceptionally high tides. Particular events, such as a severe storm, can at least temporarily convert a dissipative beach into a reflective one, while prolonged fine weather can restore a dissipative beach profile: essentially the sequence of cut and fill described previously (Section 6.10.1). The intermediate beach state may occur when a reflective beach is being modified to dissipative, with declining wave energy and the formation of bars, or when a dissipative beach is being modified to reflective, with increasing energy and bars losing sediment shoreward to the beach.

On crenulate sandy beaches of the kind seen in embayments on the New South Wales coast all three types exist. The relatively straight northern coasts, receiving relatively unrefracted ocean swell, are reflective, the slightly curved central sectors intermediate, with nearshore bars dissected by rip current channels, and the sharply curved southern shores dissipative, with waves much diminished by refraction over multiple sand bars, some of which are exposed at low tide. Within such an embayment it is possible for a moderate long swell to cut back reflective beaches towards the northern end at the same time as promoting accretion on the southern part. There is no simple relationship between the three beach states and long term trends of erosion or accretion, for changes from one beach state to another occur during short term cycles (e.g. cut and fill, Section 6.10.1) on beaches that may show long term progradation, stability or erosion.

Classification into reflective, dissipative and intermediate beach states was originally devised on microtidal coasts. As tide range increases wave action is dispersed over wider vertical and horizontal zones and tidal currents interact with incoming waves, reducing transverse gradients, smoothing bar topography and modifying rip currents. Beach profiles exposed at low tide show wider and flatter bars and troughs or low tide terraces.

Masselink and Short (1993) examined features of beach morphology associated with the interaction of wave height and tide range, and introduced the concept of relative tide range (RTR), where actual tide range (TR) is divided by breaker height (H_b):

$$RTR = TR/H_b$$

to distinguish categories of wave-dominated beach morphology as relative tide range increases. On swash-dominated beaches these range from reflective ($\Omega < 2$ and RTR < 3) to low tide terraces with rip channels ($\Omega < 2$, RTR $= 3$–7) then low tide terraces without rip channels ($\Omega < 2$, RTR > 7); from intermediate ($\Omega = 2$–5, RTR < 7) to low tide bar and rip channels ($\Omega = 2$–5, RTR > 7) and from barred dissipative ($\Omega > 5$ and RTR < 3) to non-barred dissipative ($\Omega > 5$ and RTR > 7) and ultra-dissipative ($\Omega > 2$ and RTR > 7), with multiple lines of breakers moving in over a very wide low gradient profile, as on Sarina Beach in Queensland and Pendine Sands in Wales. There are further complications on drift-dominated beaches, where profiles vary laterally with such features as migrating cusps and lobes, and bars formed at varying angles to the coastlines moving alongshore.

There are three categories of beach morphology on mesotidal and macrotidal coasts: high wave energy concave beaches with moderate gradient (1–3°), moderate wave energy gentler beaches (0.5°) with multiple bars and low wave energy backshore beaches behind wide tidal flats (Short, 1999). Sometimes there is a steep, reflective beach behind a foreshore with sediment grading finer seaward. The classification of reflective, intermediate and dissipative beaches has so far been little used in Britain, possibly because shingle beaches are normally reflective, storm waves are commoner than long swells and tide ranges are relatively large.

6.12 Use of models

Another approach to the study of beach morphodynamics is to set up scale models in tanks, in which waves and currents are generated and the rise and fall of tides simulated, in order to test their effects on beach morphology and sedimentation. Such models can be used to demonstrate cut-and-fill sequences in response to changing wave conditions, the effects of wave refraction on beach outlines and longshore drifting by waves arriving at an angle to the shore. Coastal and nearshore topography can be reduced to scale, as can waves and currents, but there are difficulties in scaling down natural sediment calibre and characteristics to fit model conditions. Sand grains can be used in a model to represent a shingle beach, but representation of a sandy beach to scale would require the use of silt, which has different physical properties and may not give the correct response of sand to wave and current action. Bearing in mind this limitation, scale models have been used to test harbour design and to test the effects of introducing artificial structures, such as breakwaters and marinas, to the coast.

An alternative has been to develop mathematical models that provide computerised simulations of the response of coastal features to changing tide, wave and current conditions, using information obtained from surveys. Mathematical modelling is useful as a means of exploring process–response relationships, but coastal systems are complex and predictions can prove unreliable. Monitoring of coastal changes is needed to check predictions and obtain further data for refinement of models.

There is a trend towards field experiments, using temporary structures, as a prelude to more permanent coastal engineering works, and there are advocates of a trial-and-error approach in beach nourishment projects.

6.13 Beach compartments

Many beaches occupy distinct compartments or cells on the coast, separated by rocky reefs or protruding headlands, particularly those that end in deep water or are defined by river mouths or the heads of submarine canyons close inshore. Some have been artificially delimited by structures such as breakwaters. Beach compartments have also been described as sediment cells or coastal sediment compartments (Davies, 1974). Some contain relict beaches (no longer receiving sediment), and some show cyclic changes related to onshore and offshore movements of beach material. Others are less clearly defined, in the sense that beach material may drift in and out, round bordering rocky headlands (in nearshore water up to 20 m deep), by-passing river mouths, or lost into submarine canyons. There may be sediment cells between boundaries such as major promontories that prevent any longshore movement of sediment and sub-cells within embayments between minor promontories that can be by-passed by drifting sediment under unusual conditions such as severe storms.

Preliminary identification of beach compartments can be made from maps, charts and air photographs, but it is necessary to study patterns of coastal sediment flow to determine compartment boundaries. In Nova Scotia cliffed headlands cut in glacial eskers are sources of sand and gravel delivered to beaches in intervening sediment cells (Carter et al., 1990). There are also sediment cells with fine grained material, silt and clay, usually supplied by a river and occupying the coast on either side of the river mouth. The term sector has been used in this book to describe a portion of coastline between any two points – it is not necessarily a cell, subcell or compartment.

There can be striking differences in the nature of beaches in adjacent bays, as on the steep coasts of SW England, where the grain size or mineral composition of beaches varies from one cove to another, the beach compartments being separated by rocky promontories. Within each compartment wave action can move the confined beach material to and fro along the shore, such alternations of longshore drifting building up first one side then the other, or change the transverse profile by alternations of shoreward and seaward movements of sediment. Sand also moves from beaches to backshore dunes, and from dunes back to beaches.

There are gains and losses from a beach compartment if the intervening headlands are small enough for sand and gravel to be carried past them on the sea floor by waves and currents. Such natural by-passing often takes place predominantly in one direction as a response to prevailing longshore wave and current action, particularly during storms. Beach sand may also be delivered to dunes that spill over headlands on to the next beach (Section 6.4.4). Some beach compartments are delimited by river mouths or tidal entrances, where transverse ebb and flow currents impede longshore drifting in much the same way as headlands or solid breakwaters. This can result in beach accretion on the updrift side, and erosion downdrift, but waves and associated currents may carry sand and gravel across the sea floor, by-passing the mouths of rivers and tidal entrances, sometimes with the formation of longshore bars.

On the south coast of England Lizard Point, Dodman Point, Start Point and Portland Bill are major compartment boundaries, but even these are by-passed by sand and gravel moving eastward along the sea floor. Bray, Carter and Hooke (1995) defined littoral cells on the central south coast of England, identifying Christchurch Bay as a sediment transport unit. Well defined beach compartments occur within coves on the coasts of Devon and Cornwall, but Lyme Bay can be subdivided into several compartments. Some (like Tor Bay) are delimited by major headlands ending in deep water, others (between Lyme Regis and West Bay) are separated by rocky ebbs around which sediment may drift along the sea floor. Chesil Beach occupies a compartment that once extended eastward from Eype, but since 1866 it has been artificially delimited by the stone breakwaters built at the entrance to West Bay harbour.

There are beach compartments in southern California where sand delivered to the shore mainly by rivers (with some from eroding cliffs, and possibly some from the sea floor) drifts southward along the coast past headlands (such as Point Dume) and through successive bays until it is lost into submarine canyons that run out from the southern ends of embayments. A similar situation exists on fluvially fed gravel beaches in Georgia, on the Caucasian Black Sea coast. These are unusual situations, however, for most beaches receive sediment from several sources, including rivers, eroding cliffs and the sea floor, and very few lose sediment into submarine canyons.

At Wewak on the north coast of New Guinea tectonic uplift led to emergence of fringing reefs as promontories that have segregated a beach-fringed coastline into a series of bays and cut off the former supply of eastward-drifting sand delivered by rivers draining to the coast to the west. Thus deprived, the bay beaches become relict, finer in texture and eroding.

The management of renourished or artificial beaches (Section 7.4) may be facilitated by inserting groynes to delimit artificial beach compartments that can be treated as sedimentary units. There are problems where beach sediment is withdrawn seaward during storms and later carried back onshore beyond the limiting groynes. Problems also arise when oblique waves build up sediment alongside a groyne until it spills round it and alongshore. In both cases longer groynes will reduce the problem.

6.14 Beach budgets

Coastal sediment budgets deal with the volumes of sediment supplied to a particular sector by onshore and longshore drifting and yields from the hinterland and the volumes of sediment lost offshore, alongshore or landward over a specific period. Beach budgets show a net gain or a net loss in beach volume, which can be determined by making repeated surveys along and across a beach, using conventional methods with instruments such as a level or theodolite, to measure variations in the plan and profile of the beach. Beach profiles are generally surveyed at right angles to the coastline, from backshore datum points (marked so that they can be found for subsequent surveys) down to the low tide line, and some distance seaward in shallow water. They can be used to monitor the advance or retreat of the coastline, supplementing information from series of dated air photographs. When they are linked by alongshore surveys gains and losses of land area on the coast and in the intertidal zone can also be measured. Within defined beach compartments it is possible to compute beach budgets by multiplying the mean cross-sectional area of neighbouring profiles (using arbitrarily defined basal and rear planes) by the intervening longshore distance to obtain volumetric changes. Pierce (1969) used these methods to assess a sediment budget for the beaches between Cape Hatteras and Cape Lookout on the Atlantic coast of the United States, identifying gains from shoreward and longshore drifting and losses through tidal entrances and by wind and wave action over intervening barrier islands.

If the volumes of sediment gained and lost on each sector within a beach compartment over particular periods are calculated, these changes can be expressed in terms of a beach budget for that compartment. Beach budgets can also be established for a coastline that includes more than one beach compartment, or for any other defined sector, such as a seaside resort waterfront. On the Natal coast in South Africa surveys showed that 380 000 m^3 of sand moved NE past Durban each year, and that with inputs from sand-yielding rivers to the north this increased to 500 000 m^3/yr at Richards Bay.

Alternatively, it is possible to make successive contour maps of the beach surface, and to determine patterns and volumes of gain or loss between each survey from these. Various tracking meters and vehicles have been developed to accelerate beach surveying, and use has been made of the satellite-based GPS (global positioning system), particularly on beaches firm enough for vehicle-based surveys, as on the Gulf Coast in Texas (Morton *et al.*, 1993). On the Adelaide coast, in South Australia, GISs (geographical information systems) have been used to generate serial contour maps showing the pattern of beach surface gains and losses as a basis for calculating beach budgets and planning beach nourishment (Fotheringham and Goodwins, 1990).

Changes on beaches have been surveyed three dimensionally in the United States using LIDAR (Section 6.1).

6.15 Tracing beach sediment flow

Patterns of longshore drifting on beaches can be deduced from accretion alongside headlands, groynes, breakwaters or landslides, migration of beach lobes, deflection of river mouths and lagoon outlets or growth of spits. Some of these indications of the direction of longshore drifting may be misleading where patterns of beach accretion result partly from sediment movement in from the sea floor, rather than alongshore. Shoreward drifting of sand and gravel by waves arriving parallel to the coastline can deposit beaches in patterns that result in river mouths or lagoon outlets migrating to sectors of lower wave energy, and accretion alongside headlands can be the outcome of shoreward drifting of sediment from a nearby shoal. Paired spits (Section 8.2.1) may result from a convergence of longshore drifting, but can also be formed by the breaching of a barrier or by wave refraction into a maintained river outlet or lagoon entrance. Usually at least some of the longshore drifting of sand and gravel takes place in the nearshore zone, or by way of exchanges between the beach and the nearshore zone, with subsequent delivery to the beach downdrift.

Pebbles of an unusual rock type or specific mineral sands may act as natural tracers, indicating longshore drifting from a source area such as a cliff outcrop or river mouth. Patterns of beach sediment movement can also be determined by introducing and following tracers. These are materials that move in the same way as the natural beach sediment, consisting of particles similar in size and shape, with similar hardness and the same specific gravity as the sediment already present on the beach. Tracer material is deposited on a beach or in the nearshore zone at a particular point or along a selected profile, and surveys are made

Figure 6.23 Pathways of mineral tracer sand movement on to beaches in Discovery Bay and Portland Bay, Australia, after tracer injection at Whites Beach, Lawrence Rocks and Narrawong. The aim was to determine whether there would be significant sand movement into the harbour that was subsequently built at Portland (H). Accretion has occurred in this harbour, but it consists of fine sand washed up from the nearby sea floor rather than coarse beach sand moving along the coast

subsequently to see where it has gone. It must be readily identifiable after it has moved along a beach or across the sea floor. Some may become buried within a beach or lost seaward from the nearshore zone, but usually a proportion remains on the beach face to indicate the direction in which sediment has moved.

Some tracer projects have used sand minerals that were not naturally present on the beach. This mineralogical method depends on a complete preliminary inventory of the minerals already present in the beach to be sure that the introduced material is indeed alien, and will be identifiable downdrift. Baker (1956) introduced alien minerals, such as iron pyrites (actually of higher specific gravity than the natural beach sand) to beaches of quartz and calcarenite sand

on the west coast of Victoria, Australia, as a prelude to surveys of patterns of sand movement alongshore and across Portland Bay, where a harbour was to be constructed (Figure 6.23). The project indicated that most of the sand moving along the coast was passing out across the sea floor, and was unlikely to accumulate in the harbour, but there has been some harbour accretion as the result of inwashing of sand from the adjacent sea floor. There are risks in using tracer of higher specific gravity than the natural beach sediment, because the introduced minerals may not reproduce the behaviour of the natural beach sediment. Lighter grains are moved more frequently and further than heavy grains, and their net long term drift may not be in the same direction.

Sand or gravel that is naturally or artificially coloured can be used as tracer, but there are difficulties in observing coloured grains when they form only a small proportion of the sediment on a beach. More effective is the use of natural or artificial sand or gravel coated or imbued with a colloidal substance containing a fluorescent dye (Ingle, 1966). Sand or gravel labelled in this way is dumped on a beach, and its subsequent movement traced by locating the dyed material at night, using an ultraviolet lamp, or sand samples taken from the beach can be examined under ultraviolet light in a darkroom, where the fluorescent stain stands out brightly. Fluorescent quartz has been used to trace beach sand movement along the shores of the Nile delta (Badr and Lotfy, 1999) (Section 12.6), and Rink (1999) has explored possibilities of using quartz luminescence in tracing beach and nearshore sand movements.

Another method of tracing sediment flow depends on the use of radioactive materials, which can be followed by means of a detection meter (Geiger counter). Artificial sand can be made from soda glass containing scandium oxide, which is ground down to the appropriate size and shape for use as a tracer. It is then placed in a nuclear reactor to acquire the radioactive isotope scandium 46, which has a half-life (i.e. time taken for the radioactivity to diminish to half its original strength) of 85 days, which means that it can still be detected by Geiger counter three or four months after it has been introduced. Gravels can be traced by using granules, pebbles or cobbles that have been taken from the beach and labelled with a radioactive substance, or artificial materials such as concrete in which radioactive material has been embedded. After the tracer has been deposited, surveys are made by carrying a detection meter over a beach, or dragging it across the sea floor mounted on a sledge. The location and intensity of radioactivity can thus be mapped, and paths of sediment flow deduced.

Radioactive tracers are expensive and a health hazard, but they provide a good means of tracing sediment movement, for the tracer can be detected even when it has been buried in a beach, where coloured or fluorescent tracer would not be visible. However, fluorescent tracers are cheaper and safer, and often more durable for long term projects, whereas radioactive tracer permits only one project, and cannot be used again until all the tracer used in the first project has disappeared. Different colours of paint or fluorescent dye can be used to trace sediment movement from several places at the same time. An alternative on gravel beaches is to use radio transmitters embedded in artificial pebbles introduced to the beach to register their patterns of movement.

Tracers have been used to estimate rates and quantities of beach sediment movement. On Sandy Hook, New Jersey, Yasso (1965) introduced four grain size classes of fluorescent sand, and subsequent sampling on a profile 100 feet downdrift showed that the smallest particles (0.59–0.70 mm) arrived first, and the next class (0.70–0.84 mm) soon afterwards, the maximum rates of flow being 61 and 79 cm per minute respectively.

If the predominant direction of longshore drifting is known, it is possible to estimate the volume of sediment moving along the beach by introducing a standard quantity of tracer at regular intervals to a particular injection point, and taking samples at another point downdrift to measure the concentration of tracer passing by. As the concentration of tracer downdrift is proportional to the rate at which the sediment is moving, the quantities in transit can be measured. Jolliffe (1961) applied this tracer concentration method successfully to the measurement of rates of longshore drifting on shingle beaches on the south coast of England, using pebbles coated with a fluorescent substance. It is more difficult to trace sand movement by this method because of the vast number of grains involved, but sand movement on the Caucasian Black Sea coast has been measured using fluorescent tracer sand down to a dilution of one grain in 10 million.

6.16 Lateral grading

Some beaches show lateral grading from fine to coarse sediment along the shore. Grading is a condition, whereas sorting and attrition are processes that may achieve it. Grain size composition of beach material may vary in the vicinity of source areas such as eroding headlands, where the proportion of locally derived coarse material is high, or river mouths, where a larger proportion of coarse fluvial sediment is likely to be present. There is no single, simple explanation for lateral grading of beaches, and several hypotheses have been put forward. One is that longshore drifting acts selectively on beaches with particles of various sizes, the finer grains being moved further because they are more easily mobilised by waves and transported by associated currents. This is supported by Yasso's (1965) measurements at Sandy Hook, New Jersey.

However, the reverse is found on some beaches, the larger particles having been differentially moved further downdrift. McCave (1978) explained downdrift coarsening of sand to shingle beaches on the Norfolk coast as the result of preferential removal of sand to offshore bars as longshore drifting progressed. At Hampton, on the shores of Port Phillip Bay, Australia, a beach renourished with sediment dredged from the sea floor, a mixture of sand and shelly gravel, was depleted by longshore drifting, the shelly gravel moving downdrift farther and faster than the sand, and so that shells dominated the beach downdrift. Alternatively, finer grains may be selectively removed from the beach, either because they are blown offshore or onshore by wind action, or because they are withdrawn seaward by

backwash, the extent of such removal diminishing alongshore (Komar, 1976).

Chesil Beach, a swash-dominated beach in Lyme Bay (Figure 6.24) in southern England, is subject to longshore drifting when waves arrive obliquely to the shore. It consists mainly of flint pebbles, and grades from granules on the beach at the western end, near Bridport, through to pebbles and cobbles (mainly of locally derived limestone) towards the SE end, near Portland Bill. Grading is correlated with a lateral increase in wave energy, with larger waves coming in through deeper water to build the higher and coarser beach at the more exposed SE end. However, it is not clear how this contrast in wave energy has achieved lateral sorting: perhaps there has been longshore edging of the finer particles from higher to lower wave energy sectors. There is no doubt that Chesil Beach used to receive shingle from the beaches of Lyme Bay to the west, but the dominant eastward longshore drifting of shingle has been interrupted by landslide lobes and bouldery ebbs. The harbour breakwaters at West Bay now form the western limit of a beach compartment in which the shingle deposits are now essentially relict (Bray, 1997). The beach is still receiving small amounts of shingle from cliff erosion to the SE, where Portland limestone yields cobbles and pebbles (mainly from quarries). These become mixed with the flint shingle, and it seems that pebbles of particular dimensions then drift quickly along the shore to their appropriate size sectors, and are there retained, while pebbles that are larger or smaller move away. It has also been suggested that as Chesil Beach was rolled obliquely landward by storm surges it became modified to a laterally graded beach by attrition, the farther travelled western sector being reduced to finer particles than the less travelled eastern sector.

Bascom (1951) related similar grading in beach sand particle size on the shores of Half Moon Bay, California, to variation in degree of exposure to refracted ocean swell coming in

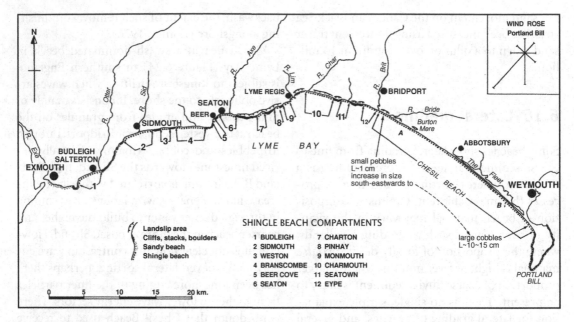

Figure 6.24 Shingle beach compartments in Lyme Bay, Dorset. Chesil Beach is a laterally graded shingle barrier extending out to Portland Bill and enclosing a lagoon (The Fleet). *A* and *B* indicate locations of photographs shown in Figure 6.13

from the NW, coarse sand on the exposed sector grading to fine sand in the lee of a headland at the northern end. A similar pattern is seen in Wineglass Bay, on the east coast of Tasmania, where a curving sandy beach is steeper, narrower and coarser on the more exposed northern sector and becomes flatter, wider and finer towards the more sheltered southern end. In both cases there is an implication that lateral sorting developed as sand was edged from the higher to the lower wave energy sector.

Another possible explanation of lateral grading on beaches is where longshore drifting is stronger in one direction than the other, as the result of contrasts in the height or frequency of incident waves. Larger waves carry the whole range of available coarse to fine particles in one direction, but smaller waves take back only the finer material. Alternations of longshore drifting can thus result in separation of sand from shingle. Chesil Beach could have become laterally

sorted in this way (Jolliffe, 1964), and the other beaches of Lyme Bay are also finer to the west and coarser eastward in response to similar alternations of stronger SW wave action and weaker SE wave action (Bird, 1989). Farther east, between Weymouth and Osmington (in the lee of Portland Bill), the beaches are graded in the opposite direction because SW wave action is much reduced, and SE wave action becomes dominant.

On drift-dominated beaches lateral grading may be the outcome of wearing and attrition of sediment delivered to a particular sector of the coast as longshore drifting carries it away from the source. This may explain the graded beach on the coast of Hawke Bay, New Zealand (Figure 6.25), which is supplied with gravelly material largely from the Mohaka River, and shows a gradual reduction of mean particle size from pebbles to granules, then coarse to fine sand, along the shore eastward from the mouth of the river.

Figure 6.25 The coastline of Hawke Bay, New Zealand, showing lateral grading of the beach east from the mouth of Mohaka River (A) to Waitaniwha (F), as indicated by modal grain size graphs. The wind rose shows the prevalence of SW waves (X) and SE waves (Y), but there is also a southerly swell (W). Predominant longshore drifting is eastward, and grading has been attributed to attrition of pebbles to sand as they move along the shore. There are local variations alongshore because sand and gravel is added from eroding cliff sectors. It is also likely that some sand and gravel have been swept in from the sea floor

In the Bay of Anges, between Nice and Antibes in southern France, pebbles delivered by the River Var became reduced in size (and more rounded) as they moved along the 15 km shore. The mean diameter of pebbles was 9.7 cm at the river mouth, 6.8 cm in the middle of the bay and 4.3 cm at Antibes. The pebbles were large and ovoid at the Var mouth while at Antibes, smaller, flatter and rounded pebbles predominated. However, it is possible that longshore sorting has contributed to this grading,

the initially smaller pebbles travelling farther (Cailleux, 1948). Landon (1930) described how angular gravel from a cliff on the west coast of Lake Michigan became rounded and reduced in size as it drifted southward along the shore, and similar longshore attrition may explain the reduction of cobble size that occurs northward along the spit at Westward Ho! in Devon. Flint and chalk fragments eroded from the Chalk cliffs south of Studland in Dorset become rounded as they drift northward along the beach, the chalk particles being reduced much more quickly than the harder flints, and soon disappearing along the shore. Where lateral grading is the outcome of more rapid attrition on a beach sector subject to strong wave action, the finer sediment will be found in the high energy sector.

By contrast, there is often a correlation between higher wave energy and coarser sediment passing laterally to lower wave energy and finer sediment (King, 1972), the implication being that there has been little or no attrition. As has been noted, there may have been lateral sorting as finer sediment drifted away from the higher wave energy sector. Grain size of beach sediments is not necessarily well correlated with wave energy sectors, for although storm-piled cobble and shingle beaches are more likely to be found on high wave energy coasts many oceanic coasts have sandy beaches. On the other hand, beaches on low wave energy coasts may be gravelly if there is a local source of coarse material, even if wave action is rarely strong enough to mobilise it.

Lateral variation in the mineral composition of drifting sediment may reflect the pattern of sources. In Encounter Bay on the South Australian coast the SE end of the beach, near Kingston, consists almost entirely of calcareous sand (almost 90 per cent calcium carbonate) washed up from the sea floor, but the proportion of quartz sand increases north-westwards to Goolwa, at the mouth of the Murray (calcium carbonate content less than 10 per cent). This lateral variation is the outcome of a pattern of earlier deposition of fluvial quartzose material on the sea floor off the mouth of the Murray during Pleistocene low sea level phases, so that the sand swept onshore is there dominated by quartz rather than the calcareous sand of marine origin further along the bay shore (Sprigg, 1959).

As some beaches show lateral grading, it is worth asking why others do not. Beaches still receiving sediment of various sizes from various sources are less likely to be graded than relict beaches and swash-dominated beaches are less likely to be graded than drift-dominated beaches. The swash-dominated cobble beach at Newgale in Pembrokeshire is composed of gravel that has been washed in from the floor of St Brides Bay, and is not laterally graded. Intricate local patterns of grading are found on beaches behind intermittent reefs or shoals, as in the shingle beach in Ringstead Bay, Dorset, where cobbles are concentrated in the lee of limestone reefs and smaller pebbles in intervening small bays.

6.17 Prograding beaches

Beaches that show net accretion, receiving more sediment from various sources than they lose onshore, offshore or alongshore, become higher and wider, prograding seaward. Their transverse profiles may be convex, or in the form of a terrace ending in a seaward slope. The dimensions of a prograding beach are determined by the balance of gains from fluvial sources, cliff and rocky foreshore yields, the sea floor or dunes blown from the hinterland against losses alongshore and offshore, the removal of sand by wind to build landward dunes and the washing of sediment into estuaries and tidal inlets. Beach progradation may be induced by human activities, notably where the sediment supply has increased as the result of coastal quarrying or where fluvial sediment yields have been augmented by hinterland deforestation or mining (Section 6.4.1).

Progradation of a beach is often indicated by the formation of successive beach ridges of sand or shingle, built above high tide level (Section 6.18). Successively formed foredunes mark stages in the seaward advance of a sandy coastline, as on Winterton Ness in Norfolk and on the shores of Carmarthen Bay in south Wales, where sand is being washed and blown onshore from extensive shoals exposed at low tide and the coastal dunes at Pembrey are extending seaward.

Some beaches are prograding as the result of shoreward drifting of sand or shingle from the sea floor, notably where there are bars or shoals in the nearshore area. On the west coast of Wales there is northward longshore drifting from Cardigan Bay and eastward longshore drifting along the Lleyn Peninsula, converging in Tremadoc Bay, where there are extensive sandy shoals and intertidal flats washed into the Afon Glaslyn estuary and built up as bordering beaches, spits and dunes. In SE Australia beaches are receiving sediment washed in from shoals in Streaky Bay, South Australia (Figure 6.26), off Hunter Island in NW Tasmania and off

Figure 6.26 Prograding beaches supplied by shoreward drifting from sand shoals in Streaky Bay, South Australia. Beach accretion has been shaped by convergence of SW waves refracted by the linear shoals, and in the lee of Eba Island

Corner Inlet, east of Wilson's Promontory, Victoria. The sand shoals east of Vansittart Strait, between Flinders Island and Lady Barron Island off NE Tasmania, are a graphic example of sea floor sand drifting that in this case has passed through a strait instead of accumulating on west-facing beaches on these islands.

Beach progradation on the Danish island of Kyholm resulted from the disappearance of sea-grass meadows in the surrounding waters and the ensuing shoreward movement of fine sand that had previously been retained on the sea floor by this vegetation (Christiansen *et al.*, 1981). Progradation is more likely to take place on coasts where emergence is in progress, stimulating shoreward drifting of nearshore sediment, as at Kalajoki on the Finnish coast of the Gulf of Bothnia. In northern Britain, uplift is continuing as the result of postglacial isostatic recovery, and is likely to have contributed to the progradation of wide sandy beaches near Holy Island in NE England. The sand has been swept shoreward from sea floor deposits of glacial drift, probably including eskers similar to those seen inland in the Tweed valley.

Progradation continues on the ends of spits, as on Hurst Castle spit in Hampshire and at the western and of Blakeney Point in Norfolk (Section 8.2), and on the shores of cuspate forelands, as on the eastern shore of Dungeness in Kent and the north shore of Benacre Ness in Suffolk. Beaches have also prograded on coasts where longshore drifting has been intercepted by a protruding headland, so that there is accretion on the updrift side, as at Point Dume in California, Apam in Ghana and Cape Wom on the north coast of New Guinea. Such beaches will continue to prograde until sand or gravel by-pass the promontory to reach the downdrift shore.

Beach sediment drifting alongshore is intercepted where jetties or breakwaters have been constructed, usually to stabilise a river mouth or lagoon outlet, or shelter a harbour from winds and waves. This results in progradation updrift of the intercepting structure. In SE England, longshore drifting is generally southward on the East Anglian coast, where beaches have prograded on the northern side of harbour break-waters at Yarmouth, Lowestoft and Southwold, while on the Channel coast eastward longshore drifting has led to accumulations on the western side of breakwaters at Black Rock marina at Brighton, and at Newhaven, Rye and Folkestone. At Rye the breakwaters built at the mouth of the Rother River have intercepted eastward-drifting shingle to form a prograded shingle beach on the western side. To the east are Camber Sands, a pebble-free sector, with shingle reviving towards Lydd and Dungeness. Thus the Rother break-waters have created a break in the shingle beach that was migrating alongshore. At Studland in Dorset the beach has prograded south of a training wall built in 1924 to prevent the northward drifting of nearshore sand from shoaling the entrance to Poole Harbour.

In the United States beaches have prograded alongside breakwaters built in 1930 at Santa Barbara and at Redondo in California and South Lake Worth in Florida, with concomitant erosion downdrift. Similar features are seen at Sochi on the Black Sea coast, Lagos in Nigeria, Durban in South Africa and Madras in India. Among many other examples are the interception of southward-drifting sand by the breakwaters at Praia da Barra near the Aveiro Lagoon and at Figuera da Foz, both in Portugal, and eastward-drifting sand alongside the jetty built at the mouth of the Vridi Canal on the Ivory Coast when it was cut in 1950. In Australia, northward-drifting sand has accumulated along the southern sides of breakwaters at Port Adelaide and at Tweed Heads in northern New South Wales.

In some places, progradation has occurred on both sides of protruding breakwaters, indicating either an alternation of longshore drifting or a predominance of shoreward drifting on to beaches on either side of the outflow. There has

Figure 6.27 The artificial entrance to the Gippsland Lakes, cut through the outer barrier in 1889 and bordered by stone breakwaters. Sand accretion on either side of the breakwaters has formed coastal forelands with successive parallel foredunes, and wave patterns show the location of the looped offshore bar. Crown Copyright Reserved

been sand accretion on both sides of the breakwaters built at Lakes Entrance in SE Australia to stabilise the artificial outlet from the Gippsland Lakes cut in 1889 (Figure 6.27). Similar accretion has taken place on both sides of paired breakwaters at Onslow in NW Australia, Rogue River and Siuslaw River on the Oregon coast, Newport in California, Ijmuiden on the Dutch coast and the Swina Inlet in Poland.

Beach accretion can occur in the lee of a harbour breakwater, as at Warrnambool in Victoria, Australia, where shoreward drifting of fine sand occurred from the sea floor into the harbour area, which became more sheltered after the breakwater was constructed, leading to beach progradation. After an offshore breakwater was built in 1934 to protect the pier at Santa Monica, California, a sandy foreland developed during the ensuing 15 years, and dredging became necessary to maintain access to the pier. Similar local progradation has occurred behind offshore breakwaters built to protect Italian beaches, notably between Rimini and Venice on the Adriatic coastline. Progradation of a beach can also be induced in the lee of a wrecked ship close to the shore, as at Sukhumi on the Caucasian Black Sea coast (Zenkovich, 1967).

Accretion has taken place alongside some tidal inlets (lagoon entrances) where no breakwaters have been built, the transverse ebb and flow of currents having had the same effect as an intercepting breakwater. There is usually some deflection of an unprotected outflow channel by longshore drifting, and if this continues and the gap becomes sealed off by deposition the accreted foreland may become a lobe, which then moves on along the shore (Figure 6.33).

On the NE coast of Port Phillip Bay, Australia, seasonal patterns of longshore drifting occur. At Sandringham, near Melbourne, breakwaters built to form a boat harbour have intercepted sand drifting south in winter in a situation whence it cannot return northward in summer. The beach has prograded within the harbour because the breakwater unexpectedly formed a sand trap (Figure 6.28).

6.18 Beach ridges

Beach ridges have formed where sand or shingle has been banked up above high tide level by wave swash on a prograding coastline. Sandy beach ridges may originate as berms built by

Figure 6.28   The natural configuration of the coast at Sandringham, Port Phillip Bay, Australia (a) was modified after the construction of a breakwater (1949–54) to shelter Sandringham Harbour. As at nearby Black Rock (Figure 6.15), the sandy beach is modified by northward drifting in summer and southward drifting in winter. The breakwater created a trap in which southward drifting sand has accumulated (b). Depletion of Hampton Beach led to beach nourishment in 1987 north of the New Street groyne, and in 1997 to the south. The nourished beaches have been losing sand northward in summer and southward in winter (c)

constructive wave action, whereas shingle beach ridges have been piled up by storm waves. Sandy beach ridges are typically 5–50 m wide; shingle beach ridges generally less than 20 m wide. The persistence of a beach ridge depends on overall progradation of the coastline and its separation from earlier ridges by swales is often the outcome of a phase of erosion. In Tasmania, Davies (1957) described the evolution of parallel beach ridges on successive wave-built berms, the outcome of alternations of cut and fill on a prograding coast. In New South Wales, McKenzie (1958) found that foredunes were initiated on successive strandlines of seed-bearing vegetation on a prograding beach. Beach ridge and foredune ini-

tiation have been much discussed, and exemplify 'multiple causality' (Hesp, 1984, 1988; Bird and Jones, 1988; Taylor and Stone, 1996; Sanderson, Eliot and Fuller, 1998).

Sandy beach ridges can be built up along the shore by storm swash where the nearshore area is shallow and incoming waves become constructive. Psuty (1965) described parallel beach ridge formation by this process on the prograding shores of the deltaic coast of Tabasco, Mexico, where fluvially supplied summer sand accretion has been built into successive ridges (a beach ridge plain) by winter storm swash on a gently shelving shore. Sandy beach ridges are found on the east coast of Peninsular Malaysia, where they are known as permatang (Teh Tiong Sa, 1992).

The height and spacing of parallel beach ridges is determined by the rate of progradation (i.e. the rate of supply of sand and shingle, depending on available sources and patterns of sediment flow), the incidence of cut and fill and the upper swash limit of the waves that built them, which is modified by changes in the relative levels of land and sea. There are numerous roughly parallel shingle beach ridges on the cuspate foreland at Dungeness, each marking a former coastline, and stages in the evolution of this structure have been deduced from the ridge pattern (Lewis, 1932). Similar shingle beach ridges have been used to trace stages in the progradation and southward growth of Orfordness, and the parallel beach ridges at Hoed in Denmark (Figure 6.11).

The height of the crest of a beach ridge is also related to the beach profile preceding the storm. Swash limits are higher on steep beaches than on those of gentle gradient. Moreover, the amount of beach material available to be storm-piled affects beach ridge height. If there is too little storm swash overtops the beach, and if there is too much storm wave energy is expended on the beach profile rather than on ridge building.

Variations in the crest levels of successive beach ridges are likely to reflect variations in the upper limit of storm swash, but they could also be related to changing sea levels. Lewis and Balchin (1940) found that the crests of some of the inner and older shingle ridges on Dungeness were 2–3 m lower than those formed more recently along the eastern shore, and deduced that sea level had been rising. On the other hand, some Scottish shingle beach ridge systems have ridge crests declining seaward, as in Spey Bay and on the west coast of Jura, and similar features are seen where eskers have been submerged, reworked by wave action and built into successive ridges on an emerging coast (Figure 6.29).

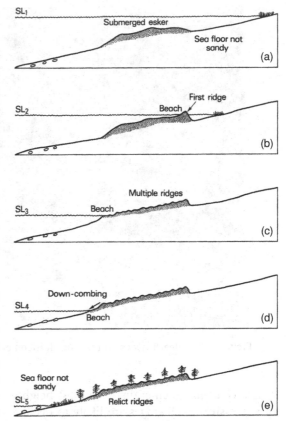

Figure 6.29 As sea level falls in the Gulf of Bothnia submerged eskers are reworked by wave action to form beaches (a), (b), and as emergence proceeds multiple beach ridges have formed on a seaward slope (c). As the sea falls below the limit of the former esker beach ridge formation comes to an end (d), (e)

Examples of beach ridges that have formed in this way are seen near Hove in southern Sweden and around the Gulf of Bothnia, as on the island of Hailuoto in Finland (Figure 6.30). These are coasts that have been emerging as the result of postglacial isostatic recovery, and the older beach ridges formed when the sea stood at higher levels. The separation of emerging beach ridges may be due to minor oscillations of sea level, or to the effects of recurrent storms.

Figure 6.30 Beach ridges on the Finnish island of Hailuoto, formed in the way shown in Figure 6.29

If a coastline is emerging, the crest heights of successive beach ridges are likely to decline seaward, but there are also variations associated with differing upper swash limits in each constructional phase. On St Vincent Island, Florida, Tanner (1995) considered that prograded beach ridge plains were the outcome of recurrent sea level fluctuations of between 5 and 20 cm during the past 4000 years.

A seaward decline in the crest heights of beach ridges is not necessarily an indication of emergence, because it could have been produced by a sequence of diminishing upper swash limits (for example if the nearshore area were shallowing as the result of accretion, and incident wave heights were consequently being reduced). In a similar way a seaward rise in the crest levels of shingle ridges could result from increasing storm surge levels rather than a rising sea level.

On sandy beaches the formation of a succession of parallel beach ridges is often accompanied by the evolution of successive foredunes. Backshore vegetation spreading forward on to the beach indicates that progradation has been taking place (Figure 6.31), especially where the vegetation canopy declines smoothly seaward from trees through shrubs to grassy communities, a zonation that indicates a plant succession is accompanying the deposition of new terrain.

Beach ridges can also be formed by the successive addition of spits that have grown parallel to the coast. Historical maps of South Haven Peninsula on the Dorset coast indicate stages in the formation of three broad sandy beach ridges, surmounted by dunes, and the peninsula at Falsterbo, in SW Sweden, also prograded in this way. Beach ridges of a different kind are formed on cold coasts in high latitudes, where the sea freezes and wave and nearshore processes are halted, at least in winter. When storm waves break up a winter ice fringe and drive it on to the shore, ice-pushed beach ridges are

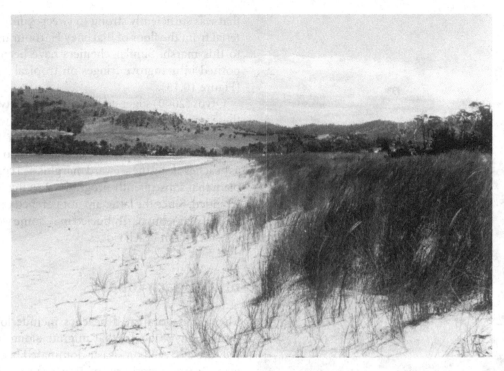

Figure 6.31 Backshore dune vegetation spreading behind a prograding beach in Frederick Henry Bay, Tasmania

formed. These have been described from arctic coast beaches in Alaska, northern Canada and Siberia and from Antarctica. In summer melt water from glaciers cuts channels across these ridges on the backshore.

6.19 Cheniers

Cheniers are wave-built ridges of sand, shelly sand and gravel that have been deposited on coastal plains of fine grained sediment (silt, clay or peat), particularly in deltas and marshlands. They are typically up to 3 m high and 40–300 m wide. Cheniers were originally described on the deltaic plains of Louisiana, where they were called cheniers because oak trees (*chênes*) grow on them (Russell and Howe, 1935). They are parallel to the coastline and are usually separated by broad flat swampy areas. They were emplaced by

occasional storm surges as the low-lying terrain prograded, and subsidence of deltas and coastal plains results in the older ridges disappearing landward beneath younger alluvium.

There are numerous cheniers on the deltas of NE China, each deposited when shelly material and fine sand were driven landward by storm swash on to a prograding marshy plain (Liu Cangzi and Walker, 1989). Similar features are seen on the coastal plains of northern Australia, notably to the east of Darwin, where sandy ridges rest upon, and are separated by, flat swampy deltaic lowlands. They are bordered seaward by a mangrove fringe with only minor beaches of sand with coral and shell gravel (as at Point Stuart), and it appears that sand has been washed up through the mangroves from the floor of Van Diemen's Gulf during storm surges (tropical cyclones) on a generally low wave energy coast. They are emphasised by the growth of

Figure 6.32 Cheniers of shelly sand on a mangrove-fringed (m) coastal plain south of Van Diemen Gulf, northern Australia, dissected by tidal creeks

pandanus palm trees, the intervening areas being bare of vegetation, apart from some sparse grassland (Figure 6.32). This is a chenier plain, a coastal plain with several cheniers formed parallel or sub-parallel to the coastline (Augustinius, 1989).

Ridges of shelly sand driven landward on to salt marshes in East Anglia, notably on the Dengie Peninsula in Essex, have also been termed cheniers, and are the outcome of swash action in successive storm surges. A shelly chenier runs east–west across the salt marsh at Morston in north Norfolk, in the lee of the Blakeney Point spit, emplaced during a storm surge

that was sufficiently strong to sweep sandy material from the floor of Blakeney Harbour up on to this marsh. Similar cheniers have been deposited in mangrove fringes on tropical coasts (Figure 10.13).

Otvos (2000) suggested a distinction between these transgressive cheniers, formed when waves or surges sweep beach sediment in on to a coastal plain, and regressive cheniers, deposited as beaches along the seaward margins of a coastal plain that subsequently progrades, leaving them stranded. Since the latter are not perched on the coastal plain, but partly buried in it, some would consider them beach ridges.

6.20 Beach lobes

Many drift-dominated beaches include lobate protrusions (salients) that migrate alongshore; they are also seen on swash-dominated beaches where some longshore drifting occurs. These beach lobes may be formed by the local convergence of longshore drifting as the result of waves approaching obliquely, first from one direction then the other. Alternatively, they may be the result of local and temporary progradation beside a river mouth or lagoon outlet, moving on when the gap becomes sealed by deposition. Some lobes form during occasional storms when waves arriving obliquely are strong enough to drive beach material round a headland to be deposited as a lobe on the lee side. Others take shape when shoals of sediment swept out of a river mouth by floodwaters are washed up on to the shore downdrift. Some form as the result of accretion in a zone of wave convergence in the lee of a shoal, reef or islet and remain in position; others (particularly on drift-dominated beaches) may move intermittently alongshore until they are intercepted beside a headland or breakwater or added to the end of a spit. Some are asymmetrical, with shorter updrift and longer downdrift flanks.

Figure 6.33 The formation of a beach lobe (A) as the result of accretion updrift of a creek outlet that is deflected by breaching to form a new outlet with the lobe downdrift (B). Beach erosion and accretion accompany the subsequent migration of the lobe (C) alongshore

Features here described as beach lobes were termed accretion and erosion waves by Inman and Jenkins (2005). These migrate in the direction of longshore drifting, and there are examples on the coasts of California and Assateague Island, Maryland. Unfortunately, this terminology risks confusion with the wave processes that cause accretion and erosion on beaches, and with such features as sand bars and troughs.

A succession of sandy beach lobes has formed and migrated along the coast between Somers and Sandy Point in Westernport Bay, Australia,

as the result of longshore drifting by waves from the SW. The evolution of one of these began with the formation of a longshore spit that deflected the outlet from a creek (Figure 6.33, A), but was breached to release a lobe of sand (Figure 6.33, B) that has subsequently drifted alongshore to the east (Figure 6.33, C). The passage of such a lobe is marked by progradation, followed by erosion as the lobe moves on. Similar lobes have migrated eastward along the Ninety Mile Beach in SE Australia, and southward on the east and west shores of Lingayen Gulf in the Philippines. Some lobes grow by the addition of beach ridges, and become travelling forelands, with beach ridges truncated on their western flanks and added successively to their eastern flanks. Beach lobes are not found on swash-dominated high wave energy coasts.

6.21 Summary

Beaches consist of sediment ranging from very fine sand to boulders, and can be described in terms of the relative proportions of these grain sizes. They have been emplaced within the Holocene still-stand (generally since 6000 years BP) and derived from various sources including eroding cliffs and foreshores, rivers and glaciers, and the sea floor (including shelly and calcareous beaches). They often include artificial materials, such as glass and pottery fragments. Some beaches are relict, no longer receiving sediment. Beach sediments are reduced by weathering and abrasion, including frost and ice action on cold coasts, but they can also be cemented by precipitation, notably of carbonates, to form beach rock.

Beaches are shaped largely by incident wave action, which also generates longshore, onshore and offshore drifting. Outlines in plan and profile are determined by incident wave patterns and influenced by other coastline features (e.g. headlands, estuaries, deltas), and there may be a tendency towards equilibrium between form and process. There are short term changes caused by alternations of erosive and constructive waves (cut and fill), with gains and losses measured within a sweep zone. Beach gradients are related primarily to sediment size, and sorting by wave action may develop contrasting upper and lower beaches. Stratification within a beach results from alternations of these processes.

Beach face features shaped by wave action include beach berms and cusps, and are influenced by wind action, downwashing and tidal oscillations. Eroding beaches generally have concave profiles, often with microcliffs, whereas accreting beaches are typically convex in profile. Beach morphodynamics (mutual adjustments between morphology and processes) can be characterised by reflective, dissipative and intermediate beach states identifiable by selected parameters and demonstrable in models. Beach compartments, typically delimited by headlands or breakwaters, contain sediment cells, but there may be by-passing laterally. Beach budgets assess gains, losses and longshore movements of beach sediment, which can also be followed using tracers. Some beaches show lateral (longshore) grading in response to sorting or attrition. Beaches gaining sediment are built up and prograde seaward, sometimes in stages marked by beach or dune ridges. Cheniers are beaches swept on to coastal plains by storm surges. Beach lobes are local accretions that form and may migrate alongshore.

7
Beach erosion

7.1 Introduction

Beaches are eroded when they lose more sediment alongshore, offshore or to the hinterland than they receive from the various sources
shown in Figure 6.2. Processes that lead to
beach erosion include destructive wave action in
stormy periods and the depletion of beach sediments by weathering and winnowing, as well
as a reduction in inputs from rivers, cliff and
shore erosion, spilling dunes and drifting from
the sea floor. As the volume of beach material diminishes the beach face is lowered and
cut back. A convex profile typical of prograding beaches is replaced by a concave profile,
often backed by a low erosional scarp (Figure
6.21). Indications of beach erosion are shown in
Panel 7.1.

Beaches that line cliffed or rocky coasts become narrower, thinner and discontinuous as
losses proceed, and may eventually disappear,
while on depositional coasts the previously deposited landforms are cut back. The rate of retreat of the high tide shoreline, which is often
also the seaward boundary of terrestrial vegetation communities, can be measured by comparing dated sequences of maps and charts, or air
and ground photographs. Mean annual recession rates have been generally small (less than a
metre per year), but there are records of beaches
retreating by up to 40 m/yr, for example on the
Nile delta (Section 12.6).

Beach erosion has been reported from many
coasts. Seaside resorts such as Deauville in
France, Miami in Florida and Surfers Paradise
in Australia are among the many that have suf-
fered when storms have removed the beaches
that attracted their visitors. In such places cut-
and-fill cycles are incomplete, only part of the
sand removed by storm waves being returned
to the beach in calmer weather. There has been
retreat of sandy coastlines in many parts of the
world during the past several decades, and in
many places for a much longer period. Recession is in progress even on the shores of coasts
that were previously prograded by deposition,
for example along the Ninety Mile Beach in SE
Australia, where progradation during the past
century has been confined to sectors on either
side of the Lakes Entrance harbour breakwaters
(Figure 6.27). Around the world there are relatively few sectors of naturally prograding sandy
beach, whereas receding sandy coastlines are extensive. Seaside resorts fortunate enough to have
had their beaches substantially widened by natural accretion include Seaside in Oregon, Riumar
on the Ebro delta in Spain and Malindi in Kenya
(Figure 6.4).

Coastal Geomorphology: An Introduction, Second Edition. Eric Bird.
© 2008 by John Wiley & Sons, Ltd

Panel 7.1 Indications of beach erosion

1. A concave beach profile, especially with a microcliff.

2. Cliffed backshore dunes (whereas prograding beaches are backed by beach ridges and incipient foredunes).

3. Truncated vegetation zones (whereas prograding beaches are backed by tree canopies descending to beach level, or by shrub and grass zones on recently formed sandy terrain).

4. Patches of sand adhering to rock outcrops indicating the previous higher level of the beach.

5. Exposures of beach rock that formed within the former beach.

6. Exposure of rocky or muddy substrate.

Panel 7.2 The causes of beach erosion

1. Submergence and increased wave attack

2. Reduction of fluvial sediment supply

3. Reduction in sediment supply from cliffs

4. Reduction of sand supply from inland dunes

5. Reduction of sediment supply from the sea floor

6. Extraction of sand and shingle from the beach

7. Increased wave energy

8. Interception of sediment supply by longshore drifting

9. A change in the angle of incidence of waves

10. Intensification of obliquely incident wave attack

11. Increased losses of beach sediment to the backshore

12. Increased storminess

13. Attrition of beach material

14. Beach weathering

15. Increased scour by wave reflection from a sea wall

16. Migration of beach lobes

17. A rise in the beach water table

18. Removal of beach material by runoff

19. Diminished tide range

20. Abrasion by driftwood

21. Removal of a sea ice fringe

The causes of beach erosion listed will now be examined in greater detail.

Between 1976 and 1984 the Commission on the Coastal Environment (International Geographical Union) assembled evidence of coastline changes around the world over the preceding century, and found that beach erosion had become widespread. More than 70 per cent by length of beach-fringed coastlines had retreated over this period, less than 10 per cent having advanced (prograded), the balance having either remained stable or shown alternations with no net gain or loss (Bird, 1985a).

Investigating the causes of beach erosion, the Commission on the Coastal Environment found that there was no single, simple and universal explanation. Some 21 factors have been identified as having initiated or accelerated beach erosion, their relative importance varying from one coast to another (Panel 7.2). Usually more than one of these factors has contributed to the onset or intensification of erosion on a particular beach. Attempts to explain the erosion of a beach should consider each of the possible causal factors, and those found to be applicable should be ranked in importance.

7.1.1 Submergence and increased wave attack

Coastal submergence may be due to a rise of sea level, coastal or nearshore land subsidence, or some combination of land and sea movement that results in the sea standing higher relative

Figure 7.1 The Bruun rule states that a phase of sea level rise on a sedimentary coast will be accompanied by recession (R) of the beach as a volume of sediment ($V_1 = V_2$) is transferred from the backshore to the nearshore zone in such a way as to restore the transverse profile landward from D, the initial seaward boundary of nearshore sand deposits

to the land. It is likely to lead to beach erosion, for a sea level rise usually results in the deepening of nearshore waters, allowing larger waves to reach the shore and erode the beach face, thus withdrawing sand or gravel to the sea floor. Bruun (1962) suggested that a sea level rise on beaches that had attained a profile of equilibrium (which he took as a condition where they were neither gaining nor losing sediment over a specified period) would cause erosion of the upper beach and transference of sand or gravel from the beach to the adjacent sea floor. This would in due course restore the previous transverse profile in relation to the higher sea level (Figure 7.1). The transverse profile would thus migrate upward and landward, the coastline retreating further than it would have done with an equivalent sea level rise on a rocky shore of similar profile.

This has become known as the Bruun rule. It has been reproduced in wave tank experiments (Schwartz, 1965) and is supported by observations of the backshore erosion and nearshore deposition that have taken place on beaches around the Great Lakes during and after each episode of

rising water level, and on Cape Cod beaches between low neap and low spring tides (Schwartz, 1967), but it is a two-dimensional model that has limitations in practice.

Beach erosion has become widespread on coasts where the sea has been rising because land subsidence is in progress (Figure 3.12), as on the Gulf and Atlantic coasts of the United States. Land subsidence resulting from extraction of groundwater has accelerated beach erosion on the northern Adriatic coast of Italy, as on the Lidi of Venice and at Ravenna, and removal of oil and water from the Gippsland Basin may have contributed to erosion on the Ninety Mile Beach in Australia. Beaches were cut back suddenly on sectors of the Alaskan coastline that subsided during the 1964 earthquake.

Evidence from tide gauge records suggests that there has been a world-wide sea level rise of 1–2 mm/yr during the past few decades, offset on some coasts by equal or greater land uplift, and varying also in relation to the geophysical factors that complicate the surface topography of the oceans. If there has indeed been a sea level rise the Bruun rule may provide at least a partial explanation for the modern prevalence of beach erosion.

7.1.2 Reduction of fluvial sediment supply

On coasts where beaches have been supplied with sediment carried down to the coast by rivers, erosion is likely to follow any reduction in sediment yield to river mouths as a result of reduced runoff. Such a reduction may be due to diminished rainfall or less melting of snow and ice within the river catchment, but is more often a sequel to the building of dams to impound water upstream. These have intercepted the discharge of fluvial sediment, cutting off the supply of sand and gravel to beaches at and near the river mouth. The result is the onset of erosion on beaches that were formerly maintained or prograded by the arrival of this sediment. Erosion will develop more quickly, and become more severe, on coasts where there is also strong longshore drifting of sediment away from river mouths.

The best known example of such erosion is on the shores of the Nile delta (Section 12.6), where sandy beaches had been prograding for many centuries as the result of the longshore spread of sandy sediment delivered to the mouths of Nile distributaries. Erosion of beaches near the mouths of the Rosetta and Damietta distributaries was first noticed early in the present century, soon after barrage construction began upstream in 1902. It became much more rapid and extensive after the completion of the Aswan High Dam in 1964, which resulted in large scale sediment entrapment in the Lake Nasser reservoir, and during the next few years, beach erosion on parts of the deltaic coastline attained annual rates of up to 120 m (Sestini, 1992). Some of the sediment removed from these beaches drifted eastward along the coast towards Port Said, but much has been lost offshore (Lotfy and Frihy, 1993).

Similar beach erosion has occurred on other deltaic coasts following dam construction upstream. Examples include the Rhône delta in France, the Dnieper and Dniester deltas in the Ukraine, the Citarum delta in Indonesia and the Barron delta in NE Australia. On the coast of southern California, where relatively high wave energy has largely suppressed delta development, beaches nourished by longshore distribution of sand and gravel delivered by rivers have become depleted since dam construction reduced fluvial sediment yields. Barrages built on the Tijuana River in Mexico have retained about 600 000 m^3 of sediment annually, and this has deprived beaches north of the river mouth of a sediment supply, so erosion has ensued, particularly on Imperial Beach near San Diego. Beach erosion on the Caucasian Black Sea coast has been attributed to the damming of rivers. In northern Sweden, where most rivers have been harnessed for hydroelectric power production, the necessary dams have so diminished fluvial sediment yields that deltaic coasts such as that of the Indal River are eroding despite continuing isostatic uplift of the land.

Beach erosion near river mouths has followed the dredging of sand from river channels, as in the Tenryu River in Japan, and the cessation of hinterland mining that had previously augmented the fluvial sediment supply to the beach. Examples of this can be seen at Par and Pentewan in Cornwall, where beaches prograded with convex profiles during the period when sand and gravel waste from tin and copper mining was carried down to the coast by rivers between 1750 and 1950. When mining ceased this sand supply declined, and the beaches were reshaped and reduced by erosion and sediment attrition to concave profiles.

Excavation of sand from the Rhine during the 1940s resulted in river-bed hollows, and while these were filling a diminished sand supply to the delta region downstream contributed to the onset of erosion of beaches along the delta coastline.

Another cause of reduction of fluvial sediment yield has been successful soil conservation works (slope terracing, runoff management, reafforestation) in the hinterland, as exemplified by the rivers draining to the Gulf of Taranto in southern Italy. Diminished river flow during prolonged droughts has had similar consequences. In southern California, Orme (1985) found that beach erosion occurred during dry periods, especially in 1939–1968, the beaches being restored during wet years such as 1969, 1978, 1980 and 1983, when the fluvial sediment supply revived. In Queensland, Australia, the Burdekin delivered abundant sand to its mouths during floods, to drift northward and be deposited as beaches and spits. The 1991–1995 drought diminished fluvial runoff and sand supply to this coast, and beaches north of the river mouths were eroded as SE waves continue to move sand away northward.

The importance of a river maintaining a sediment supply to a beach is also illustrated where there have been natural or artificial diversions of the river mouth. Beach erosion has occurred on the abandoned deltas of rivers that have been diverted to other parts of the coastline, where a new delta is forming, as on the Hwang Ho delta in China after the river diverted naturally during an 1852 flood. On the coasts of Greece and Turkey beach erosion has resulted from a diminished sediment yield from rivers because of long-continued soil erosion in the river catchments, where increasing areas have been stripped of all unconsolidated material, leaving bare rock widely exposed.

On the world scale much beach erosion has been caused by diminished fluvial sediment yields, but there has also been extensive erosion on beaches remote from river mouths,

and for this some other explanation must be sought.

7.1.3 Reduction in sediment supply from cliffs

The supply of sediment from erosion of cliffs and foreshore rock outcrops may diminish for various reasons. Cliff erosion is often preceded and accompanied by subaerial weathering, and runoff after heavy rainfall may cut ravines or gullies in the cliff face, as at Black Rock Point on the shores of Port Phillip Bay, Australia. Sediment slumps or is washed down from the cliff on to adjacent beaches in rainy periods. If annual rainfall diminishes, or runoff is controlled (e.g. by the insertion of drains along the cliff crest to intercept the water flow and prevent it flowing down the cliff face), the supply of sediment to nearby beaches will diminish, and they will be depleted.

Stabilisation of coastal landslides has a similar effect, as has the cessation of coastal quarrying, halting the supply of waste material to the beach. Thus beaches at Porthallow and Porthoustock on the coast of SW England were augmented by quarry waste between 1878 and 1985, but diminished after the quarrying ceased (Figure 6.12). It remains to be seen whether the recent revival of quarrying here will restore them.

A decline in the strength and frequency of wave attack on cliffs may occur as a result of a climatic change leading to calmer conditions in coastal waters, or when waves are reduced by nearshore shoaling or the growth of reefs. Zenkovich (1967) put forward the theory that if sea level remains unchanged the rate of cliff recession will eventually decline as the result of the widening of the shore platform cut in front of them, across which wave action gradually weakens, and this implies a diminishing sediment yield from cliffs to beaches.

The commonest cause of a reduced sediment yield from cliffs to beaches is the construction of sea walls along the cliff base. Beach erosion at Bournemouth in England has occurred during the progressive extension since 1900 of a promenade on the seafront. This was intended to halt cliff recession and preserve coastal properties, but it also cut off the sand supply that had come from the cliffs, and with little (if any) replenishment of the sand and gravel carried away by longshore drifting the Bournemouth beach gradually diminished.

On the shores of Port Phillip Bay near Melbourne, Australia, similar beach erosion is a sequel to the walling and stabilisation of formerly eroding cliffs of Tertiary sandstone. Other examples of beaches depleted as the result of the building of sea walls on adjacent cliffed sectors include Ediz Hook, in the state of Washington, USA (Schwartz and Terich, 1985), and Byobugaura in Japan (Sunamura, 1992).

In recent decades erosion has become more severe and extensive, on the southern shore of Hurst Castle Spit, where reduction of the supply of sand and shingle because of stabilisation of Barton Cliffs to the west accelerated beach erosion. On the Norfolk coast Clayton (1989) found that erosion of 33 km of cliffs in Norfolk, averaging 25 m in height, had under natural conditions yielded 500 000 m³/yr of sediment, about two-thirds of which was sand and gravel supplied to beaches extending for more than 60 km downdrift. With artificial stabilisation of 70 per cent of these cliffs during the past century (many of the coast protection works have proved to be only partially effective in halting cliff erosion), the sediment yield has been reduced by 25–30 per cent, and beach erosion is spreading downdrift.

In situations where cliff outcrops of sandstone or conglomerate, yielding sand or gravel to nearby beaches, are backed by contrasted rock types, such as a massive resistant formation, or soft silts and clays that do not yield beach-forming sediment, cliff recession may eventually expose these. Beaches formerly derived from the sandstone or conglomerate may then no longer be maintained.

7.1.4 Reduction of sand supply from inland dunes

Beaches that have been supplied with sand from dunes spilling from the land on to the shore may start to erode if the sand supply is reduced or terminated because the dunes have become stabilised. This may result from the natural spread of vegetation, or from conservation works such as the planting of grasses or shrubs, the spraying of chemicals such as bitumen or rubber compounds, or paving, road-making and building over the dune surface. Alternatively, the sand supply may run out because the whole of an available dune has moved on to the shore.

Examples of beach erosion resulting from diminished aeolian sand supply have occurred on the south-facing Cape Coast of South Africa, where the prevailing westerly winds are driving dunes along the coast, as at Sundays River. On Phillip Island, Australia, the partial stabilisation of dunes that had been spilling eastward across the Woolamai isthmus to nourish the beach in Cleeland Bight (Figure 2.5) was followed by beach erosion.

7.1.5 Reduction of sediment supply from the sea floor

There is much evidence, particularly on oceanic coasts, that sea floor sand or gravel drifted shoreward during and after the Holocene marine transgression. As sea level rose across the continental shelf waves reworked sediment that had previously been deposited by rivers or wind action, and eroded material from weathered rock outcrops. The coarser components were swept shoreward, and deposited as sand or gravel beaches. As the marine transgression came to an end, continuing shoreward drifting prograded the beaches, often forming successive backing beach ridges and parallel dunes (Figure 6.5(a)–(c)).

Shoreward movement of sand is still continuing where sediment is being washed in by waves and currents from nearshore shoals, but on many coasts this supply of sediment has declined as the transverse nearshore profile has became smooth and concave and beaches have no longer received sediment from the sea floor. If there is no compensating input of sediment from other sources (such as rivers) progradation comes to an end, and with continued input of wave energy this transverse nearshore profile migrates landward, so the beaches are progressively consumed. Many beaches that prograded earlier in Holocene times have passed into this condition, and are now being cut back by erosion, continuing wave action driving the transverse concave profile landward (Figure 6.5(d)). This may be the explanation for beach erosion on the shores of St. Ives Bay, at Newquay on the north coast of Cornwall and at Braunton Burrows in Devon.

The onset of erosion has come at different times in different places because the development of the concave profile and the cessation of shoreward drifting of sediment from the sea floor have not occurred simultaneously along the coastline. Tanner and Stapor (1972) described this change on beaches as a transition from an economy of abundance to an economy of scarcity of sediment supply, and the sequence has been described as a maturing of the system. They illustrated the sequence of events by reference to the erosion that had developed on the beach ridge plain at Cabo Rojo, SE of Tampico in Mexico. It also provides an explanation for erosion on the Ninety Mile Beach in SE Australia, which borders a sandy coast formerly prograded by accretion of sand supplied from the adjacent

floor of Bass Strait, and is now being cut back by marine erosion (Figure 7.2). There is still plenty of sand in the nearshore area, but the transverse profile has become smooth and concave, and there is no longer shoreward drifting of sand to this beach.

Beach erosion could also result from diminished production of shelly deposits washed in from the sea floor because of ecological changes, such as the destruction of shell fauna by pollution. It is possible that this has reduced shelly beaches fringing the shores of Corio Bay, near Geelong, Australia. The supply of sand and shingle from the sea floor has diminished where increased growth of sea grasses or other marine vegetation has impeded shoreward drifting and trapped the sediment offshore.

7.1.6 Extraction of sand and shingle from the beach

Where sand or gravel is removed from a beach for use in road and building construction the shore profile is artificially lowered, allowing larger waves to attack the beach more strongly during storms. This has increased beach erosion at Klim, on the north coast of Jutland, Denmark. Similar depletion of beaches has occurred on the south coast of England, where pebbles have been quarried from the western end of Chesil Beach at West Bay on the Dorset coast for use in industrial filters, and from nearby Seatown for ornamental purposes. Coastal erosion at Hallsands in South Devon followed the extraction of gravel from Start Bay in the 1890s for use in the building of Plymouth Dockyard (Section 4.9). Beaches in Jersey (notably St Ouen's) and Guernsey were much reduced by the extraction of sand from beaches by the German occupying forces during the Second World War for use to build bunkers and gun emplacements, and increased backshore erosion has led to the construction of massive sea walls. Beach and backshore

Figure 7.2 Erosion (arrowed) of the Ninety Mile Beach, on the outer barrier of the Gippsland Lakes in SE Australia, with accretion (+) on either side of the Lakes Entrance breakwaters (see Figure 6.27)

erosion resulting from sand extraction has occurred on Casuarina Beach, near Darwin in northern Australia. Beach quarrying may be less damaging where the sand or gravel is extracted from a sector that is prograding, notably as the result of updrift accretion alongside a breakwater, as at Timaru in New Zealand.

Beaches have been depleted by the extraction of shell sand and gravel for agricultural use as lime on farmland in Cornwall, where this is permitted below high tide level by an Act of Parliament dating from 1609. Such extraction has traditionally been on a small scale (50 to 100 tonnes/yr) from several beaches, notably at Bude, where Summerleaze Beach has been depleted. Recent use of bulldozers to take large

quantities from the beach at Poldhu, near Mullion in Cornwall, has severely depleted this beach and called the historical right into question. Erosion has taken place where shelly material has been taken for aggregate and lime making from beaches on islands in the Hebrides and at Kinnego Bay in County Donegal, Ireland. It is sometimes assumed that because shelly material is of marine origin it will be naturally replenished, but it is necessary to investigate the relative rates of extraction and biogenic replenishment to be sure that extracted material will indeed by naturally replaced.

Some beaches have been depleted by the extraction of mineral sands, such as rutile, tin or gold, or of diamonds, as in Namibia. In general beach quarrying has led to instability, but the effects depend on the rate of extraction and the size of the beach compartment. On some beaches the effects are almost instantaneous, on others they may take several years to become obvious. Harvesting of seaweed from beaches results in losses of adhering sand and pebbles, as when kelp is extracted from beaches on the coast of King Island in Tasmania and in California.

Intensively used beaches at seaside resorts gradually lose sand as it is removed by visitors to the beach, adhering to their skin, clothes or towels, or trapped in their shoes. The quantities are small, but the losses are cumulative and no one brings sand to the beach. Pebbles and shells are also carried away as souvenirs by beach visitors. Regular beach cleaning operations, when bulldozers or tractors scrape or sweep seaweed and litter from the beach, lower the beach profile by compaction and also remove sediment, particularly sand. Samples taken from heaps of bulldozed seaweed on the beach at Beaumaris in Victoria, Australia, were found to contain up to 20 per cent (dry weight) of sand, and it was estimated that annual beach cleaning was removing up to 0.5 m^3 of sand from each 100 m^2 of beach surface.

7.1.7 Increased wave energy

Increased wave attack resulting from the deepening of nearshore water because of a rise in sea level relative to the land (or tectonic subsidence of the sea floor) has already been discussed, but nearshore water can also deepen when a shoal is removed. Such shoal migration has caused beach erosion on Benacre Ness on the Suffolk Coast (Section 6.9) and the same effect was observed on Matakawa Island, New Zealand, when beach erosion followed the shoreward movement of a tidal channel in the approaches to Tauranga Harbour (Healy, 1977). Similar features have been noted in French Guiana and Surinam. Deep-

ening of the nearshore zone off Rhode Island, in the United States, during the 1976 hurricane permitted larger waves to reach the coastline, accelerating subsequent beach erosion (Fisher, 1980).

Wave energy increases when nearshore sea grass disappears, and the sediment it had retained is dispersed, so that the water deepens. Nearshore dredging also deepens the water, so that larger waves reach the shore (Hesp and Hilton, 1996). In Botany Bay, Australia, beach erosion accelerated at Brighton-le-Sands after the bay floor was dredged to provide material for the extension of a runway at Sydney International Airport. The cutting of a trench across the sea floor in St Ives Bay in Cornwall in 1994 to lay a sewage outfall was followed by erosion on nearby Porthminster Beach. On the arctic coast of Russia increased beach erosion has been attributed to larger waves arriving as the result of nearshore deepening due to downwarping of the adjacent sea floor.

Destruction of nearshore and fringing coral reefs may initially increase the sediment supply to adjacent beaches, so that a phase of progradation ensues, but as the reefs disintegrate the nearshore water deepens and increasing wave action leads to beach erosion. Beach erosion at Colombo in Sri Lanka was partly due to greater exposure to wave attack following the decay of an old reef a short distance offshore, and there is erosion on beaches backing damaged fringing coral reefs on the Perhentian Islands off NE Malaysia, and on the east coast of Lombok, Indonesia, where reefs have been destroyed by illegal use of explosives by fishermen. Reduction of coral reefs by dredging has resulted in increased wave energy on some oceanic islands. The Indian Ocean tsunami in 2004 (Section 2.5) changed the nearshore profile on some coastal sectors by sweeping sea floor sediment landward or seaward, and where this resulted in deepening larger waves now move on to the shore.

7.1.8 Interception of longshore drifting by breakwaters

Where breakwaters have been built to stabilise river mouths or lagoon entrances and improve their navigability, or create boat harbours, sand or gravel drifting alongshore have been intercepted on the updrift side, and there is beach erosion on the downdrift side. There are many examples of this. At South Lake Worth, Florida, breakwaters have intercepted longshore drifting of sand from the north to prograde the beach updrift while to the south beaches deprived of their longshore sediment supply were eroded. At Tillamook Bay, Oregon, a breakwater built north of the inlet has trapped southward-drifting sand, and erosion has become severe downdrift on Bayocean Spit.

At Lagos in Nigeria breakwaters built in 1912 to maintain a navigable entrance to Lagos Lagoon intercepted eastward-drifting sand on Lighthouse Beach, to the west, and resulted in erosion on Victoria Beach, downdrift of the harbour. By 1975 the coastline had advanced more than 1300 m seaward alongside the breakwater, but Victoria Beach had retreated by up to 1300 m (despite nourishment) (Figure 7.3). Eventually sand accretion will extend out to the end of the breakwater, whereupon the natural eastward drifting to Victoria Beach will resume, but there will then be problems in maintaining a navigable entrance to the port of Lagos.

Until about a century ago there was a shingle beach, maintained by eastward drifting, beneath the chalk cliffs between Dover and Deal but this has almost disappeared as the result of the interception of shingle drifting alongshore by the breakwaters at Dover Harbour. Protruding areas of land claimed from the sea can have a similar effect to breakwaters. At Map Ta Phut, near Rayong on the coast of Thailand, a wide protrusion of land claimed for port and industrial development has led to updrift accretion of beach sand and erosion downdrift. Interception of drifting shingle by groynes to retain a beach for the sea-

side resort of Brighton in England was followed by beach depletion downdrift to the east. The longshore supply of sand and shingle can also be interrupted by the growth of a fringing coral reef or some other depositional feature, as at Wewak in New Guinea (Section 6.13). At Channel Islands Harbour in California an offshore breakwater induced accretion on the beach in its lee, but reduced the supply of sand and gravel to downdrift shores, resulting in beach erosion. On the Chalk coast of southern England shingle drifting alongshore has been intercepted by landslide lobes and rock falls, which act as breakwaters and cause depletion of the beach downdrift.

7.1.9 A change in the angle of incidence of waves

Beach erosion can be initiated by a change in the angle of approach of the dominant waves, either because of breakwater construction or because of growth of nearshore reefs or islands or the formation or removal of shoals. A change in wave incidence following the construction of the Portland Harbour breakwater in Victoria, Australia, may have contributed to the onset of beach erosion at adjacent Dutton Way. Beach erosion began here in 1959 when the harbour breakwater was constructed, and then spread eastward along the coast of Portland Bay (Figure 6.23).

A change from a largely swash-dominated beach alignment to a drift-dominated beach alignment accelerated longshore drifting at Kunduchi in Tanzania, where the sandy coastline had prograded under the dominance of easterly swell, but the growth of reefs and shoals reduced the effectiveness of these ocean waves, and allowed locally generated and previously subdominant SE waves to supervene, resulting in severe erosion.

Extension of the harbour breakwater at Albissola on the Ligurian coast of Italy modified the angle of wave approach in such a way that

Figure 7.3 The building of breakwaters alongside the entrance to Lagos Harbour, Nigeria, interrupted the eastward longshore drifting of sand in such a way as to cause extensive accretion updrift on Lighthouse Beach and erosion downdrift on Victoria Beach

the adjacent sandy beach at Albissola Marina was reshaped, the eastern part being eroded as the western part widened by accretion (Piccazzo *et al.*, 1992). At Sorrento on Port Phillip Bay, Australia, Point King Beach has been reshaped, with erosion at the eastern end and accretion at the western end, as the result of oblique waves (boat swash) generated by passing ships.

7.1.10 Intensification of obliquely incident wave attack

Wave attack on one sector of a beach may intensify as a result of the lowering of the beach profile on an adjacent sector, allowing stronger waves to arrive obliquely, and thus accelerate

beach erosion. This often occurs after sea wall construction, the lowering of the beach profile by reflected storm waves allowing larger oblique waves to attack the neighbouring coastline. The outcome is that beach erosion spreads alongshore, and if sea walls are extended laterally to counter this a 'domino sequence' of cumulative beach erosion is produced, with each new sector of sea wall on a set-back alignment. This has happened at Point Lonsdale in Victoria, Australia.

7.1.11 Increased losses of beach sediment to the backshore

Losses of sediment from the beach face to the hinterland occur when onshore winds blow sand

Figure 7.4 Sand is moving inland from a beach as drifting (transgressive) dunes on the shore of Encounter Bay, South Australia. The beach is dissipative, with waves losing energy as they break across a shallow nearshore zone. Beach cusps are present

inland (Figure 7.4), or when storm surges wash beach sand and gravel on to the backshore, or over into lagoons, swales or swamps. Alternatively, sediment brought alongshore by winds, waves and currents may be swept into lagoon entrances or river mouths. If losses from the beach face are not compensated by the arrival of fresh supplies of beach sediment the profile is lowered and the coastline recedes.

In South Australia, beach erosion accelerated after landward movement of backshore dunes followed reduction of their vegetation cover by burning, grazing and trampling. The lowered backshore was then cut back more quickly because of the diminished volume of sand to be removed by wave attack. Overwash during storm surges has eroded beaches by driving back sandy barrier islands on the Atlantic coast of the United States, and shingle formations at Blakeney Point and Chesil Beach in England.

7.1.12 Increased storminess

An increase in the frequency and severity of storms in coastal waters may result in the erosion of beaches that were previously stable or prograding. Beach profiles are cut back and steepened by storm waves until they attain a concave form adjusted to the augmented wave energy. A series of storms in quick succession is particularly destructive because the second and subsequent events occur on beaches already reduced to a concave eroded profile. Worsening beach erosion on shores of the North Sea and on the Atlantic coast of the United States in recent decades may be partly due to an increasing frequency of storms, but detailed long-term weather records are necessary to demonstrate that storminess has indeed increased, and it is difficult to separate this factor from other causes of beach erosion. Tsunamis may have a similar effect to storms, deepening nearshore water and allowing larger waves to erode the beach.

On the coast of Estonia the climate has become stormier during the past few decades, with sea level frequently raised by storm surges, and also milder, with a sharp reduction in the extent and duration of shore ice cover and frozen shore sediments, allowing wave action to continue to attack the coast and shape beaches even in winter (Orviku *et al.*, 2003). As a result, coastline erosion has become more rapid and more extensive. There has been continued recession on the west coast of Saaremaa Island, and eroded material has been added to spits such as Cape Kelba.

7.1.13 Attrition of beach material

On relict beaches (no longer receiving a sediment supply) agitation of the beach by wave action leads to gradual attrition, and so to a reduction in volume of beach sediment. Erosion of Four Mile Beach, North Queensland, Australia, occurred after the fluvial sand supply (from the Mowbray River) was cut off by coral reef growth, and the relict beach sediment has been reduced to very fine sand by attrition. The beach has become compacted and is now firm enough to land an aircraft, drive a bus or car or ride a bicycle. As sediment calibre is reduced, such beaches are more likely to lose the increasingly fine sediment by winnowing and removal, either landward into backshore dunes or seaward to bars and bottom deposits. The gradual lowering and flattening of the profile of Four Mile Beach has been accompanied by increased penetration by waves, and the onset of erosion along the seaward margin of backshore dunes. On the Hebridean island of Barra in Scotland the volume of a shelly beach was reduced because it was crushed and compacted by vehicles driven along it. As it became lower and flatter, the upper beach was eroded by increased wave scour (Figure 6.8).

7.1.14 Beach weathering

Beach erosion can result from a reduction of beach volume by chemical weathering, including the decay and removal of ferromagnesian minerals and the dissolving of carbonate beach sand grains in rainwater, stream seepage or sea spray. Solution proceeds more rapidly on calcareous beaches in high latitudes because cold water is less saturated with carbonates, and thus more corrosive. As beach volume diminishes the beach profile is lowered and larger waves attack the backshore. This factor has probably contributed to the beach erosion at Port Douglas, mentioned in the previous section.

7.1.15 Increased scour by wave reflection from a sea wall

Waves breaking against a solid structure, such as a sea wall built of concrete, stone blocks, steel sheeting or timber, are reflected, and form seaward currents that carry sediment away from the foot of the wall (Figure 7.5).

Depletion of beaches has been observed on many coasts where sea walls have been built behind a beach to halt cliff recession (Figure 7.6) (Kraus and Pilkey, 1988; Tait and Griggs, 1990). Storm waves at high tide then overwash the beach to splash against the wall, and their reflection causes further beach erosion and the eventual undermining and destruction of the sea wall. The outcome is usually the building of a new set-back sea wall, but at Porthcawl in Wales in 1934 a new sea wall was set forward of one that had been ruined. Boulder ramparts and gabions are less reflective, some of the wave energy being absorbed by percolation between the boulders, but beaches in front of them are nevertheless depleted.

On the Queensland coast in Australia the sandy beach at the resort of Surfers Paradise

Figure 7.5 A receding cliff supplying sand to a beach is stabilised as a bluff after a sea wall is built to halt coastline recession. Reflection of waves from the sea wall then causes beach erosion

Figure 7.6   A sea wall and stabilised bluff at Black Rock in Port Phillip Bay, Australia, where the beach has disappeared as the result of reflection scour (as in Figure 7.5)

was reduced by a series of cyclones in the 1960s and a large boulder wall was built to safeguard the coastal hotels and apartments. The boulder wall then caused scour by reflecting the waves, which lowered the beach further, and prevented the natural restoration that had occurred after previous similar erosion by cyclones before the boulder wall was inserted.

Reflection scour following the construction of sea walls has lowered beaches in Jersey in the Channel Islands. On the west coast, St Ouen's Bay had a wide sandy beach, backed by dunes, formed as the result of shoreward drifting of sand during the Holocene marine transgression (Figure 6.5(a)–(c)). By the 19th century it was eroding as the result of a diminution in sea floor sediment supply (Figure 6.5(d)), partly because of offshore dredging of shelly sand. The building

of a sea wall to halt coastline recession resulted in further lowering and flattening of the beach (Figure 6.5(e)), which is now submerged at high tide and remains wet with groundwater seepage at low tide. Similar changes have taken place on the shingle beach east of Porthleven in Cornwall, which has been lowered following construction of massive sea walls to halt cliff erosion and protect buildings.

Where beaches persist in front of sea walls it is generally because they are being maintained by longshore drifting, and are essentially migrating trains of beach sediment. An example of this is seen at the wide Paris Plage fronting the sea wall at Le Touquet in France. Alternatively, gentle wave action in calm weather may succeed in building a beach in front of a sea wall, but it is usually short lived, disappearing when storm waves break over it and are reflected from the wall.

7.1.16 Migration of beach lobes

Where beach lobes (Figure 6.33) form and migrate downdrift along the coast, there is accretion as each lobe arrives, but erosion as it moves on. At Somers, on the coast of Westernport Bay, Australia, a yacht clubhouse was built on one such lobe in the 1970s, and is threatened by the beach erosion that has developed as that lobe moved on.

7.1.17 A rise in the beach water table

It has long been known that a wet sandy beach is eroded more rapidly by wave action than a dry one. Wet sand is more coherent, like a soft sandstone, whereas dry sand is dispersed by swash. Analysis of changes over 95 years at Stanwell Park Beach, near Sydney, Australia, by Bryant (1985) identified rises in the level of the beach water table as a contributory cause of beach erosion. Such a rise in beach water table may be due to the ponding or diversion of river or lagoon outlets, to unusually heavy or prolonged rainfall or to increased discharge following land use changes in the hinterland.

Attempts have been made to reduce erosion and stabilise beaches by pumping water out of them, but the effectiveness of dewatering has yet to be convincingly demonstrated (Turner and Leatherman, 1997). Towan Beach, at Newquay in Cornwall, had been lowered by reflection scour from a sea wall, but an attempt to restore it by dewatering was ineffective because the beach is totally submerged when the water rises above mid-tide level: the pumps were then merely withdrawing sea water that was immediately replaced. In Estonia sandy beaches lowered to the water table have been invaded by reed-swamp vegetation.

7.1.18 Removal of beach material by runoff

Beach erosion can occur as the result of runoff during a period of heavy rain, or the melting of snow or ice, particularly from a backing cliff or steep slope. Beach sediment can be swept into the sea by strong runoff issuing from a stream or drain. These effects are stronger on sandy beaches, especially if they are already wet, than on gravel, where runoff disappears more quickly by percolation. The seepage mentioned in the previous section also contributes to beach erosion by washing sediment seaward as the tide ebbs. Increased runoff is often due to urbanisation and the construction of roads and other sealed surfaces from which water runs off quickly, instead of percolating into the subsoil, as it did before these structures were built.

7.1.19 Diminished tide range

Erosion by waves is more effective where their energy is concentrated at a particular level, rather than dispersed by the rise and fall of a substantial tide. It follows that a diminution of tide range will increase the effectiveness of wave action, initiating or accelerating beach erosion. Examples may be found on the shores of inlets, estuaries or lagoons that have been partly or wholly cut off from the open sea by the growth of spits, or the building of structures such as weirs or barrages, so that tidal ventilation is impeded, or excluded altogether, and wave action is intensified at a particular level. This has been observed on beaches fringing coastal lagoons at the mouth of the Murray River in South Australia, which were formerly estuarine and tidal, but were separated from the sea by the construction of barrages in 1940 (Figure 11.12).

7.1.20 Abrasion by driftwood

Some beaches are littered with branches and trunks of undercut trees, or lumber piled on the shore. Driftwood is extensive where hinterlands are forested, as in Canada and on the Pacific coast of the United States, but it also occurs where trees have been undercut along the coast, and have fallen to the beach or floated along to nearby beaches. Driftwood piled parallel to the coastline can act protectively by impeding wave scour but where fallen trees and branches are jostled by wave action, and act as levers or battering rams, they contribute to the erosion of sand or gravel from the beach.

7.1.21 Removal of a sea ice fringe

On cold coasts, as in Alaska, northern Canada and Siberia, beaches are protected in the winter by the formation of a fringe of shore ice, and are subject to wave action only during the brief summer thaw. Beach erosion will increase if the climate becomes warmer, and the summer lengthens, permitting waves to reach these shores for a longer period. Many arctic and antarctic beaches rest on ice, or have interbedded ice (permafrost), and when this melts collapse depressions form in the beach surface and the beach sediment is loosened. With a warming climate, beaches that were formerly cemented as frozen ground are mobilised by wave action, and may be removed.

7.2 The multiple causes of beach erosion

No single explanation can account for the modern prevalence of erosion of the world's beaches, or indeed for the onset or acceleration of erosion on any particular beach. It is not simply the outcome of human activities, artificial structures, a

sea level rise, an increase in storminess of coastal waters or the maturing of the system as the sediment supply from the sea floor dwindled during the Holocene still-stand. Each of these 21 factors may have contributed to beach erosion, to an extent that differs from place to place.

An example of multiple causality in beach erosion is seen on the Lido di Jesolo, the beach fronting the Lagoon of Venice on the Adriatic coast. Since this island was developed as a resort in the 1950s beach erosion has been severe. The causes include coastal subsidence, augmenting a rising sea level in the Adriatic, nearshore deepening and changes in current flow and wind and wave regimes, all of which have curtailed sandy supply from the sea floor and favoured increasingly energetic wave attack. In addition, hinterland reafforestation, reservoir construction and excavation of sand and gravel from river channels have diminished the former sediment yield from the Piave River to this coast. Artificial structures include groynes, which have reduced longshore drifting, and concrete and boulder sea walls, which have increased reflection scour (Figure 7.5).

The task of ranking the relevant factors and apportioning their contribution to beach erosion on a particular coast requires investigation of past and present patterns and rates of change on beaches and the process systems operating in coastal waters. An example of quantitative assessment of beach erosion was provided by Fisher (1980) from Rhode Island, USA (Figure 7.7). He found that between 1939 and 1975 the beach-fringed coast had retreated at an average rate of 0.2 m/yr, in a period when sea level rise averaged 0.3 cm/yr, and calculated that 35 per cent of linear beach recession over this period had been due to the washing of sand into tidal inlets and 26 per cent to losses of sand washed or blown over the barrier islands to form migrating dunes and washover fans. This left 15 per cent of the beach retreat accountable as the direct result of submergence, and 24 per cent lost by

Figure 7.7 The distribution of sediment losses accompanying recession of sandy barriers in Rhode Island, United States, based on Fisher (1980)

transference of beach sand seaward. Movement of sand over and between these barrier islands indicates that they were continuing the long term landward migration, accompanied by the transgression of backing lagoons and marshes, that has characterised much of the Gulf and Atlantic coastline during the Holocene. A similar sequence of evolution is now being observed on the submerging coasts of the Caspian Sea.

The modern prevalence of beach erosion calls into question the idea that beaches are naturally stable, having attained some kind of equilibrium with the processes at work on them. This is difficult to sustain from geological evidence, for beaches form only a very minor proportion of the sedimentary formations found in the geological column. They have been preserved in Tertiary deposits in Texas, but they are not prominent in most depositional sequences. This is because they have been transient features, formed and reworked by waves on coasts destined to retreat as marine planation proceeds (reducing land masses ultimately to surfaces planed down to the limit of wave erosion). Beaches have thus been eroded and dispersed. It could be argued that beach erosion is natural, and that the anomaly is the widespread progradation that occurred in Holocene times, related to the unusual

conditions of a still-stand following a major marine transgression over a sand-strewn sea floor (Clayton, 1989). The attempt to preserve existing beaches should be considered in terms of this geological perspective.

7.3 Effects of artificial structures on beaches

A common response to beach erosion, especially where it threatens to undermine and destroy developed property such as roads or buildings, has been to construct, extend and elaborate sea walls. These are intended to prevent wave attack on the eroding coast, usually a receding cliff, an undermined and slumping bluff, or a truncated dune, sometimes fronted by a beach that is insufficiently high and wide to prevent waves reaching the back of the shore. Often they are initially banks of earth excavated from a parallel ditch, but when these are damaged by storm waves they are replaced by more solid stone or concrete structures. Solid walls that are designed to withstand the force of the breaking waves inevitably reflect these seaward, as backwash that scours away the beach. Boulder ramparts, also known as revetments or riprap, or artificial structures such as tetrapods, made of reinforced concrete, are less reflective than solid sea walls, but still cause backwash erosion on their seaward sides. Artificial structures designed to protect the coastline form an increasing proportion of the world's coastline (Walker, 1988). In Japan long sectors of the cyclone-prone east coast of Honshu are now protected by large sea walls and concrete tetrapods weighing up to 50 tonnes each.

Demands for the halting of coastal erosion have led to littering of the world's coastline with an array of artificial structures of various kinds. Some have been successful, but many have failed and are derelict. Some have helped to protect and maintain beaches, but many have resulted in further beach erosion. In recent decades attention has been given to artificial beach nourishment, particularly in the United States, Western Europe and Australia, as a means of countering beach erosion.

7.4 Beach nourishment

Beaches that have been depleted by erosion can be restored by dumping sand or gravel on the shore (Schwartz and Bird, 1990). Beach nourishment (replenishment, restoration, recharge, reconstruction or fill) is artificial in the sense that the sediment has been brought to the shore by engineers. Beach renourishment is the maintenance or replenishment by deposition of suitable sediment on a beach that has been previously nourished, and the term artificial beach should only be used where there was previously no natural beach, as at Ibiza in the Balearic Islands and Praia da Rocha in Portugal (Psuty and Moreira, 1990).

Beach nourishment aims to restore beaches that have been depleted by erosion, and to create a beach formation that will protect the coastline and persist in the face of wave action. Many beach nourishment projects have been at seaside resorts that had eroding beaches, and wanted them restored for recreational use, but in recent years increasing awareness of the importance of beaches in absorbing wave energy has resulted in the use of nourished beaches to protect the coastline and prevent further cliff erosion or damage to coastal property.

Prior understanding of coastal geomorphology is essential in planning a beach nourishment project. Preliminary research is necessary on the movement of sand and gravel in relation to incident wave regimes and the effects of any artificial structures on the shore sector to be treated. It is necessary to know why the natural beach was eroded and where the sediment has gone: landward, seaward or alongshore. Modelling of beach forms and processes can guide a beach nourishment project but the complexity and variability of coastal systems are such that an experimental approach, based on accumulated experience, may be more realistic than theoretical modelling (Pilkey and Clayton, 1989). It is generally easier to renourish and maintain a beach in a bay than on a straight or salient coastline.

Sediment used for beach nourishment can be brought from inland, alongshore or offshore sources and deposited mechanically or hydraulically on the shore (US Army Corps of Engineers, 1984). Almost any kind of sediment of suitable grain size can be used for beach nourishment, but it should be at least as coarse as the natural beach sediment (engineers would prefer coarser sediment; ecologists would like it to be of the same type and size as the pre-existing natural beach). On the east coast of Jersey the beach at Anne Port was nourished with matching grey gravel obtained from quarries near Cork in southern Ireland. The dumped sediment should also be durable, not quickly reduced by weathering or abrasion. Restoration of a beach near Aberystwyth with gravel from a nearby quarry failed because the material was too friable, and rapidly disintegrated into fine

Figure 7.8   Beach nourishment in progress at Mentone, Port Phillip Bay, Australia

sediment that was quickly washed away (So, 1974).

Most beach nourishment projects begin by depositing sediment to form a beach terrace (Figure 7.8), which is then reshaped by waves and currents towards a natural profile, often with sand bars just offshore. It is necessary to deposit more beach material than is required to restore a beach to its natural dimensions (i.e. to form a terrace that is higher and wider than the natural beach), in order to allow for expected losses onshore, offshore or alongshore. A nourished beach may be held in place by building one or more retaining breakwaters or a series of groynes; alternatively, a nearshore breakwater can be inserted to protect artificial beach from strong waves, as at Niigaata in Japan.

It may be possible to nourish a beach by dumping sediment in places where longshore or shoreward drifting will deliver it to the shore. This requires knowledge of the direction and rate of longshore and shoreward drifting, taking account of variations in coastal aspect. Offshore breakwaters can be used to create a pattern of refracted waves that will concentrate sand deposition and prograde the beach in their lee, as at Port Hueneme in California. A floating breakwater anchored off successive sectors of the shore can be used to induce local accretion of sand and gravel by shoreward drifting of sediment to nourish a beach in stages along the coast. As the breakwater is moved away the accreted sand will begin to erode, and so a cyclic system of offshore breakwater placement is necessary to maintain such a beach.

Sand or gravel carried away by longshore drifting can be intercepted and brought back, a process known as recycling or recharging. At Rye in SE England the Kent River Board has been trucking loads of shingle taken from alongside the breakwater at the mouth of the River Rother back round to Cliff End and dumping them to restore the depleted beach. Sediment that has been withdrawn from the beach to the nearshore zone, particularly during stormy phases, can be dredged and brought back to the shore, a procedure known as backpassing.

Some nourished beaches persist for many years; others diminish rapidly. It is generally acknowledged that nourished beaches will be eroded by the same processes that depleted preceding natural beaches, and that they will have to be replaced at intervals. Monitoring of changes on and around nourished beaches provides an understanding the processes that erode and distribute emplaced beach material, and can guide further beach management procedures, including the insertion of groynes, the introduction of regular renourishment updrift or the repeated restoration of the profile of a beach that loses sediment offshore. Experience gained from one beach nourishment project can be applied to another, providing the geomorphological situation is comparable.

By contrast, it has been found necessary to remove sand from a prograding beach at La Pineda, south of Barcelona in Spain, to prevent it being blown into a backing marina.

Examples of beach nourishment projects were provided by Bird (1996a), together with a review of the principles and problems of beach nourishment. The more technical details are dealt with in the Shore Protection Manual produced by the US Army Corps of Engineers (1984) and the Delft Hydraulics Laboratory (1987) manual on beach nourishment.

7.5 Summary

Beach erosion occurs when losses of beach sediment exceed gains. The beach is then lowered and cut back, developing a concave profile. Beach erosion has become widespread around the world's coasts, and is obvious where seaside resort beaches have been depleted. A set of 21 factors has been identified as contributing to beach erosion (Panel 7.2), but only a few of them are responsible for the onset of erosion on any particular beach. Prominent among them are a rising sea level with increased wave attack (which will become the dominant factor as the forecast global sea level rise proceeds), reduction in fluvial sediment supply (e.g. as a consequence of dam construction), reduction in sediment supply from stabilised cliffs and a decline in sediment supply from the sea floor. Sea walls, often built to halt beach erosion, may cause reflection scour, worsening the problem. Beach nourishment, bringing sand or gravel in from alongshore, offshore or hinterland sources, has been widely used to counter beach erosion.

8
Spits, barriers and bars

8.1 Introduction

Depositional features closely related to beaches, and shaped by similar processes, include spits of various kinds, barriers built offshore, or across inlets and embayments to enclose lagoons and swamps, and bars in the intertidal and nearshore zones. The shaping of these various features will now be discussed.

8.2 Spits

Spits are beaches built up above high tide level and diverging from the coast, usually ending in one or more landward hooks or recurves (Schwartz, 1972). They have grown in the predominant direction of longshore drifting by waves arriving obliquely to the shore, and their outlines have been shaped largely by dominant patterns of wave action. Some longshore (barrier) spits are almost straight, like the southern part of Orfordness on the east coast of England, where the mouth of the River Alde has been deflected about 18 km to the south and SW, but most end in recurves, representing earlier terminations, as at Hurst Castle spit (Figure 8.1).

Recurves can be formed where two sets of waves arrive from different directions, as on Hurst Castle spit, or where one set of waves is refracted in deep water around the distal end of the spit, as at Cape Henlopen on the east coast of the United States, where northward-drifting sand has been built into a recurved spit shaped by SE Atlantic swell refracted into the mouth of Delaware Bay. Traces of older recurves on the landward side mark former terminations of a spit that has grown intermittently, a feature well shown by Blakeney Point on the Norfolk coast (Figure 8.2). There have been phases of prolongation, possibly during storm surges, between the formation of each recurve. Salt marshes or mangrove swamps often develop on the sheltered landward side of spits, notably between the recurves. The growth of a spit results in coastal progradation and the protection of a pre-existing coastline, where reduced wave energy may permit the formation of marshland, as at Morston to the south of Blakeney Point.

Sediment deposited to form a spit has usually come from an alongshore source, but shoreward drift may also have contributed. Blakeney Point is built of sand and gravel largely derived from cliffs cut in glacial drift in the Sheringham district to the west, but also includes beach material washed in from glacial drift on the adjacent sea floor. A glacial moraine deposited at the margin of the Last Glacial ice sheet, which crossed the

Coastal Geomorphology: An Introduction, Second Edition. Eric Bird.
© 2008 by John Wiley & Sons, Ltd

Figure 8.1 Shaping of a recurved spit where sand or shingle delivered by longshore drifting, resulting from oblique wave action (A) is shaped by waves from direction B, with recurves formed by waves from direction C. Based on the shingle spit at Hurst Castle spit in Hampshire, which is also being driven landward by overwashing storm waves (broad arrow)

Norfolk coastline in this area, has been sorted and rearranged in the course of the Holocene marine transgression and ensuing wave and tidal current action, and built into a spit. The westward growth of Blakeney Point is indicated by successive recurves, shaped by westerly and NW wave action, and the main shingle bank has been driven landward by North Sea storm surges, athwart the successive recurves. It terminates in Far Point, the most recent recurve, added after a storm surge in 1978 (Figure 8.3).

Similar features are seen on Dungeness Spit in Washington State, USA, formed downdrift from eroding cliffs of glacial sand and gravel, diverging from the coastline and backed by recurves (Schwartz, Fabbri and Scott Wallace, 1987). It is narrow where it diverges from the land, and widens towards the distal (far) end, a pattern that may indicate a reduced rate of sediment supply.

Figure 8.2 Blakeney Point, a recurved spit on the Norfolk coast. The recurves mark stages in westward growth, the most recent (Far Point) having been added in the 1970s. The spit shelters an estuarine harbour, with salt marsh on the southern side bearing sandy cheniers emplaced by waves during storm surges

Figure 8.3 Changes in the configuration of Far Point, the western termination of Blakeney Point, Norfolk, between 1886 and 1979. A new recurve was added after a storm surge in 1978

Figure 8.4 shows the patterns of longshore drifting on this spit.

Spurn Head is a 4.5 km long sand and gravel spit at the mouth of the Humber estuary that has grown southward by longshore drifting produced by NE wave action, and been driven westward by North Sea storm surges. Much of its sediment has been carried southward from the eroding glacial drift cliffs of Holderness, but some may have been washed in from the adjacent sea floor.

Spits have formed on the coast of Burghead Bay in NE Scotland where gravel supplied by the River Findhorn has drifted westward along

Figure 8.4 Dungeness Spit, on the coast of Washington State, USA, with arrows indicating the directions of longshore drifting, determined by local exposure to wind-generated waves

the shore as the result of waves arriving from the north and NE to form spits with recurves at their western ends. To the west, Whiteness Head on the southern shores of Moray Firth is another recurved spit of well rounded shingle, derived from glacial moraine deposits, and carried westward by longshore drifting.

Hurst Castle spit in Hampshire has been supplied with sand and shingle derived from cliff and shore erosion in Christchurch Bay and carried eastward along the coast by longshore drifting; again, some sediment may have been washed in from the adjacent sea floor. The shaping of this recurved spit in relation to the direction of approach of dominant waves was demonstrated by Lewis (1938). The main shingle bank has been driven landward by storm surges. King and McCullach (1971) used a computer model (SPITSIM) to simulate the processes at work here and examine the relative significance of waves approaching from various directions, the effects of wave refraction and the influence of

submarine topography, and then trace probable stages in its evolution to the present outline. When the simulation was extrapolated into the future it generated an additional elongated recurve extending north-westwards, but this is unlikely to form because of the presence of marshland in this direction. The value of this computer simulation was in identifying constraints that had not been considered, such as the morphology of the backing areas into which recurves might grow and the ways in which shoals, marshes or the backing coast can influence the shaping of a spit built by longshore drifting.

Some spits have been widened by the accretion of beach material, usually forming successively built ridges on the seaward side, and stages in their growth can be deduced from the ridge pattern. In a study of Cape Cod, Davis (1896) showed that sand eroded from cliffs of glacial drift had been built into a spit, with beach ridges marking stages in progradation at Provincetown

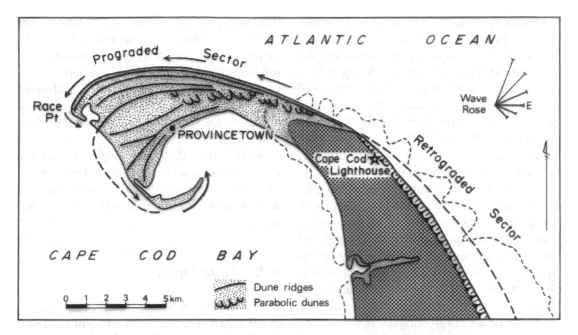

Figure 8.5 The spit at Cape Cod, Massachusetts, United States, has grown out from a peninsula of glacial drift, cliffed on its seaward side. Sand and shingle derived from the cliffs has drifted westward to form a broad spit with beach ridges marking stages in progradation. These have been truncated by wave action on the shores of Cape Cod Bay, with the formation of minor spits at Race Point and south of Provincetown

to the west (Figure 8.5). The beach ridge pattern showed that the spit had been reshaped, partial truncation of the formerly prograded sector by marine erosion having released sand that formed smaller spits on the shores of Cape Cod Bay.

Many spits have beach ridge patterns indicating that they have been eroded along their outer shores, yielding sediment that has drifted alongshore to prograde their distal ends. Examples include Pointe de la Coubre and Pointe d'Arcay on the French Atlantic coast, and Farewell Spit, which has grown out from the northern end of South Island, New Zealand, in this way, nourished by northward drift of fluvially supplied sediment along the west coast of South Island, and by shoreward drift of sand from the sea floor (McLean, 1978). It is now more than 30 km long and up to 1.5 km wide at high tide, with a wide beach backed by dunes on the seaward side and

extensive intertidal sand shoals on the more sheltered landward side.

Spits have formed on the shores of deltas where sandy sediment delivered to a river mouth drifts alongshore, as on the flanks of the Ebro delta in Spain (Figure 12.5). Sand spits on the western shore of the Mississippi delta have been derived from sectors of eroding deltaic coastline where sand has been sorted by wave action from finer material.

Evidence of the evolution of a spit can be obtained from sequences of historical maps and air photographs. Maps made since about 1530 A.D. suggest that the shingle spit at Orfordness has grown about 6 km during the past four centuries, and stages in the evolution of Penouille spit, Gaspé, Quebec between 1765 and 1981 were traced from maps and air photographs by Fox, Haney and Curran (1995). Historical information of this kind is rarely available outside Europe and North America, but eventually it will be possible to use successive surveys and series of dated air and satellite photographs to trace the evolution of spits. Carr (1965) used this kind of evidence to demonstrate short-term changes at the southern end of the Orfordness spit between 1945 and 1962.

The shape of a spit is influenced by the space available for its growth and the formation of recurves, and by the adjacent sea floor topography, as demonstrated by Schou (1945) in the Danish archipelago. Spits grow more rapidly across shallow nearshore areas than into deep water, and their configuration may be related to variations in fetch (exposure to wave action) determined by nearby headlands, islands or reefs. The size and shape of recurves depends on the space available on the inner side of a growing spit. The evolution of many spits has been modified, or even halted, by the addition of artificial structures. Dawlish Warren in Devon and Sandy Hook in New Jersey are examples of spits with complex histories that have been armoured by the building of sea walls intended to stabilise their present configuration.

Some of the best examples of spits are found on the shores of landlocked seas, lakes and coastal lagoons, where sand and shingle carried along the shore have been deposited as spits where the orientation of the coastline changes, in forms related to prevailing wave conditions. Examples of this are found in the Danish archipelago, around Puget Sound and on the New England coast, where in each case cliffing of glacial drift deposits has yielded sand and gravel for longshore drifting to nourish nearby spit structures. In the Rade de Brest, a Brittany ria (Section 11.2), wave-built sand and shingle spits built in tributary bays incorporate sediment eroded from the periglacial drift deposits (head) that mantle bordering coastal slopes that have been basally cliffed (Section 4.3.3). Their outlines have been shaped largely by wave patterns related to the variations in fetch and wave

incidence resulting from the configuration of the ria, but some have been truncated by strong tidal currents (Guilcher *et al.*, 1957). Truncation of spits by strong tidal currents occurs in narrow straits between the islands of Puget Sound on the NW coast of North America, as at West Point, bordering Deception Pass in Washington (Section 2.3.1).

Some features that look like spits on maps and charts may actually be moraines, eskers or reefs. The Ness of Portnacutter in Dornoch Firth is a narrow strip of glacial drift extending out into the sea.

8.2.1 Paired spits

On some coasts there are spits that have grown in different directions at different times. On Rattray Head in Scotland there is evidence of spit growth first to the SE, and later to the NW, implying a reversal of longshore drifting, but it is possible that sediment came in from the sea floor and that the spits were largely swash built. A similar alternation is seen at Sandwich in Kent, where the mouth of the River Stour was first deflected southward by the growth of the Stonar spit, then northward by a younger spit built along the shore of Sandwich Bay. This alternation may have resulted from a change in the pattern of dominant refracted waves from NE to SE as the result of migration of the intertidal Goodwin Sands offshore.

Paired spits often border river mouths and lagoon entrances. They have formed either as the result of convergent longshore drifting or the breaching of a former coastal barrier. Spits of this kind border the entrances to tidal estuaries at Poole, Christchurch and Pagham on the south coast of England (Robinson, 1955) and Braunton Burrows beside the Taw–Torridge estuary in North Devon (Kidson, 1963). These spits have been supplied with sediment from the sea floor, as well as from adjacent sectors of eroding coastline, and have been shaped by incident wave patterns. The paired spits bordering Pagham Harbour attained their present form as the result of the breaching of the shingle barrier during a storm in 1910. Paired spits are well developed alongside the tidal entrances between barrier islands, as at the tidal entrances to Corner Inlet in Australia. Paired spits formed between islands, as at Mawar on the SE coast of Malaysia, may grow to unite as a tombolo (Section 8.3).

8.2.2 Trailing and flying spits

Trailing (arrow or comet-tail) spits form in the lee of islands, as at the Plage des Grands Sables on the eastern end of the Ile de Groix off the Brittany coast and the similar feature on neighbouring Belle Ile. The high island of Ailsa Craig in Scotland has a triangular lee spit, and there are various spits of coral sand and gravel trailing leeward of high islands off the North Queensland coast (Figure 8.6) (Hopley, 1971). These spits consist mainly of coralline beach material that has drifted from reefs on the high wave energy coast on the windward end of the island along the bordering shores to accumulate at the leeward end.

Where the islands are of soft material, such as glacial drift, trailing spits include sand and gravel derived from eroding cliffs on their coasts. Elongated spits trail from several islands in the Danish archipelago (Figure 8.7), and from cliffed drumlin islands (glacial drift) in Donegal Harbour and Clew Bay, NW Ireland. On a smaller scale, trailing spits may form on the downdrift side of prominent shore rock outcrop or boulder, or in the lee of an isolated mangrove tree.

Where an island has been completely destroyed by erosion the depositional trail may persist as a flying spit, aligned at right angles to the predominant waves. Examples are found off the Boston coast and in the Strait of Georgia,

western Canada, in both cases derived from former islands of glacial drift.

8.3 Tombolos

Tombolos (Figure 8.8) are wave-built ridges of beach material that link islands, or attach a stack or island to the mainland. The term comes from the west coast of Italy, where a double tombolo is well developed at Orbetello, enclosing a lagoon. They are also found on coasts cut into glacial drift deposits, as at Nantucket in New England, and in SW Finland. Chesil Beach on the south coast of England is the western part of a double tombolo attaching the Isle of Portland to the Dorset mainland and in Australia the Yanakie isthmus ties the granitic upland of Wilson's Promontory to mainland Victoria in a similar way. Some tombolos have formed by leeshore deposition of sediment by waves refracted round a nearshore island or stack to produce paired spits that grow into a linking isthmus, as at Cape Verde on the Senegal coast in West Africa. The tombolo at Tyre in the Lebanon is thought to have been formed by sand deposition on either side of a causeway built out to an offshore island by Alexander the Great in 332 B.C.

Stages in the evolution of tombolos are seen in the Isles of Scilly, several of which have become linked by depositional banks of sand or shingle. On Samson a sandy isthmus links two former islands, and a shingle bank ties Gugh to St Agnes. The Ayres of Swinister in the Orkney Islands have an indented coastline with shingle tombolos and in the Shetlands the large sandy St. Ninian's tombolo (known locally as an ayre) links St Ninian's Isle to the SW coast of Mainland, and encloses a lagoon. Small tombolos have formed as the result of the growth of spits in the lee of offshore breakwaters, as at Rimini on the Adriatic coast of Italy. In the East Coast Park, Singapore, erosion between longshore breakwaters built at intervals along the coast to protect a

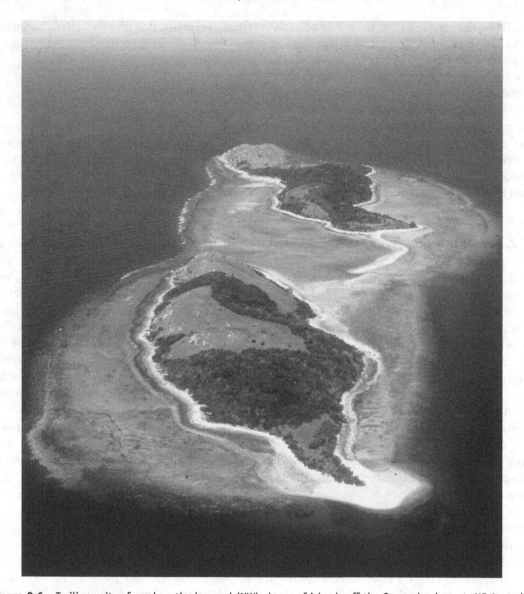

Figure 8.6 Trailing spits of sand on the leeward (NW) shores of islands off the Queensland coast, NE Australia

plain of sand deposited for reclamation has resulted in the shaping of erosional tombolos in the lee of each breakwater.

A tombolo that is partly or wholly submerged by the sea at high tide is known as a tie-bar. The Ilha Porcat in Santos Bay in Brazil is attached to the mainland by a tombolo, whereas adjacent Uruboqueçaba is linked by an intertidal tie-bar.

8.4 Cuspate and lobate spits

Cuspate spits form where beach sediment is deposited as a protruding, more or less symmetrical structure, formed where waves approaching at an angle to the shore from either direction are stronger and more frequent than those coming directly onshore.

Figure 8.7   A long trailing spit on the island of Lindholm in Denmark

Figure 8.8   A tombolo linking north and south Bruny Island, SE Tasmania

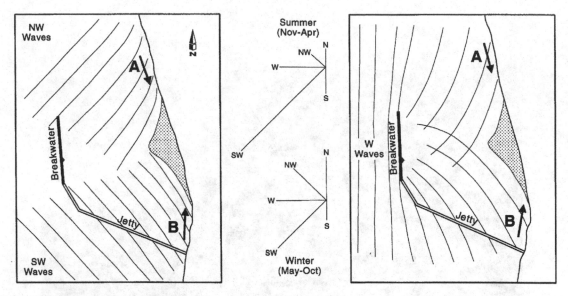

Figure 8.9 The building of a breakwater at Middle Brighton harbour, Port Phillip Bay, in 1954 was followed by the growth of a cuspate spit in its lee, supplied with sand that has drifted in from beaches to north (A) and south (B) and shaped by waves from the SW and NW (left) and westerly waves refracted round the offshore breakwater (right)

Such conditions are common on the shores of narrow straits and coastal lagoons where the transverse fetch is small. The waves generate a convergence of longshore drifting, so that sediment is supplied from both directions. The sediment has usually been derived from erosion in adjacent bays, but may come from rivers updrift, or from the sea floor. Cuspate spits of this kind are seen on the shores of fiords in NW Iceland, as at Isafjördur. Typically a cuspate spit ends in a sharp point projecting seaward, but where the outline is more rounded they may be called lobate spits.

Cuspate and lobate spits are also found where wave refraction has concentrated beach material in the lee of an island, stack, breakwater, reef or shoal. Where chains of nearshore breakwaters have been built parallel to the coastline, cuspate spits have formed in their lee, and some of these have grown into tombolos, attaching each breakwater to the mainland. Examples are seen on the Mediterranean coast, as at Sitges in Spain and Rimini in Italy.

The Anse Vata lobate spit near Nouméa in New Caledonia consists of coral sand and gravel that has been shaped by waves refracted across a fringing coral reef. Lobate spits have formed as the result of convergent drifting by waves refracted round offshore breakwaters in several harbours around Port Phillip Bay, Australia, notably at Middle Brighton (Figure 8.9).

Cuspate spits have formed as the result of the breaching of a former tombolo, as at Gabo Island in SE Australia. Some cuspate or lobate spits grow to enclose a lagoon or swamp. At Ebeys Landing on Whidbey Island in the State of Washington a looped barrier (not a bar because it extends above high tide level) diverges from the coastline and curves back in to enclose a small lagoon.

The asymmetrical spits on the north coast of the Sea of Azov (Figure 8.10) are shelly sand spits

Figure 8.10 Sand spits on the north coast of the Sea of Azov are migrating westward. A curving sandy barrier has formed on the western coast, and erosion has been prevalent on the generally cliffed southern coast

that have grown out at an angle to the coast as the result of longshore drifting, usually ending in a slight recurve. They occur in a regular sequence on the Osipenko coast, where they are migrating westward. Secondary accumulation of sand on the leeward side forms a series of curved beach ridges that may be truncated by wave erosion on the eastern side (Zenkovich, 1967).

Cuspate and lobate forelands may also form as the result of deposition at stream mouths (i.e. deltas) or of volcanic material, as on the lava salient at Budarhaun in NW Iceland or the pyroclastic deposits at South Point on Montserrat Island in the Caribbean, which came from the erupting Soufrière volcano.

8.5 Cuspate and lobate forelands

Cuspate and lobate forelands are similar to cuspate and lobate spits, but have been enlarged by the accretion of beach ridges parallel to their shores. They are known as nesses in Britain and, like cuspate spits, are found on the shores of narrow straits or in the lee of islands or shoals. Stages in their evolution can be deduced from the patterns of beach ridges. Some cuspate forelands have remained stationary and grown symmetrically, while others have been eroded on one side and built up on the other, so that the cuspate foreland has migrated along the coast as a travelling foreland.

This is true of Dungeness, a massive shingle cuspate foreland shaped by SW waves from the English Channel and easterly waves arriving through the Strait of Dover. Stages in its evolution can be traced with reference to the patterns of shingle beach ridges, each of which indicates an earlier position of the coastline. A mass of shingle that accumulated off Rye a few thousand years ago has been reshaped and sharpened into the cuspate foreland (Figure 8.11). Beach ridges have been truncated along the southern

coastline, and shingle eroded from here has drifted round the point, to be added as successive beach ridges on the prograding eastern shore. The point has thus migrated eastward. Similar features are seen on Presque Isle on the south coast of Lake Erie in the United States, which has also travelled eastward, and Sandy Point in Westernport Bay, Australia (Figure 6.19).

Coastal lowlands of cuspate or lobate outline have not necessarily been formed by beach ridge accretion. On the coast of Cardigan Bay in Wales, Morfa Dyffryn and Morfa Harlech are somewhat asymmetrical cuspate lowlands. It has been suggested that they originated as spits of sand and shingle that grew out NW at an angle to a former steep or cliffed coast, then curved round to the NE, enclosing an area of sandy and marshy flats. However, both incorporate glacial or glacifluvial deposits in front of an abandoned Pleistocene cliffed coastline, and the beach and dune fringe on their shores is the outcome of erosion and reworking of the margins of glacial drift forelands during and since the Holocene marine transgression. Gravelly cuspate lowlands on the shores of fiords in NW Iceland (e.g. Patreksfjördur) and on the Alaskan coast at Point Barrow originated as glacifluvial deltas.

Winterton Ness and Benacre Ness (Section 6.9) on the East Anglian coast are cuspate forelands that have migrated (southward and northward respectively) in response to changing wave and current patterns related to longshore movement of shoals in coastal waters. Cuspate forelands and tombolos on the coast of Western Australia have formed in the lee of rocky islands or reef segments (Sanderson and Eliot, 1996). Capes Hatteras, Lookout, Fear and Kennedy are large cuspate forelands on the Atlantic coast of the United States, each consisting of sand delivered by longshore drifting and retained on a sector of convergent wave refraction behind nearshore shoals. They show erosion of the northern or eastern flank and progradation to the south. Cape San Blas in Florida

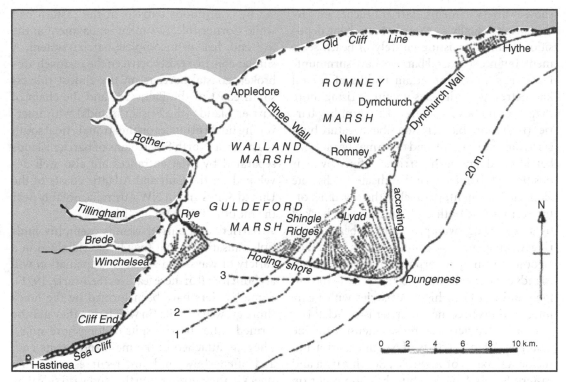

Figure 8.11 Shingle ridges (dotted) show stages in the growth of the cuspate foreland at Dungeness in Kent, formed in front of a former cliffed embayment now occupied by drained marshland. The ridge pattern indicates that the cuspate foreland grew seaward (1, 2, 3), the point migrating eastward as the southern shore was truncated by erosion and shingle drifted round to the prograding eastern shore

is a cuspate foreland that has been nourished with sand derived from nearshore shoals deposited by the Apalachicola River. The Darss Foreland on the German Baltic coast has grown out from a promontory of glacial drift, truncated in cliffs on the western shore, with multiple sandy beach and dune ridges that indicate northward progradation during the past 3500 years (Sterr, Schwarzer and Kliewe, 1998).

8.6 Coastal barriers and barrier islands

Coastal barriers and barrier islands are elongated landforms that have been formed by the deposition of beach material offshore, or across the mouths of inlets or embayments. They extend above the normal level of highest tides (Shepard, 1952) and may partly or wholly enclose lagoons and swamps (Schwartz, 1973). They result in coastal progradation, and are so called because they stand as a barrier between the open sea and an earlier coastline. Barriers, thus defined, are distinct from bars, which are submerged for at least part of the tidal cycle (Section 8.8), and from reefs of biogenic origin, built by coral and associated organisms (Chapter 13).

Coastal barriers and barrier islands fringe about one-eighth of the world's coastline (Leontiev and Nikiforov, 1965) They may be up to 100 km long, and often more than 1 km wide. They show a variety of forms. Barrier beaches are narrow strips (usually < 200 m) of low-lying depositional land consisting entirely of beach sediment, but most coastal barriers have surmounting dunes and some attain widths of several kilometres, with dunes sometimes rising more than 100 m above sea level. The term bay barrier (baymouth barrier) describes a feature built across an embayment, and there may be mid-bay barriers and bay-head barriers, defined by their position. A barrier island indicates a discrete elongated segment, parallel to the coastline, often recurved at both ends, and usually backed by a lagoon or swamp. It usually bears beach ridges, dunes and associated swamps and minor lagoons and may incorporate recurves. Barrier islands are typically 0.5–5.0 km wide, 1–100 km long and 6–100 m high. A barrier with many interruptions becomes a barrier island chain.

Coastal barriers are most extensive on the Gulf and Atlantic coasts of North America and the ocean coasts of Australia, South Africa and eastern South America, but they also occur on a smaller scale elsewhere, notably in Sri Lanka and New Zealand.

Coastal barriers consist of sand or gravel (occasionally boulders) delivered by longshore drifting or carried in from the sea floor and deposited by wave action as beaches, behind which there may be wind-emplaced dunes. Barriers are best developed on coasts where the tide range is small, whereas barrier islands are usually formed where there is a large tide range generating strong ebb and flow currents that maintain gaps between them, preventing wave action from depositing sand or shingle to seal the intervening inlets. On the Gippsland coast in Australia spring tide range increases from about 1 m at the NE end of the Ninety Mile Beach, where there are no natural tidal entrances through the coastal barrier, to 2.5 m at the SW end, where tidal currents have maintained channels between a chain of barrier islands. This is why the Gippsland Lakes are almost completely sealed off from the

sea by continuous barriers at the eastern end while Corner Inlet, a similar embayment at the SW end, has an incomplete barrier system. A similar contrast exists between the relatively unbroken coastal barriers on the almost tideless south coast of the Baltic Sea and the chain of barrier islands (the Frisian Islands) with intervening tidal entrances on the strongly tidal south coast of the North Sea. Chains of barrier islands separated by tidal entrances are also well developed on the Gulf and Atlantic coasts of the United States and in NW Australia, notably near the mouth of the De Grey River.

Barriers and barrier islands exemplify multiple causality in that they have originated in a variety of ways, and no single explanation will account for all of these features (Schwartz, 1971). Some barriers have been formed by the longshore growth of spits (in which case they may be termed either barrier spits or longshore spits). They are attached to the mainland at one end, and often show landward recurves indicating stages in their growth. On the landward side they are backed by a bay, lagoon or swamps. An example is seen on the coast north of the Columbia River in Washington State, where Long Beach is a barrier spit partly enclosing Willapa Bay.

Other barriers result from the emergence of an offshore bar during a phase of sea level lowering, or the partial submergence of a pre-existing coastal sand ridge during a phase of sea level rise. Some are the result of shoreward sweeping of sand or gravel during and since the Holocene marine transgression. Many barriers have had a composite origin (Roy et al., 1994). Longshore spits, for example, can also be driven landward as transgressive barriers.

On the West African coast Guilcher and Nicolas (1954) described an elongated barrier in the form of a longshore spit, the Langue de Barbarie, which grew as the result of southward drift of beach sand in such a way as to form and prolong a barrier deflecting estuaries, notably the mouth of the Senegal River in west

Africa. Barriers that grew as longshore spits often have recurves marking stages in their growth, as on the recently formed barrier island known as the Bar on the Culbin coast, which developed during the 19th century, and grew both eastward and westward, with successive terminal recurves. Similar features are seen on Tramore spit in SE Ireland (Ruz, 1987). Stages in the growth of the longshore gravel spit at Koumac, on the west coast of New Caledonia, are also indicated by recurved ridges projecting into the backing lagoon.

Other barriers and barrier islands have formed on coasts bordered by shallow seas as the result of a relatively sudden emergence, when reduction of water depth offshore caused waves to break farther out and build a bar that emerged as it was driven shoreward. The sandy barrier island of Knotten on the SW coast of the Danish island of Laesø may have been initiated in this way. It is possible to generate a barrier island from an emerging bar experimentally in a wave tank, but it is doubtful whether it applies to many actual barriers, for the majority were initiated during the Holocene marine transgression.

Sandy barrier islands on the SE coast of the United States were probably initiated when the Holocene sea rose to partially submerge preexisting coastal dune ridges, penetrating lower land to the rear to form lagoons behind a chain of barrier islands. Many of these barrier islands are now transgressive, migrating landward as the result of recurrent washovers. Sandy barriers have formed in this way and moved landward during the past few years on the coasts of the Caspian Sea, where a marine transgression is in progress (Figure 3.11) (Kaplin and Selivanov, 1995). Transgressive barriers enclose lagoons that also migrate landward, often submerging any evidence of earlier coastlines.

Many coastal barriers are the outcome of shoreward sweeping of beach sediment during and since the Holocene marine transgression, notably where Pleistocene deposits that had been stranded on the emerging sea floor during the preceding marine regression were collected and built up, then driven landward as the marine transgression proceeded. This shoreward drift of beach sediment may have been facilitated by minor emergences during the oscillating Holocene marine transgression (Figure 3.6), which set up broomlike shoreward sweeping.

Chesil Beach on the Dorset coast (Section 6.4.3) is a shingle barrier that stands in front of a lagoon (the Fleet) and an embayed mainland coast that has escaped marine cliffing. It must have been formed and driven shoreward during the Holocene marine transgression in such a way as to protect the submerging land margin from the waves of the open sea. The barrier and lagoon are underlain by a platform cut in bedrock that descends from the inner shore of the lagoon and flattens at about 15 m below sea level. The shingle barrier is permeable, so that during storms and high tides sea water seeps through to The Fleet, washing out fans of gravel, known locally as cans, which run from the landward slopes of the barrier down into the lagoon.

Chesil Beach is still moved shoreward intermittently, when vigorous storms sweep shingle over the crest and down into the waters of the Fleet. It has overrun peaty sediments that formed in the lagoon, so that peat now outcrops locally on its seaward slope. Eventually the beach will come to rest against the mainland coast, which will thereafter be cut back as marine cliffs by storm wave activity, a stage that has already been reached at Burton Bradstock, near the western end.

The Loe Bar, near Helston in Cornwall (Figure 6.6), is a barrier 180 m wide and up to 3 m above calm weather high spring tide level. It is part of the beach that extends from Gunwalloe north to Porthleven, and runs across the mouth of a former ria, enclosing the Loe Pool. It consists largely of fine flint and chert shingle that has been washed in by wave action from Tertiary or Pleistocene gravel deposits on the sea floor offshore, with only a small proportion of quartzite and slaty material derived from bordering cliffed coasts, and some shelly sand and gravel. The presence of flint and chert shingle on this part of the coast, and its rarity on adjacent coasts, suggests that the sea floor gravels had already been concentrated into one or more beach and barrier formation segments (along coastlines now submerged) before their final shoreward movement during the Holocene marine transgression. The Loe Bar is sometimes overwashed by large waves from the SW during storms, and by occasional Atlantic tsunamis.

Contrasts in sediment size are likely between the seaward and landward flanks of a shingle barrier where overwashing is infrequent, the wave-agitated shingle on the seaward flank being reduced in calibre by attrition, the beach face pebbles being clean and actively worn, while the less disturbed shingle on the landward slope is often stained and silty, sometimes with plant growth. Where the shingle is of similar size on the beach face, barrier crest and landward slope the implication is that overwashing has been relatively frequent in relation to the rate of attrition on the beach face.

Shoreward drift of the kind that produced Chesil Beach and the Loe Bar has also been invoked to explain the barriers of sand and shingle that have developed in front of uncliffed mainland coasts in Siberia (Zenkovich, 1967), and Le Bourdiec (1958) reached the same conclusion for barriers on the Ivory Coast. Shoreward sweeping of beach sediment during the Holocene marine transgression contributed to the sandy barrier islands that border the Gulf coast of the United States (Shepard, 1973). In each case the barrier formations rest upon an older (Pleistocene) land surface submerged by the Holocene transgression.

Some coastal barriers have remained in position during Holocene times, and may have been widened seaward by progradation with the addition of successively formed beach (and dune)

ridges. On many barrier coasts this progradation has lately given place to erosion (Figure 8.12). Other barriers are (or have become) transgressive, migrating landward across lagoon and swamp deposits as the result of overwashing by storm waves (as on Chesil Beach) or landward spilling of dunes.

Chesil Beach is apparently receiving little new shingle, either from onshore or alongshore, and experiments with radioactive pebbles dumped on the sea floor failed to show any evidence of active shoreward drift to Scolt Head Island, a barrier island on the north Norfolk coast. These shingle barriers are essentially relict formations, a legacy of the Holocene marine transgression. Most British barriers are of shingle (some capped by sand dunes), but in Sandwood Bay on the west coast of Scotland there is a wide barrier beach, built of sand washed in from the sea floor, enclosing a freshwater loch at the mouth of Strath Shinary, a glaciated trough.

Other barriers are still receiving sediment derived from adjacent cliffed coasts, river discharge, and shoreward drift. The barriers that extend across embayments on the New England coast, enclosing lagoons known as ponds, are built of sediment derived from adjacent cliffs cut in glacial outwash deposits. These ponded barriers are well developed on the southern shores of Martha's Vineyard and Nantucket Island off the Massachusetts coast, and on the shores of Cape Cod. Similar spits and barriers have been derived from eroding cliffs (not in glacial drift) on the coast of Sakhalin Island in Russia. On the coast of Nova Scotia, where shingle barriers consist of gravel sorted from glacial drift deposits, Orford, Carter and Jennings (1996) related phases of shingle barrier growth, consolidation, breakdown and re-formation to sea level fluctuations. Shelly sand drifting alongshore has accumulated in the form of a curving barrier on the western coast of the Sea of Azov (Figure 8.10).

Barriers known as *nehrungen* on the south Baltic coast have also been partly derived from

Figure 8.12 Evolution of barrier coastlines where Holocene progradation formed a succession of parallel ridges, but has been followed by recession, with losses of sand landward into spilling dunes (A), as well as alongshore (from B) and seaward (C). Such a barrier can eventually become transgressive, with sand blown (or washed) over into the backing lagoon

cliffed glacial deposits and carried along the shore to enclose embayments as lagoons, and partly from sediment swept in from the sea floor. By contrast, the sand and gravel barriers on the SE coast of Iceland have been supplied largely with sediment supplied to the coast by glacifluvial streams, especially during the floods that follow the melting of hinterland ice by volcanic eruptions. Sand and gravel washed into the sea are reworked and built into barriers by the constructive action of southerly Atlantic swell. In the North Island of New Zealand sandy barriers are extensive, and mostly derived from volcanic sand carried down to the coast by rivers or eroded from cliffs of volcanic ash. In the South Island shingle barriers are commoner, derived from glacifluvial outwash deposits eroded from cliffs or swept in from the sea floor.

On some coasts there are multiple barriers with intervening tracts of lagoon and
swamp. The inner barriers are generally of Pleistocene age, having been dissected by stream
incision during the Last Glacial low sea level
phase, whereas the outer barriers are usually of
Holocene age, having come into existence during and since the Holocene marine transgression. There are many complications, for barriers
have been built, then dissected or destroyed, and
subsequently rebuilt, sometimes in overlapping
alignments, during Quaternary sea level oscillations.

On the Victorian and New South Wales coasts
in Australia the inner barriers are largely of
quartzose sand that can readily be reshaped by
wave or wind action. They have not been lithified in the manner of the calcareous sand barriers of South Australia, which are commonly
preserved as relatively durable dune calcarenite (Section 6.4.3). The numerous parallel ridges
of dune calcarenite in SE South Australia (Figure 3.4) formed successively as coastal barriers during Pleistocene times, when this part
of the coast was intermittently uplifted, and
subject to sea level oscillations. Between the
emerged dune calcarenite barriers are corridors
of swamp land that developed in former lagoons.
The Coorong, behind the outer barrier on the
shores of Encounter Bay, is the last in this sequence of inter-barrier lagoons, not yet filled by
swamp deposits. It is bordered by a Holocene
outer barrier (Younghusband Peninsula). Farther south, between Robe and Beachport, the
coast consists of a cliffed and dissected ridge
of Pleistocene dune calcarenite, there being no
Holocene outer barrier of unconsolidated sand
here.

The sandy barriers in SE Australia have their
counterparts on similarly oriented coasts in
South Africa (Natal) and South America (between Rio and Cabo Frio). Many of these barriers have been prograded by the addition of
successive parallel beach ridges on alignments
that, like the curved outline of the present beach,
result from the refracted patterns of dominant
swell approaching the coast. Often the beach
ridges are surmounted by parallel dunes, developed successively during progradation, as on
the barrier islands in SW Florida, where beach
ridge formation has been correlated with fluctuating Holocene sea levels (Stapor, Mathews and
Lindfors-Kearns, 1991).

The pattern of beach ridges and dunes can
be used to decipher the history of barrier evolution, particularly if there are relics of former recurves indicative of stages in longshore
growth. The outer barrier on the East Gippsland
coast (Figure 8.13) originated as a chain of barrier islands that grew to the NE by longshore
drifting, deflecting several outlets from coastal
lagoons (the Gippsland Lakes) and sealing off
most of them. This barrier was subsequently
widened by progradation, with successive parallel ridges added on the seaward side, until
it took up the alignment of the Ninety Mile
Beach along the present coast. Radiocarbon dating of shelly material obtained from drilling in
these sand barriers was used by Thom (1984)
to establish stages in the Holocene progradation, but the barriers have been truncated
along their seaward margins by ensuing marine
erosion.

By contrast, some of the barriers and barrier islands of the Gulf and Atlantic coasts of
the United States are transgressive. The Outer
Banks of North Carolina north of Cape Hatteras
have been migrating landward as the result of
spillover dunes and washover sand invading
Pamlico Sound, and accompanying recession of
the seaward margin. On the Texas coast deep
transverse channels cut by storm surge overwash
are known as bogues, as on Andros Island. Others have prograded, with the addition of successive beach ridges, as on Seabrook and Folly
Islands in South Carolina. The barrier on the
shores of Encounter Bay in Australia is becoming
transgressive, with numerous blowouts spilling

Figure 8.13 The Gippsland Lakes in SE Australia are enclosed by a Inner Barrier and an Outer Barrier (the Ninety Mile Beach), separated by lagoons and swamps, and with relics of an earlier, Prior Barrier that pre-dates the enclosure of the existing lakes. Sections through the outer barrier on the East Gippsland coast show that it formed in Holocene times by the deposition of beach and foredune sands over earlier lagoonal deposits (based on Thom, 1984)

sand over into the backing lagoon, the Coorong. Tidal inlets between barrier islands may migrate downdrift, become sealed off by deposition or reopen during stormy episodes. It is not surprising that borings in transgressive barrier formations reveal a complex stratigraphy, with overlapping sequences of beach, dune and lagoon sediments, as was shown in Van Straaten's (1965) study of Holocene deposits on the Dutch coast.

On some coasts there are elongated islands that look like barrier islands but are not wave-deposited features. Often they are segments of glacial moraines, as on Long Island, New York, and Walney Island in Cumbria, which is cliffed

on the seaward side. Spit growth at the limits of such islands increases their similarity to wave-shaped barrier islands. On the German North Sea island of Sylt, sediment derived from receding cliffs cut into glacial drift on the outer coast has been distributed by northward and southward longshore drifting to dune-capped recurved spits at each end.

Low sectors of a barrier through which storm waves or surges flow are called swashways, and sediment swept over a barrier or through these swashways is deposited as washover fans on the inner shore.

Several references have been made to dunes associated with beaches, spits and barriers, and the following chapter will deal with the evolution of coastal dunes.

8.7 Intertidal sandflats

As the tide falls in front of beaches, spits and barriers it exposes an intertidal zone that is generally sandy, often with some gravel. Rocky intertidal features were discussed in Chapter 5 and muddy foreshores will be considered in Chapter 10.

Sandflats have formed where waves and currents have carried sand onshore or alongshore and deposited it in intertidal areas. They are best developed where wave action has been strong enough to have prevented the deposition of silt and clay, and they may grade laterally into mudflats where wave energy diminishes alongshore.

In general sandflats are wave-dominated features, but currents (particularly tidal currents) also contribute to their morphology, notably in the formation of bars, troughs and ripples (see below). Sandflats can also be reshaped by wind action, especially in the upper intertidal zone that remains dry between the highest tides: on Morfa Harlech in North Wales the wide sandy intertidal zone grades upward into hummocky dunes across a boundary that is not distinguished on Ordnance Survey maps.

Where wave action is strong at high tide an intertidal sandflat may have only subdued topography, or be almost featureless, consisting of smooth sand packed firmly by wave action. Such areas have been used for sports such as sand yachting in the Netherlands and on the French Channel coast east of Dunkirk, and as sites for motor racing on Pendine Sands in South Wales.

8.8 Bars and troughs

A bar is an elongated ridge or bank of sand (sometimes gravel) a few metres wide, deposited and shaped by waves and associated currents, extending parallel to the coastline, and submerged at least at high tide. Between a bar and the coastline, and between nearshore bars, are elongated troughs where the water is deeper (Figure 6.16). Some bars are exposed at low tide but submerged at high tide, and the intervening troughs may contain lagoons as the tide falls. A break-point bar is a concentration of sediment parallel to the coastline, formed by breaking waves where sand or gravel carried shoreward by spilling waves meets similar sediment withdrawn from the beach by backwash. Break-point bars can be shaped in wave tank experiments (Rey, Davies and Belzons, 1995), where it is found that their size and distance from the shore are related to the dimensions of the waves, higher waves building larger bars farther offshore. Longshore bars are formed parallel to a beach, and at least partly exposed at low tide, while oblique bars run at an angle to the beach, and transverse bars at right angles to it. Bars may also occur in cuspate, lunate, looped, crescentic, reticulate or chevron patterns. Multiple parallel bars and troughs form on gently sloping sandy shores where spilling waves break and re-form. Trailing bars (banner banks) occur in the lee of headlands or islands, particularly of glacial drift, and may become flying bars if they are disconnected, or if the island is consumed by erosion. Examples of these various bars can be found in Puget Sound, Boston Harbour and the Danish archipelago. Before Shepard (1952) defined bars, such terms as offshore bar, bay bar and looped bar were often used to describe features that would now be termed barriers. Some features that are named bars (such as Loe Bar in Cornwall) are really coastal barriers, built above high tide level.

Most bars are sandy, but pebbles and shells are often present. Shingle bars are unusual, but are found off river mouths and tidal inlets that interrupt shingle beaches and barriers: a looped shingle bar is generally present off the mouth of the River Teign in South Devon.

Some sand bars may remain in position, but most bar crests move to and fro as tides rise and fall, or with alternations of calm and stormy weather. Migrating bars become asymmetrical, with steeper advancing slopes. In calm weather, when constructive swash is more effective, bars move closer to the shore and may become swash bars, flatter in profile, sometimes with a steeper shoreward advancing slope. Bars of this type are well developed along the Ninety Mile Beach in Australia, where in fine weather gentle waves wash them shoreward until they become welded on to the beach as berms (Section 6.10.6).

At high tide waves break across intertidal bars and troughs, generally as spilling breakers (Section 2.2.6). After the strong swash crosses the bar, the wave may re-form as it moves into in the deeper water of the trough, the surf falling back from the crest as a new smooth wave front emerges. Bars are often interrupted by transverse channels (gats) formed and maintained by rip currents, which complete the nearshore water circulation by carrying water driven on to the shore by breaking waves back seaward through the surf zone. Rip currents may form a lobe or fan of beach sediment at the seaward end of each channel. There are variations in nearshore topography and rip current channel patterns with alternations of storm waves and calmer conditions (Goodfellow and Stephenson, 2005).

Segmented bars may be straight, but are usually crescentic in outline: some are parallel to the coastline, others angled. On Murvagh Strand, Donegal, in NW Ireland evenly spaced lobate swash bars exposed at low tide are separated by channels that drain from an intertidal lagoon. The pattern may have developed in a similar way to beach cusps, as the result of interactions between breakers and edge waves (Section 2.2.8) as the tide ebbed.

Where the tide range is sufficient to expose a broad foreshore at low tide there may be numerous subdued bars and troughs running parallel, or at a slight angle, to the coastline (Figure 8.14).

These are known as low-and-ball or ridge-and-runnel in Britain, where they are well developed on the lower part of the beach near Formby in South Lancashire (Gresswell, 1953). Their amplitude rarely attains a metre, and the ridge crests are almost flat, and often as much as 100 m apart. They are essentially swash-built bars, formed by relatively weak wave action. Their formation as rhythmic topography is an adjustment between surface form and the oscillating turbulence produced by spilling breakers: once the bars have formed, these waves break across them, re-form as they cross the intervening swales and break again over the next bar. The number, spacing and amplitude of multiple bars varies with the height and period of breaking waves, but bar topography, in turn, influences where and how incident waves break. The profiles of bars and troughs are modified by wave action as the tide rises and falls, the bars remaining subdued as their margins show minor progradation shoreward with the rise and seaward with the ebb.

Where waves have been arriving at an angle to the shore the ridges and troughs also run obliquely, as on the south-facing shore at Porth Neigwl in North Wales, where the dominant waves come in from the SW. As the tide falls the troughs may be temporarily occupied by lagoons, which drain out by way of transverse channels: the trough sands are often rippled and the bar crests smoother. Once established, ridge-and-runnel patterns persist on sandflats through many tidal cycles, with minor modifications by wave action (particularly during storms and when the angle of incidence of waves changes) and currents that strengthen as neap tides grow into springs. On the Danish coast Aagard (1991) reported that multiple bars were shaped by low frequency edge waves (Section 2.2.8) and found that the outer bars changed only during stormy periods. On the French coast south of Arcachon Michel and Howa (1999) investigated sandy ridge-and-runnel on a shore

Figure 8.14 Multiple nearshore sand bars at Rosebud, Port Phillip Bay, Australia

where the ridge was interrupted by curving rip channels flowing out from the runnel. Sand washed in by waves was added to the ridge, while sand washed out of the runnel through rip channels by strong ebb currents was reworked by waves and returned to the ridge.

On the Australian coast parallel sand bars of this kind are found on shores where the effects of ocean swell are weak or excluded. Up to ten parallel sand bars occupy the intertidal sandflats between Rye and Rosebud on the SE coast of Port Phillip Bay (Figure 8.14), where at high tide incoming waves break, re-form and break again in such a way as to develop and maintain the rhythmic intertidal topography. Similar features are seen on sandy intertidal sandflats in the Darwin area and on the coast near Port Hedland in NW Australia.

Some bars have grown alongshore, and are essentially intertidal spits. At Gibraltar Point on the Lincolnshire coast there are parallel sand bars that look like swash-built ridge-and-runnel but have actually grown from north to south as the result of deposition by currents, and possibly waves coming in from the NE, at an angle to the coast.

Transverse or finger bars are usually orientated at a high angle or perpendicular to the coastline, extending out into the intertidal sandflat. They are common off dissipative sandy beaches (Section 6.11) on low to moderate wave energy coasts, as in Florida and parts of North Queensland. No entirely satisfactory explanation of their origin is available, but they are produced where the seaward flow of water after waves break upon the shore at high tide is more dispersed than in rip currents, and they may be produced by intersecting edge waves (Section 2.2.8) (Holman and Bowen, 1982). Once formed, incoming waves break along them in an intersecting pattern, but this is a consequence rather than a cause of transverse bar formation.

8.9 Sand shoals

Shoals form offshore in various ways, some as the result of sand deposition in slack water between tidal channels, others by fluvial sand deposition off river mouths, usually during floods, and others by the dissection of former sand bars. They persist where wave energy is too low to move or disperse them. They may be at least partly exposed at low tide, as the ebb drains into shallowing troughs and channels, but some are subtidal. Sandy shoals emerge at low tide offshore in the Goodwin Sands off East Kent and the Brake Bank off Lowestoft in eastern England, where the sand has been derived from glacial drift deposits, reworked by waves and currents on the sea floor.

Elongated linear sand shoals and swales have been formed parallel to tidal currents in the Straits of Dover. Intertidal sand shoals are found in the mouths of estuaries, as in the macrotidal gulfs of northern Australia. The three King Shoals, where the estuary of the Ord River opens to the northern end of Cambridge Gulf, are intertidal sandy (and muddy) ridges 15–20 km long and about 1 km wide, aligned parallel to the ebb and flow of tidal currents. Estuarine shoals, with intricate and variable creeks, out of which the ebb tide drains, lie behind barrier islands and spits on the North Norfolk coast and in the Wadden Sea on the coasts of Holland, Germany and Denmark. Shoals are also found between ebb and flood channels in estuaries. Occasionally wave action on sand shoals may build up islands that rise above high tide level, but these are usually only temporary, swept away during storms.

8.10 Ripples

Various kinds of ripple form on intertidal sandflats. They are typically more or less regular, at intervals of between 1 cm and a few metres, and may be formed either by wave action, or by currents, or by a combination of both processes. Wave-formed ripples are usually asymmetrical and parallel, shaped by eddies that form as a wave passes, separating, steepening and sharpening the ripple crests. Current-formed ripples are also generally asymmetrical, with a steeper face away from the current, and associations of cross-currents generated by waves and tides over sandflats at high tide can produce decussate or rhomboidal networks of transverse current ripples. In addition, there are longitudinal ripples that form parallel to a strong unidirectional current (similar to those that form on sandy river beds), and these can be complicated by interfering wave motion and cross-current patterns.

There are also giant ripples (megaripples), with crests spaced at intervals of several metres and amplitudes attaining, and sometimes exceeding, 1 m. They are formed by strong currents, and are usually aligned at right angles to the tidal flow, migrating in the direction of the current, but sometimes they run diagonally or parallel to the current flow. Megaripples are found in sandy intertidal areas where the tide range is very large, as in the Bay of Fundy in Canada and in macrotidal estuaries, such as the Rhine and Scheldt in the Netherlands, where Van Straaten (1953) identified 20 different kinds of megaripple. McCave and Geiser (1979) described megaripples 0.3–0.6 m high and 10–15 m long, formed by strong currents during flood tides on intertidal sandflats in The Wash. On outer sandy shoals they graded into wave-formed ridge-and-runnel morphology, with wave-formed ripples in the troughs, which may be lined with muddy sediment. Multiple bars are essentially wave-built megaripples.

Details of current ripples are discussed at length by Allen (1968), and a classification has been presented by Tanner (1982).

8.11 Sandstone reefs

Where outcrops of sandstone, notably beach rock and dune calcarenite, have been planed off at sea level they form reefs that are similar in many ways to the constructional reefs formed by corals and algae (Chapter 13). There are several sectors on the southern and western coasts of Australia where parallel ridges of dune calcarenite formed during Pleistocene alternations of sea level, and some of these that extended seaward from the present coastline have been planed off by marine erosion. There are good examples west of Esperance on the south coast of Western Australia, where the basal dune calcarenite is often a shelly beach rock. A flat-topped sandstone reef awash at low tide is separated from the beach by a shallow lagoon and serves to protect the beach by reducing wave impact. Similar features are seen on the so-called stone reefs off the coast of Brazil.

8.12 Summary

Spits are formed by longshore drifting of beach sediment, and often end in recurves, sometimes multiple and indicating former terminations as the spit grew. Paired spits at river mouths or lagoon entrances may result from convergent drifting or from the breaching of a barrier. Trailing spits extend downdrift from islands, and may become separated as flying spits. Tombolos are spits that link islands, or attach an island to the mainland. Spits may become cuspate or lobate in response to incident wave patterns, and spits enlarged by the accretion of beach ridges may become cuspate or lobate forelands.

Coastal barriers have formed by the deposition of beach material offshore, or across mouths of inlets or embayments, above the level of normal high tides, and may partly or wholly enclose lagoons and swamps. Some have evolved from longshore spits, others from sediment washed up from the sea floor. Barrier islands are interrupted by tidal inlets. Some barriers remain in position, and may be widened by beach progradation; others are overwashed and driven landward (transgressive barriers). Many have beach or dune ridges in patterns that indicate their history of growth.

Intertidal sandflats may front beaches or fringe tidal estuaries and lagoons. They often show bars and troughs parallel to or angled against the beach and shaped by waves. Sand shoals are less regular, and may be shaped by currents as well as waves. Ripples occur on various scales up to several metres (megaripples) and are formed by currents, including those associated with wave action.

On some coasts there are intertidal reefs of sandstone, often submerged and truncated beach rock or dune calcarenite.

9
Coastal dunes

9.1 Introduction

Coastal dunes generally form where sand on the shore has dried out and been blown to the back of the beach, to accumulate above high tide level, particularly where deposition occurs against obstacles such as driftwood, or within strand litter or vegetation. Their growth and shaping are related to a source of sand that can be moved by wind, wind flow characteristics, rates of aeolian transport and patterns of erosion and deposition. They are most extensive on windward coasts behind wide sandy beaches, notably on the Atlantic coasts of Europe and the Pacific coasts of the Americas and in SE Australia and southern Africa. Coastal dunes differ from inland (desert) dunes in that they are subject to a wider variety of processes, including wave action and vegetation, which influence their size, shape, evolution and persistence.

Backshore dune development is aided by frequent strong onshore winds and by the availability of a wide sandy beach as a source area: it is also influenced by wave processes, especially when storm waves trim back the seaward margins of coastal dunes. Coastal dunes are more often found behind wide gently sloping dissipative beaches (Section 6.11) than behind steeper, coarser and narrower reflective beaches; newly formed dunes are usually associated with prograding sandy coasts. They are well developed where the tide range is large, as on the Atlantic coasts of Britain, where prevailing westerly winds have blown sand onshore from beaches, as in the Ainsdale dunes in south Lancashire. On the North Sea coast dunes back wide beaches in the vicinity of Holy Island and along the shores of Moray Firth (Culbin Sands), and are extensive in Belgium, Holland, Germany and Denmark (Bakker *et al.*, 1990). Papers on various aspects of dune geomorphology were assembled by Nordstrom, Psuty and Carter (1990).

Even where the tide range is small, coastal dunes have formed where there has been a sufficient supply of accreting sand to be blown onshore from the beach. Sandy beaches shaped by ocean swell are generally backed by dunes on the coasts of Australia, Africa and the Americas, although some of the older dune systems (especially dune calcarenites, Section 9.10) include sand that was carried inland by wind action from the emerged sea floor during Pleistocene low sea level phases. This has contributed to the major dune systems on the SE coast of Australia, notably on Fraser and Moreton Islands in Queensland and behind Newcastle Bight and Wreck Bay in New South Wales and Discovery Bay in Victoria (Hesp and Thom, 1990).

Coastal Geomorphology: An Introduction, Second Edition.   Eric Bird.
© 2008 by John Wiley & Sons, Ltd

On the Pacific coast of North America dunes have been derived from beaches nourished by an abundance of fluvial sand supply from the Columbia and other rivers. There are dunes on the coastal barrier at Long Island, north of the mouth of the Columbia River, and Clatsop spit, to the south. Fluvial sand swept out to the sea floor, reworked by wave action and delivered to beaches, has been the source of the extensive dunes at Coos Bay in Oregon (Cooper, 1958).

Some backshore dunes have not been derived from adjacent beaches, but from sources along the coast or inland. On the south-facing Cape Coast of South Africa there are backshore dunes derived from beaches upwind and driven from west to east along the coast, spilling over headlands, as at Cape Recife near Port Elizabeth. Other backshore dunes are of desert origin, as on the coast of Mauritania, where Saharan dunes drifting down to the coast deliver sand to the beach (Vermeer, 1985). Coasts bordering the Sahara and other arid regions have areas where desert dunes meet and mix with dunes derived from beach sands.

Dune sands have similar characteristics to the beach sands from which they have been derived and generally consist of quartz, feldspar and calcareous particles (including foraminifera, bryozoa and comminuted shells and corals), sometimes with heavy minerals such as rutile and ilmenite. The white dune sands in the Isles of Scilly are of quartz and pale feldspar derived from the local granites. Where there are volcanoes or volcanic deposits on the coast or in the hinterland, dunes may be derived from deposits of volcanic ash.

Sand blown from a beach is typically fine grained (sand grains of diameter 0.1–0.3 mm, or just below 2ø to just above 3ø, are most readily moved by wind action), well sorted and well rounded. Grain size analyses show that dune sands are often (but not always) finer and better sorted than beach sands, with positively skewed grain-size distributions (Panel 6.1). Dune sands

often have highly polished grain surfaces. Their rounding may be the result of abrasion during transport by the wind.

Wind data for coastal dune studies has often been obtained from meteorological stations, which may be some distance from the dune area. This has proved useful in demonstrating relationships between dune forms and dominant winds, but for detailed process–response studies it is necessary to obtain site data using anemometers, preferably set to measure wind velocity close to the ground.

Entrainment of sand by the wind depends on near surface airflow and surface morphology. Wind moving across a sandy beach develops a shear stress, equivalent to air density (averaging about 1.22 kg/m^3, cold or dry air being denser than warm or moist air) multiplied by wind velocity, and when this exceeds the entrainment threshold value wind energy overcomes gravitational inertia and loose sand particles are mobilised. Well rounded and well sorted sands are more easily lifted by the wind than sand composed of more irregular grains. Mobilisation is impeded where there are cohesive forces that raise the entrainment threshold. Deflation of sand from beaches is inhibited by moisture. There is no doubt that wet sand is more cohesive and less readily moved by wind action across the shore, but strong onshore winds soon dry a beach surface and can drift sand even when it is raining. On some tropical beaches the formation of a surface crust by salt evaporated from sea water or spray impedes the movement of sand by wind action. Morton (1957) attributed the poor development of dunes on the coast of Ghana to the salt binding of beach sands, noting that dunes had developed at Old Ningo where the sand was unusually coarse and shelly, and not bound by salt. On some beaches a surface accumulation of pebbles or shells left where sand has been winnowed by the wind or washed away by waves prevents further sand deflation (Pye, 1983): on Magilligan Point in Northern Ireland this resulted in a reduction of sand supply to backshore dunes (Carter and Wilson, 1990). In such situations wind energy may be strong and sustained enough to move much more sand than is available in source areas.

On dry, loose sand the flow of wind over sand grains causes a pressure gradient that lifts the particles, which then travel downwind by saltation (bouncing), traction (rolling) or, if the wind is strong, suspension. Sand movement by wind action is influenced by the shape of sand grains because rounder grains roll more easily, and flatter grains, such as mica flakes, are readily deflated. Dune surfaces occasionally have ripples formed by wind action, which are similar to ripples formed by nearshore currents (Section 8.10). An analysis of sand mobilisation by wind action has been provided by Sherman and Hotta (1990).

Onshore winds sweep sand from the beach to the backshore. The sand is carried by until the wind velocity diminishes, the ground surface rises, or vegetation is encountered. Olson (1958) used anemometers to compare the wind velocity profile over bare sand with that over grassy dunes, and found that the vegetation raised the upper limit of the zone of calm air near the ground. Vegetation also reduces shear stress and presents a surface roughness that promotes sand deposition. This deposition increases the backshore gradient, and the backshore is shaped into a ridge that impedes further landward transport of blown sand. The outcome is the upward growth of a vegetated sand ridge, known as a foredune. Sometimes strong winds carry fine sand over the ridge and deposit it as a thin sheet or low hummocks on the landward side.

Sand is most readily winnowed from a beach where it is fine grained (but not necessarily well sorted) and where the grains are irregular in shape. Sand transport rates and sediment budgets on beaches and coastal dunes can be quantified by making repeated surveys of the beach face and developing dune to determine changes

in volume, by trapping blown sediment in a surveyed trench or receptacle such as a cylindrical pipe (Bauer *et al.*, 1990) or by using fluorescent tracers, as outlined by Sherman and Hotta (1990). Horikawa (1988) reviewed methods of determining beach and dune sediment budgets.

The amount of sand entrained by the wind depends on the strength of the onshore wind, and is limited by beach width. More sand can be blown from a beach to a backshore dune by onshore winds arriving obliquely than by those that arrive at right angles to the coastline, which also generate stronger wave action. On the shore of Lake Erie, Davidson-Arnott and Law (1990) found that the width needed for maximum sand transport by a wind exceeding 50 km/hr was over 40 m.

Beach–dune interactions occur when onshore winds deliver beach sand to backshore dunes and when winds blow sand from the dunes to the beach or storm waves erode sand from backshore dunes and incorporate it in the beach (Psuty, 1988). The first process tends to leave the beach sand coarser and less well sorted than the dune sand, but the second can obscure such differentiation.

On cold coasts where there is a winter snow cover, as in Korea and eastern Canada, deflation of sandy surfaces may occur as they dry out after the spring thaw, forming dunes before annual plants can grow to stabilise the surface. Dunes have formed in this way from river channel sands on the Colville delta in the Canadian arctic (Walker, 1988).

Coastal dunes have formed over a variety of timescales, but most have been shaped in Holocene times from sand supplied to beaches from the sea floor (notably during the Holocene marine transgression) and alongshore sources such as cliffs in soft sandstone or glacial drift deposits. On some coasts the Holocene dunes overlie, and are backed landward by, dune formations that originated in the Pleistocene, some of which may have formed during phases of falling sea level, when onshore winds swept sand from emerging sea floors.

Coastal dune landforms include ridges, mounds, terraces and low lying swales. Swales may be temporarily or permanently occupied by dune lakes related to the level of the water table, and wet swales (often marshy) are termed slacks.

9.2 Foredunes

Foredunes are ridges of sand built up at the back of a beach or on the crest of a sand or shingle berm, where dune grasses have colonised, and are trapping blown sand (Figure 9.1). The colonising vegetation acts as a baffle, diminishing wind velocity close to the ground and so creating a sheltered environment within which blown sand is deposited.

Foredunes become higher and wider as accretion continues, depending on rates of wind-blown sand supply and coastline progradation. Vegetation certainly traps and retains sand blown onshore from a beach, and the initiation of a foredune may occur along a high spring tide line where sand-trapping vegetation grows up from a seed-bearing strandline of plant litter on the beach face. There has been much discussion of whether such a strandline follows the crest of a wave-built berm of sand or shingle, or whether a foredune can be initiated from seed-bearing litter deposited in a line along the beach face (seaward of pre-existing vegetated dunes). Foredunes run parallel to the high tide shoreline, as do berm crests and seed-bearing litter zones, but they can also be initiated along the seaward edge of a grassy backshore terrace or an area of irregular hummocky dunes when cliffing by storm waves at high tide is accompanied by swash-borne or wind-blown sand accretion on top of the cliffed margin, forming a linear sand ridge that is colonised by sand-trapping grasses.

Tussock grasses, herbs or shrubs that form circular patches tend to build mounds or hummocks of sand, whereas the close networks of stalks in more uniform grassland is more likely to build a ridge or terrace in the backshore zone (Hesp, 1988; Bird and Jones, 1988). Backshore dune terraces form where vegetation spreads forward on to the beach, and are generally horizontal or with faint ridges parallel to the coastline, marking stages in intermittent progradation.

On the coasts of the United States foredune initiation and growth have been aided by the building of sand-trapping fences parallel to the coastline, as on Fire Island in New York State (Psuty, 1990). Sand is precipitated as wind velocity slackens across such a linear baffle.

There are geographical variations in the plant species that act as pioneer colonists and foredune builders. In the British Isles and Europe a common pioneer dune plant is marram grass (*Ammophila arenaria*), a species that colonises bare sand, and has been introduced to stabilise dunes in many other parts of the world. Lyme grass (*Elymus arenarius*) and sea wheat grass (*Agropyron junceum*) are also dune pioneers in Europe, the latter extending around the Mediterranean and Black Seas. Marram grass is the most vigorous of the dune-building plants, thriving on cool, moist coasts, where it is able to trap and grow up rapidly through accreting blown sand. It can add up to 2 m of sand to an accreting foredune in the course of a single year, but is overwhelmed if sand is supplied too rapidly. Where sand accretion is slow marram grass grows poorly, and other grasses may take its place (Ranwell, 1972).

In the United States another form of marram grass, *Ammophila brevigulata*, is a common pioneer dune species, with the tussocky grass *Panicum amarum*, with *Spartina patens* common on low moist dunes. On sandy coasts in tropical regions the vegetation is generally sparser, and dominated by creeping vines such as *Ipomoea* and *Canavalia*, rather than tussocky grasses.

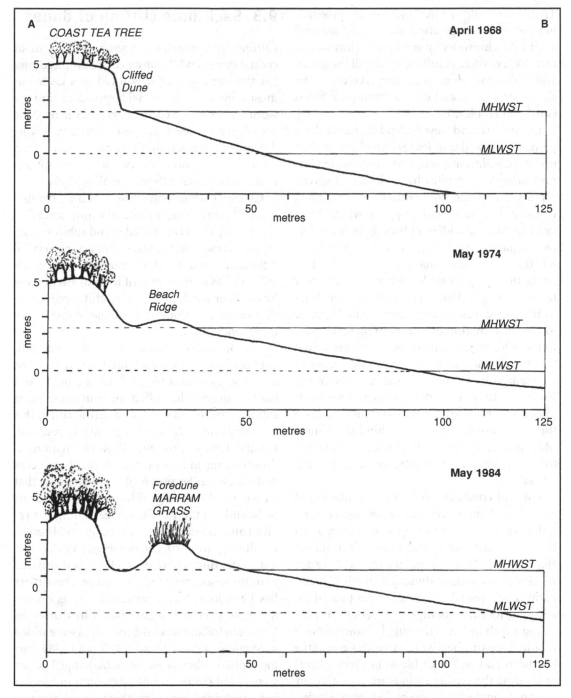

Figure 9.1 Successive surveys showing the formation of a beach ridge in front of a previously cliffed dune, and the development of a foredune when the beach ridge was colonised by marram grass

These can build a foredune, but on prograding shores this vegetation may spread seaward and build a hummocky or ridged backshore terrace. Where there is sufficient rainfall the grassy and herbaceous dune vegetation is colonised by shrubs and trees, and eventually tropical forest (Bird and Hopley, 1969).

In Australia and New Zealand the native dune pioneers are sand spinifex (*Spinifex hirsutus*), sea rocket (*Cakile maritima*) and coast fescue (*Festuca littoralis*). Spinifex has stolons that spread rapidly across the sand surface, putting down roots at intervals, and trapping wind-blown sand to form a widening backshore dune terrace (again hummocky, or with low ridges), whereas introduced marram grass forms tussocks that tend to build higher and narrower foredune ridges. However, foredunes were built by the native Australian grasses and herbs, prior to the introduction of European marram grass in the 19th century. It has been suggested that the introduction of European marram grass to Australia (often displacing native *Spinifex* and *Festuca* grasses) has altered coastal dune morphology, but foredunes were certainly built on the Boole Boole Peninsula, behind the Ninety Mile Beach, in the early Holocene, long before marram grass was introduced here (Bird, 1978a).

On arid coasts (as in Baja California and at the head of the Great Australian Bight) vegetation is too sparse to trap sand blown from beach to backshore, and dunes drift inland (Figure 7.4). Even on humid coasts there are sectors where coastal dunes are drifting inland because the available vegetation has proved inadequate for the trapping of sand, either because of a superabundant sand supply blown from a rapidly accreting beach (as on Sandy Cape at the northern end of Fraser Island in Queensland) or because the onshore winds are exceptionally strong, as on the NW coast of Tasmania (Davies, 1980).

9.3 Backshore cliffing of dunes

Cliffing (scarping) of the seaward margins of coastal dunes and the absence of new foredunes (or the formation of new foredunes lasting at most a few years) is very widespread around the world's coastline, a consequence of the modern prevalence of beach erosion (Figure 9.2). Undercut by wave action, dune cliffs recede with sand collapsing on to the beach or into the sea, to be carried away offshore or alongshore.

Clumps of dune vegetation may bind the surface sufficiently to form blocks of dune sand that break away from the cliff edge and subside as irregular terracettes. Dune cliffs may stand vertical if the sand is moist and coherent, but as it dries it falls to a basal apron that (if it is not swept away by waves or wind) grows as the cliff recedes, until it becomes a slope at the angle of rest of dry dune sand (about 32°).

On some coasts the formation of a dune cliff, cut back by storm waves, is followed by renewed accretion as sand is blown from the beach and banked against the cliffed dune in subsequent calmer weather. As vegetation spreads on to this accreting sand, the earlier profile is restored. Coastal dune margins may show alternations of dune cliffing and restoration during beach cut-and-fill cycles (Section 6.10.1). A foredune that runs parallel to the high tide line on the NE coast of Norfolk in the vicinity of Sea Palling has revived after phases of severe erosion and breaching during storm surges, notably in 1953.

Coastal dunes have generally cliffed seaward margins in southern Britain, except where there has been local beach progradation, as on the northern part of Studland beach in Dorset. In Northern Ireland several dune ridges were added to Magilligan Point between 1953 and 1983 during a phase when adjacent dune-fringed coasts were being cut back, but progradation has not been sustained here, and these dunes are now cliffed (Carter and Wilson, 1990).

Figure 9.2 Cliffed backshore dune on the beach east of Somers, Victoria, Australia

In northern Britain, where emergence is in progress as the result of continuing isostatic uplift (Section 3.3.4), there are young foredunes behind beaches that are still receiving sand swept in from sea floor glacial drift deposits in the vicinity of Holy Island in Northumbria and at Tentsmuir in Scotland, where the dune fringe has advanced about 90 m seaward since concrete tank traps were built in 1940.

During storms, the cutting of a cliff along the seaward side of a foredune may be accompanied by accretion of wind-blown and wave-washed sand on the crest and lee of the ridge, raising its altitude.

9.4 Parallel dunes

On some coasts there are multiple dune ridges, usually running parallel to the coastline, which have formed successively as foredunes behind a prograding sandy beach. They differ from pro-graded backshore terraces in that there are intervening elongated swales or troughs. These result from sequences of cut and fill on an intermittently prograded shore. Incipient dunes form as sand blown from the beach is deposited to build a new foredune seaward of earlier foredunes, which thus become parallel dune ridges separated by corridors that remain as relatively low swales. The seaward margin of a foredune is trimmed back by waves during a storm, forming a crumbling cliff of sand. Subsequently, during calmer weather, waves restore the beach and a new foredune is initiated along a high tide berm or strandline litter zone in front of, and parallel to, the trimmed margin of the earlier foredune, separated from it by a zone that remains unvegetated, and becomes a low lying swale (Figure 9.1). Often the seaward slopes of parallel dunes are steep as the consequence of the cliffing that preceded the formation of the next foredune, followed by slope degradation.

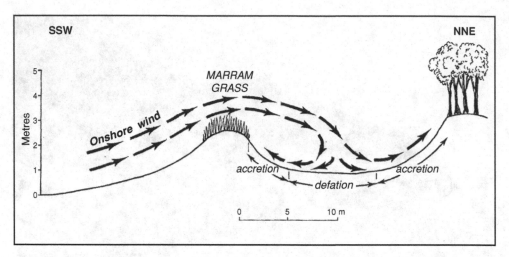

Figure 9.3 Onshore wind flowing over a grassy foredune develops eddies which may contribute to the shaping of the backing swale. Pattern determined by observing the flow of smoke from a fire on the beach

The formation of swales separating parallel foredunes may be aided by the occurrence of wind eddies in the lee of each developing ridge, the swales being zones excavated by the wind so that the root systems of dune vegetation are laid bare (Figure 9.3).

Parallel foredunes are well developed on parts of the North Sea coast of Britain, as at Winterton Ness in East Anglia, on the Oregon and Washington coasts in the United States, and in SE Australia, notably on the coastal barriers of East Gippsland.

Successive foredune formation parallel to a prograding coastline is well illustrated at Lakes Entrance in SE Australia, where sandy forelands have developed as the result of accretion of sand on either side of jetties built beside the entrance in 1889 (Figure 6.27). The pattern of beach and dune accretion in the ensuing century has been determined by ocean swell, refracted on encountering a sand bar offshore, so the coastline of each foreland is lobate in form. Three foredunes have formed successively, parallel to this prograding coastline, when beach accretion following storm scour provided foundations for wind-blown sand to be intercepted by colonis-

ing vegetation and build roughly symmetrical foredunes, each about 3 m high. The youngest of these was initiated on a sandy beach berm east of the harbour jetties in 1957 (Bird, 1978a).

The height and spacing of parallel foredunes is a function of the rate of sand supply to the shore, the history of cut and fill and the effectiveness of vegetation in binding sand and building the dunes. Where sand supply has been rapid on a prograding shore subject to frequent storms a large number of low, closely spaced parallel dunes are formed, but on a similar shore less exposed to storms the effects of backshore erosion, which is responsible for the separation of the dunes into parallel ridges, are less frequent, and there are fewer, but larger, parallel dunes. Where the sand supply has been meagre, parallel ridges are less likely to form on a stormy shore, but a few low parallel dunes may develop on more sheltered sections of the coast. The numerous precisely parallel dune ridges behind Disaster Bay (Section 6.9) on the South Coast of New South Wales have an amplitude of 1–2 m and are spaced at intervals of about 20–30 m. They consist of leached sand, bearing heathy woodland, and are bordered seaward by a higher (4–5 m) scrubby ridge that has been cliffed on the seaward side,

Attempts have been made to deduce changes in sea level from the heights of successive parallel dunes and swales. An overall seaward descent of dune crests and swales could indicate that the coast was emerging during their formation, but this pattern could also have been produced by a diminishing sand supply or increasing storm frequency, without any change in the relative levels of land and sea, and so cannot be accepted as conclusive evidence for coastal emergence (Davies, 1957). In the Netherlands, parallel dune ridges have formed during progradation on a coast where relative sea level has been rising.

Most parallel dunes have been formed by sequences of cut and fill on a prograding sandy shore, but there are exceptions. On the Falsterbo peninsula in Sweden there are parallel dune ridges that formed by wind-blown sand accretion on successively built longshore spits, and on South Haven Peninsula in Dorset three parallel dunes have developed since the 17th century on successive wave-built sand ridges with broad intervening low zones that have become occupied by lagoons and marshy swales. On the coast of Lake Michigan, Olson (1958) traced the evolution of successive parallel dune ridges to alternating phases of backshore dune development when lake level was falling and coastline recession when lake level was rising, overall progradation having accompanied these historical fluctuations in water level.

Continuing upward growth of a foredune gradually cuts off the supply of sand to its predecessor, which becomes relatively stable. The dune grasses are then invaded and replaced by scrub or heath vegetation, depending on the type of sand. On the coasts of the Bristol Channel, notably at Berrow in Somerset, recently formed calcareous grassy foredunes are backed by older dune topography with a rose and bramble thicket and buckthorn scrub (*Hippophae rhamnoides*), but where the dune sands are quartzose pedogenesis (soil formation) leads to the development of podzols. The upper layers are leached of shelly material by percolating rainwater, which also removes the yellow stain of iron oxides from the sand grains, leaving the surface sand grey or white in colour, while the lower layers are enriched by the deposition of iron oxides, together with downwashed organic matter, to form a slightly cemented red-brown sand horizon, known as humate or coffee rock (Section 9.14). The older dune ridges on the landward side show more advanced stages in vegetation succession, often with heath or heathy woodland on sand that has been deeply leached, as on the South Haven Peninsula in Dorset. In SE Australia parallel dunes of quartzose sand show a landward sequence from grassy foredunes through tea-tree (*Leptospermum laevigatum*) scrub to heathy forest, accompanying deepening podzolic soils, an association that indicates increasing age landward. Evidence from soil profiles across parallel sand ridges at Woy Woy, New South Wales, was used by Burges and Drover (1957) to assess the rate of podzol formation.

Parallel dunes may be modified by wave overwash during occasional storm surges, which damages or destroys vegetation and sweeps sand landward into fans or sheets. Ridges and mounds are shaped into more subdued forms at the same time as the seaward margins are cut back as dune cliffs. On the Gulf and Atlantic coasts of the United States dunes have been modified in this way by recurrent hurricane surges.

9.5 Blowouts and parabolic dunes

On some coasts there are unstable dunes. These include blowouts and parabolic dunes in otherwise vegetated dune topography, and unvegetated transgressive dunes moving downwind.

Mosaics of stable, vegetated dunes and unstable bare dunes may develop either as the result of partial stabilisation of drifting sand by the arrival or introduction of vegetation (notably the planting of European marram grass during the past few centuries) or as the result of disruption of formerly well vegetated dunes, as on the coast near Cape Arnhem in northern Australia, where dunes have been partially mobilised by the impact of trampling and grazing by introduced Asian buffalo.

Coastal dunes that have been stabilised by vegetation may subsequently be eroded and reshaped by the winds in areas where the vegetation cover has been weakened or removed, either naturally (for example by increasing aridity or strengthening wind action) or as the result of disturbance by human activities, so that sand is no longer held in position. Formerly vegetated dunes can thus become mobile sand sheets. Parallel dunes held by vegetation have in places been interrupted by blowouts, which are unvegetated or sparsely vegetated hollows excavated by onshore winds, with sand driven mainly landward to form a looped ridge. Local weakening or destruction of dune vegetation may be initiated by intensive and human activity, where footpaths are worn by people walking to a beach, or where trackways are by vehicles driven to and from the shore. Burning of coastal vegetation and excessive grazing by rabbits, sheep, cattle or goats can destroy the vegetation and initiate blowouts. The effects of rabbits are accentuated where they burrow into the dunes, disrupting the surface, as on Blakeney Point in Norfolk. Measurements by Rutin (1992) on Dutch dunes showed dune topography being modified by local accumulation of up to 0.5 m^3 of sand excavated from each rabbit burrow, and by subsidence as the rabbit burrows collapse. Grazing by introduced rabbits is thought to have depleted dune vegetation on the coast at the border between South Australia and Western Australia, mobilising dunes that advanced at up to 20 m/yr and overran

the township of Eucla in the 1940s (Jennings, 1963). Dunes and their vegetation are more stable where there are no rabbits, as on islands such as Flinders Island, Tasmania and the Chatham Islands, New Zealand,

Blowouts may develop naturally during a phase of aridity when the vegetation that had held them under preceding humid conditions is weakened, resulting in mobilisation of wind-blown sand. Dune erosion can also be initiated by stronger and more frequent wind scour associated with increasing storminess. Blowouts often form where the outer margin of vegetated coastal dunes is cut away by the sea during a storm, leaving an unvegetated cliff of loose sand exposed to onshore wind action. On a prograding shore rebuilding of the beach, with new berms developing into newer foredunes, prevents much erosion, but if the coastline is gradually receding, and the beach not fully replaced, such blowouts will continue to develop and grow, especially on parts of the coast that are exposed to strong winds.

A blowout that grows until its axial length is more than three times its mean width is termed a parabolic or U-dune, and has an advancing nose of loose sand (sloping at 30–33°) and trailing arms of partly fixed vegetated sand on either side of an axial corridor excavated by deflation (Figure 9.4).

Parabolic dunes have noses of bare sand that spill in the direction of the dominant wind. The

Figure 9.4 The formation of a blowout through a foredune

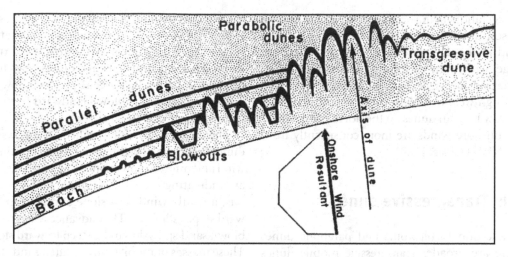

Figure 9.5   Development of blowouts in parallel foredunes and their growth into parabolic dunes and eventually transgressive dunes. The axis of a blowout or a parabolic dune runs parallel to the onshore wind resultant (Figure 6.17)

driving force of this wind shapes the parabolic form, which in turn modifies the near-surface wind flow. Analysis of wind flow patterns over a parabolic dune in the Sands of Forvie, Scotland, by Robertson-Rintoul (1990), demonstrated a crestal jet, windward and leeward eddies over the advancing nose and spiral vortices beside the trailing arms. These wind flow patterns result from parabolic dune morphology, but then shape its further evolution. Changes in mobile or partly mobile dunes thus represent an adjustment between wind flow characteristics and the surface morphology of incoherent sand.

Elongated parabolic dunes with axes trending SW–NE and W–E are also seen in Barry Links on Buddon Ness, a large cuspate foreland on the north shore of the Firth of Tay, east of Dundee in Scotland (Landsberg, 1956). On Braunton Burrows in North Devon there are blowouts and parabolic dunes, the outlines of which persist despite devegetation and damage during a phase of intensive military use.

Parabolic dunes may become stabilised beneath a vegetation cover (the term is retained when they no longer have active noses). Fixed parabolic dunes are extensive in SE Queensland,

notably on Fraser Island, where they trend SE–NW. Both active and fixed parabolic dunes are common on the SE coasts of Australia, and were described by Jennings (1957) from King Island. He showed that their movement was a response to the direction, frequency and strength of onshore winds, the axis of each parabolic dune, defined as the line bisecting the angle between the trailing arms and directed towards the advancing nose (Figure 9.5), running parallel to the resultant of onshore winds of Beaufort Scale 3 (12–19 km/hr) and over (Figure 6.17). On King Island the parabolic dunes on the west coast are moving inland eastward while those on the east coast are moving inland westward. The axial directions of numerous parabolic dunes cut through the foredunes behind the Ninety Mile Beach change as the curvature of the coastline brings in different component groups of onshore winds with different angles of onshore resultant (Bird, 1978a).

In North Queensland elongated parabolic (hairpin) dunes cut through previously fixed vegetated dunes have axes parallel to the prevailing SE winds. They continue to grow because the hollow immediately behind the nose is kept clear of vegetation, being occupied by a lake in the wet season and drying out to expose readily deflated sand in the dry season. In Tasmania there are large, wide parabolic dunes on the west coast, where onshore gales vary in direction from NW through west to SW, in contrast with the long, narrow hairpin dunes on the north coast, where the onshore winds are more consistently from the NW (Davies, 1980).

9.6 Transgressive dunes

In addition to blowouts and parabolic dunes there are broader transgressive mobile dunes that form where sand blown inland from a beach has not been retained by vegetation. Alternatively, they may form where previously vegetated coastal dunes have been disrupted by numerous blowouts until they merge into an elongated dune, spilling inland (Figure 9.5). Parabolic dunes retain their form as long as the trailing arms are held back by vegetation, but if trampling, overgrazing or fires reduce this vegetation the parabolic form evolves into transgressive sand sheets.

Transgressive dunes are extensive on desert coasts, as in Baja California, Namibia, the Atlantic coast of Morocco and along the shores of the Red Sea. They are also found in humid regions where the sand supply is rapid enough to prevent stabilisation by vegetation, as in the Slowinski National Park on the Polish Baltic coast (Borowca, 1990). At Parangtritis on the south coast of Java, where large quantities of sand arrive periodically on the beach as a sequel to eruptions of the Merapi volcano in the hinterland, dunes are blown inland by SE trade winds. On the Pacific coast of the United States large transgressive dune systems occur where there is an abundant supply of sand from beaches (particularly where the coasts swings westward, so that southward-drifting sand can accumulate on wide beaches) and frequently strong on-shore winds to blow sand inland (Cooper, 1958, 1967). These dunes are halted as precipitation ridges when they encounter forest vegetation. In SE Australia transgressive dunes may have originated during Pleistocene phases of low sea level, when sand was blown landward from the emerged sea floor.

There are also long-walled transgressive dunes that form at right angles to the dominant wind and migrate downwind. They usually have an undulating crest that is often slightly sinuous, a gentle windward slope and a steep leeward slope (slipface). They advance when wind-blown sand spills down the steep leeward slope. These masses of mobile, unvegetated sand show forms similar to dunes in arid regions, the dune crest profile at any time being a response to preceding wind conditions. A strong wind in one direction drives the dune crest one way, with a lee slope of up to 32°, but subsequent winds from other directions then modify the outline. Mobile dunes of this kind are seen on the Cape Coast in South Africa (Figure 9.6), where the predominant westerly winds are driving them eastward until they spill into Sundays River. Similar active dunes are seen on the coast of Uruguay near Montevideo and at Cape Howe in SE Australia. Dunes that drifted across headlands such as Cape Otway, Cape Patterson and Cape Liptrap in Victoria, Australia, in Pleistocene times are now largely stabilised by vegetation.

Sequences of transgressive dunes are found on the west coast of Auckland, New Zealand (Schofield, 1975). In Spain, successive waves of transgressive dunes are being driven inland by the prevailing SW winds on the shores of the Gulf of Cadiz (Vanney, Menanteau and Zazo, 1979). The Pyla dune, in SW France, is a huge mass of sand moving from the eroding shore south of the Bassin d'Arcachon, and spilling inland over the Landes pine forests. It was initiated during the 18th century, and has now attained an altitude of more than 100 m, with a steep wall (32°) of spilling sand. A dune cliff cut by tidal scour

Figure 9.6 Mobile coastal dunes drifting along the coast of Algoa Bay, east of Port Elizabeth, South Africa

on the estuary shore, resulting from deflection of the ebb channel by the southward growth of Cap Ferret spit, maintains the sand supply to this transgressive dune.

Extensive unvegetated sand areas may include barchans, which are sand mounds with lateral arms trailing downwind, similar to those that form inland in deserts. They are typically up to 30 m high, with a steep advancing leeward slope, and the trailing arms are up to 350 m apart. On the arid coasts of Baja California and SW Peru barchans move downwind at up to 30 m/yr (Goldsmith, 1985). Barchans are also found on coasts in humid regions where there are extensive areas of bare, drifting sand, as at Yanakie, near Wilson's Promontory, Australia, where the climate is cool and mean annual rainfall about 900 mm. Small barchans sometimes form during gales on wide sandy beaches and unvegetated backshore sands, migrating downwind (onshore, seaward or alongshore). They have been described from the south coast of

Honshu in Japan and the west coast of Schleswig-Holstein in Germany and observed forming on Tatlow Beach at Stanley in Tasmania during a gale.

It is sometimes possible to distinguish successive waves of transgressive dunes that have migrated inland from a shore. In Newborough Warren in North Wales three parallel transgressive sand ridges have drifted inland, each backed by low lying sandy terrain on which zones of vegetation of increasing age upwind represent stages in plant succession (Ranwell, 1972). Similar vegetation zones mark stages in the migration of Seagrens Dune, an elongated parabolic dune behind Cape Bedford in NE Queensland.

On the Sands of Forvie in Scotland seven roughly parallel sand ridges up to 30 m high have migrated to the NE, one of them having buried Forvie chapel in the 15th century. They originated from an area of bare sand at the mouth of the River Ythan, but it is not clear how the successive waves were initiated (Ritchie, 1983).

Successively formed dune ridges, separated by rather broad low areas (i.e. not just narrow swales) have migrated inland behind Newcastle Bight and several other bay beaches along the New South Wales coast, and similar sequences have been described from Morro Bay, California, by Orme (1990). Broad low lying plains can be formed within dune terrain by deflation of sand, usually to the level of the water table, sometimes with minor dunes (residual or newly formed) over the moist sand, or strewn with coarse residual lag deposits.

On some coasts dunes formerly transgressive dunes have been stabilised by vegetation, particularly marram grass planted during the past century. During phases when larger quantities of sand were arriving, or when the vegetation was sparser, dunes were bare and mobile, drifting inland, and up and over headlands. On the north coast of Cornwall some dune systems extend below present sea level, and the wind-blown sand was derived from the emerged Atlantic sea floor during Pleistocene times, before and during the Holocene marine transgression. Dunes then spilled inland, across what has become the present coastline. On the north coast of Devon and Cornwall formerly transgressive dune topography extends up to 2 km inland from the high tide line at Braunton Burrows and Gwithian Sands. Subsequently, drifting dunes buried farms, villages and churches behind St Ives Bay and Perran Sands in Cornwall and at Kenfig Burrows in South Wales, where there is archaeological and documentary evidence that sand dunes were advancing inland in mediaeval times. After earlier stability, sand drifting was unusually active in these areas in the 13th and 14th centuries because of a phase of stormier climate or a minor sea level oscillation that sharply augmented the coastal sand supply. Earlier dune instability is indicated at Skara Brae, in the Orkney Islands, where dunes overran an Iron Age settlement 4 500 years ago. Mobilisation of extensive eastward-drifting dunes at Culbin Sands,

NE Scotland, was attributed to the uprooting of marram grass for roof thatch, baring the surface for massive deflation in a 1694 storm. Marram grass removal was then halted, and the dunes were stabilised by afforestation with pine trees, which began in 1839. In 1922 the Forestry Commission laid birch litter and planted Scots and Corsican pines to create Culbin Forest on the stabilised dune topography (Steers, 1973).

The very high dunes of the coast of SE Queensland, on Fraser Island, Stradbroke Island and Moreton Island (where at one point they exceed 275 m above sea level), are not piled over rocky foundations. They consist entirely of wind-blown sand stabilised by vegetation in parabolic dune patterns, and borings have shown similar aeolian sand extending well below sea level. There are several sets of transgressive dunes, arranged in sequence parallel to the ocean coast, each partly overlapping its predecessor. These have advanced away from the shore during phases of instability and dune migration related to Pleistocene phases of aridity, strong SE wind action and a superabundant sand supply from the emerged sea floor during low sea level phases. They are now stabilised beneath a cover of scrub and forest, and show a landward sequence of deepening podzolic soils with increasing age, similar to that seen across parallel dunes (Section 9.4). The Cooloola dunes south of Fraser Island also show this sequence, with older subdued dunes on the landward side dissected by rain and rivers to hilly topography (Thompson, 1983).

On the New South Wales coast there are several sectors where transgressive dunes of mobile, unvegetated sand are migrating inland and burying older dune topography with a scrub or forest cover. Some of these active dunes were initiated by human activities, notably the grazing of stock and the burning of vegetation, liberating unconsolidated dune sand on the coastal margin. Some blowouts and transgressive dunes originated prior to European colonisation, for Captain Cook observed active dunes on the New South Wales coast and the Queensland coast north as far as Fraser Island in 1770. Aborigines may have initiated these when they repeatedly set fire to vegetation in order to hunt animals, and this weakening of the vegetation cover assisted the formation of blowouts.

Dissection of previously vegetated dune topography can result in the formation of residual mounds (knobs) of vegetated sand, crowned by a clump of grass or shrubs, within areas of bare and drifting sand, as in the dunes behind Discovery Bay in SE Australia. Lee or shadow dunes may form downwind of such remnant knobs, or vegetated mounds. Goldsmith (1985) mentioned a distinctive type of unvegetated dune, known as a medano, which is steep sided and tens to hundreds of metres high, formed where winds blow from several directions, moving sand towards the summit. Such a dune changes in shape in response to differing wind directions, but shows little if any migration. Examples have been described from Coos Bay, Oregon and SE Lake Michigan. If colonising vegetation prevents the arrival of sand from one direction the medano is shaped into an elongated ridge.

The effects of human activities on mobile coastal dunes include accretion induced by sand fence construction or the planting of vegetation and erosion resulting from trampling, grazing, burning and sand mining. In the United States dune areas have been stabilised by road construction and urban development, as on the New Jersey coast, where Gares (1990) compared processes and morphology on urbanised and undeveloped dunes, and found that wind flow patterns and the net landward movement of sand were much reduced by obstructions such as houses, roads and boardwalks, as well as by sand-trapping fences. Such urbanisation may stabilise dune topography, but it can be threatened by the recession of bordering coastlines accompanying beach erosion.

9.7 Cliff-top dunes

On some cliffed coasts onshore winds have blown sand from the beach and piled it against the cliff as a climbing dune. At Foreness Point in Kent wind-blown sand has been banked against the Thanet Chalk cliffs during the past few decades. Such dunes may grow to spill inland over cliffed headlands, and where there is no longer a source of sand they become relict cliff-top dunes, as at Church Cove and Gwithian Towans in Cornwall.

On parts of the Australian coast dunes have climbed up and over coastal promontories, as on Cape Bridgewater and Cape Paterson on the Victorian coast, where they form cliff-top dunes on the lee side. Jennings (1967) showed that some cliff-top dunes formed when the sea stood at a higher level while others (notably on dune calcarenite coasts) originated as transgressive dunes moving in from the emerged sea floor during Pleistocene phases of lower sea level on to a coast that later became cliffed (Figure 9.7).

Dunes found on the top of a cliff may not have come from nearby beach sands. In arid regions some cliffs have been cut back into terrain that carries desert dunes formed by deflation of unvegetated or sparsely vegetated landscapes. There are cliff cappings of red (pindan) desert sands in NW Australia. On the coast of Port Phillip Bay in SE Australia, the grey dunes that cap the cliffs in the bayside suburbs of Melbourne are part of a series of elongated parallel dune ridges that formed on the coastal plateau as desert dunes during a Pleistocene arid phase.

9.8 Dunes on shingle

Similar explanations may apply where are dunes on shingle beach ridges with no existing link with a sand supply from an adjacent beach. Sand may have arrived as a transgressive dune spilling from a nearby beach, possibly when sea level

Figure 9.7 Cliff-top dunes may form (a) where a transgressive dune moves up a cliff, and is later truncated by marine erosion, (b) where a dune formed at a higher sea level is stranded by cliff recession after emergence, (c) where a dune formed when sea level was lower is truncated by marine erosion as the sea rises and (d) where the dune has moved from inland to a position on the cliff top. After Jennings (1967)

was higher or lower than it is now. The grassy Northam dunes on the northern end of the Westward Ho! cobble spit probably formed when there was a contiguous sandy beach from which westerly winds supplied sand, but the sand supply from the beach diminished, the connection was broken and the dunes are now relict. Similar relict dunes stand on shingle beach ridges at Blakeney Point and Scolt Head Island in Norfolk, the link with the beach sand source having been broken by the accretion of younger shingle ridges (Steers, 1960). At West Beach, Littlehampton, sand is winnowed from low tide sandflats, but observations during an onshore gale showed

that it accumulated along the base of the shingle beach, and did not reach the grassy backshore dunes. Sand could not drift up across the shingle until it had filled the cavities between the pebbles.

9.9 Rates of dune movement

Dune movement is often impeded by vegetation or obstacles, including natural rock outcrops, but where there is unimpeded drifting measurements of dune movement have indicated that unvegetated dune fronts 10–20 m high can

Table 9.1 Components of dune calcarenite

---

1. Dune sandstone, often slightly hardened in biscuit-like layers that follow the dune bedding.

2. Calcrete, a generally undulating layer of hard white limestone that formed on the surface of a stabilised dune, and was later buried by younger dunes.

3. Palaeosols, associated soil horizons of sandy clay up to 3 m thick, also buried by younger dunes.

4. Concretionary structures indurated by the precipitation of carbonates from groundwater drawn upward by plant roots and capillary action. These include columns of cemented dune sand (calcrete) formed around roots (rhizoconcretions) or stems (phytoconcretions) and vases formed by sand cementation beneath trees or shrubs. Subsequent erosion may expose these as 'petrified forests'.

5. Gravel formed by disintegration of calcretes (hamada), locally cemented into breccias.

6. Unconsolidated sand, usually in horizons capped by a calcrete layer.

---

advance downwind at rates of 1–10 m/yr. The 100 m high dune at Pyla in SW France is said to be advancing inland at an average rate of 1 m/yr. Measurements on dune fronts at Cronulla, New South Wales, showed that a dune 20 m high advanced 19 m, and smaller dunes as much as 50 m, in a year (Chapman *et al.*, 1982).

Rapid rates of dune advance occurred in mediaeval times in Europe, possibly because of increased storminess, but perhaps because of increased clearance and impoverishment of dune vegetation. Records of advancing dunes burying buildings, villages and farmland imply that the sand advanced rapidly over a few decades.

9.10 Dune calcarenite

Where coastal dune sands are calcareous the older dunes may become lithified by internal precipitation of calcium carbonate from percolating water, to form dune calcarenite, which preserves the dune topography in solid limestone. It was originally observed in 1791 from King George Sound in Western Australia by Captain Vancouver, who thought it a coralline formation, but when Charles Darwin visited here in 1844 he correctly reported that it consisted of wind-blown sand and associated limestone.

It was later described and named (as eolianite, the American spelling) in Bermuda by Sayles (1931). The term calcarenite is strictly applied when the carbonate content of the sand exceeds 50 per cent, but dune sands can become coherent and lithified when the carbonate content is as low as 10 per cent.

Components of dune calcarenite are listed in Table 9.1. Dune calcarenite is usually of Pleistocene age, and may be overlain by unconsolidated Holocene dunes. These formations are often exposed in coastal cliffs cut in dune calcarenite, as at Jubilee Point, Victoria, Australia (Figure 5.12).

Dune calcarenite may be formed from dunes built behind beaches of calcareous sand by onshore winds, but it often extends below sea level, and includes dunes derived from biogenic sands deposited by wind action on the emerged sea floor during low sea level phases of the Pleistocene. Transgressive dunes then drifted in over what is now the coastline and spilled some way inland.

Some calcareous dunes remain relatively unconsolidated, and can be mobilised by wind action if surface vegetation (or an overlying calcrete crust) is removed. Others become consolidated as calcareous sandstone. Dune calcarenites vary in degree of lithification by precipitated carbonates, which in turn depends on leaching of carbonates from overlying calcareous sands in wet phases alternating with periods dry enough to precipitate them. Some carbonates may be brought up in rising groundwater, especially where high evaporation at the surface promotes upward capillary movement between the sand grains, and this may result in the formation of calcrete by precipitation on the dune surface.

Dune calcarenite topography is found on the coasts of the Mediterranean Sea, in Morocco, on the Red Sea and southern Arabian coast, in western India, around the Caribbean, on the Brazilian coast, in South Africa and extensively on the western and southern coasts of Australia. It also occurs on oceanic islands such as Mauritius and Hawaii. Often the dune calcarenite has been deposited in front of and over much older rock formations, such as the granites of western and southern Australia, which have been exposed locally by subsequent erosion. Residual patches of dune calcarenite may be left adhering to older coastal rock outcrops.

On the Atlantic coasts of Britain Holocene dunes derived from beaches of calcareous sand contain lightly cemented horizons, but true dune calcarenite has been observed on Balta Island, a small island off the east coast of Unst in the Shetland Islands, and behind Portsalon Bay in County Donegal, NE Ireland. There is a useful review by Kaye (1959).

Although dune calcarenites originally formed as dunes behind beaches of calcareous sand, many are no longer beach fringed, and may even have been cut into cliffs, bordered by shore platforms (Section 5.2.4). The dune calcarenites of the Nepean Peninsula in Victoria, Australia, comprise a series of superimposed Pleistocene and earlier dune formations that extend more than 140 m below, and up to 60 m above, present sea level, a notable example of coastal dune stratigraphy. They are largely the product of winnowing of sand from the emerged floor of Bass Strait during Pleistocene low sea level phases, and of shoreward drifting of sand during the Holocene marine transgression. This coastline now consists of cliffs and shore platforms cut into the consolidated Pleistocene dune calcarenites, and there are unconsolidated cliff-top Holocene dunes (Figure 9.8).

The contrast between generally calcareous beach sands on the western and southern coasts of Australia and predominantly quartzose beach sands in the SE is reflected in coastal dune morphology. There is relatively stable Pleistocene dune calcarenite topography on the western and southern coasts of Australia, but on the SE coast, where the Pleistocene dune sands are quartzose, this kind of lithification has not taken place and the dunes remain unconsolidated, either active and mobile or retained by a vegetation cover. In South Australia there is a longshore transition in the composition of dune sands SE from the Lower Murray, the proportion of terrigenous quartzose sand diminishing as the proportion of marine calcareous sand increases to the SE. The strongly calcareous dune ridges to the SE (Figure 3.4) are preserved in solid dune calcarenite, but as the sand becomes less calcareous lithification diminishes and the limestone ridges grade NW into more irregular and disrupted dune topography (Sprigg, 1959). On the eastern coast of the Mediterranean there is a similar transition, the calcarenite dune topography of the Israel coast passing southward into more irregular, mobile dunes on the coast of the Gaza Strip, where the sands have an increasing content of terrigenous (Nile-derived) quartzose sediment.

The submergence and dissection of Pleistocene dune calcarenite topography formed during low sea level phases is well known off the coast of Western Australia, where the Holocene marine transgression isolated Rottnest Island and Garden Island as it invaded Cockburn Sound and submerged the lower part of the Swan River valley. Some submerged ridges have been planed off by wave action on the sea floor, but

Figure 9.8 The evolution of dune calcarenite cliffs and shore platforms: (a) when sea level is falling dunes are stranded on the emerged sea floor in front of an earlier dune calcarenite coast; (b) when sea level rises these dunes are submerged and eroded, and sand is blown landward across the present coastline; (c) when the sea has reached its present level it cuts a cliff and shore platform in the dune calcarenite, capped by Holocene dunes. The dune calcarenites were formed in earlier such cycles as sea level oscillated in Pleistocene times

detailed soundings have shown that segments of dune topography survive off the southern and western coasts of Australia. There are also submerged dune calcarenite ridges off the Bahamas, NE Brazil and Sri Lanka, and some of these have become reefs with coral and algal crusts.

9.11 Machair

On the coasts of Scotland and Ireland there are areas of almost featureless low lying calcareous sandy terrain known as machair (Ritchie and Mather, 1984). Typically these bear herbaceous pastures, and lie behind a vegetated coastal dune fringe, as at Eoligarry on the northern part of the Hebridean island of Barra (Figure 6.8). Morrich More in Easter Ross is a low lying sandy foreland on the southern shores of Dornoch Firth, with developing foredunes behind Whiteness Sands, backed by fixed and mobile parabolic dunes, then by machair.

Machair is generally of Holocene age, much of it formed between 6000 and 2000 years ago (Ritchie, 1977) although there are sites where it may still be developing. It may be associated with areas of Holocene coastal emergence, for although it is extensive in Scotland and Ireland there is no equivalent behind the calcareous dunes of SW England, although thin splays of fine wind-blown sand are sometimes seem in the lee of such dunes, as at Bantham in South Devon. Machair can form where a coastal dune has been removed down to the level of the water table, as at Trawenagh Bay in County Donegal, Ireland, or where sand deflated from coastal dunes is deposited in lagoons or marshes. This is possible behind Machir Bay on Islay, and behind Traigh na Berie on the island of Lewis, where the beach is backed by dunes, which impound two small lochans, towards which the machair becomes marshy. Often machair is undergoing dissection by blowouts and gullies, as on Gualann Island on the west coast of South Uist and Balta Island in the Shetlands and behind Torrisdale Bay in NW Scotland (Gimingham *et al.*, 1989).

9.12 Coastal dunes in the humid tropics

Coastal dunes are best developed on coasts in the temperate and arid tropical zones. Jennings (1964, 1965) noted that in the humid tropics they are of limited and local extent, sandy coastal topography consisting of low beach ridges with little transgressive dune development. It has been suggested that the prevailing dampness of beach sands on humid tropical coasts impedes deflation and backshore dune development, particularly on microtidal sectors where the beaches are narrow. Another suggestion is that rapid colonisation of coastal sand accumulations by luxuriant vegetation prevents dune development in the humid tropics, but it is doubtful whether the vegetation has much effect: if aeolian sand were arriving at the backshore, dense vegetation would simply trap it and facilitate the building of a high foredune. Coconut palms and *Casuarina* trees often grow immediately behind the beach, forming a wall of vegetation, but this is partly because of the lack of strong onshore winds, and it is this lack, rather than the presence of vegetation, that explains the absence of coastal dunes (Davies, 1980).

The paucity of dune development in the humid tropics is thus a consequence of a relatively meagre sand supply to the backshore, largely because winds strong enough to deflate beach sand and build up dunes are rare in comparison with other climatic zones. The occasional violence of winds in tropical cyclones is usually accompanied by torrential rainfall that saturates the beach surface and impedes sand transportation by the temporarily strong wind action. The formation of salt crusts on beaches in Ghana has impeded the development of coastal dunes.

Nevertheless, dunes are found on humid tropical coasts where there is a sustained sand supply and prevailing onshore winds, as at Parangtritis in southern Java. In Fiji coastal dunes are poorly developed except in the SW, near Singatoka, where black magnetite sands brought down by the Singatoka River have been washed up by the strong SW swell to form beaches that are a source of backshore dunes. Local development of dunes occurs where there is a dry season, as on parts of the Malaysian and Sri Lankan coasts, and dunes persist on humid tropical coasts where they have been inherited from earlier drier phases, as in NE Queensland.

9.13 Old and new dunes

Some parts of the coast show evidence of distinct phases of coastal dune accumulation, with old stabilised dunes on the landward side and new dunes, either stable or mobile, bordering the coast. This pattern is clearly developed where the dunes are built of quartzose sand, as on King Island in Bass Strait, Australia (Jennings, 1957), where there are contrasts between the topography, soils and vegetation of old and new dunes. Old quartzose dunes (usually of Pleistocene age) have a comparatively subdued topography, and have been leached of shelly fragments and other calcareous sediment by percolating rainwater, sometimes to a depth of a metre or more, as podzol profiles develop. They have sometimes been called grey dunes because of the colour of the leached surface sand, in contrast with the yellow new dunes, where the surface sand still retains a colouration with sand grains coated by iron oxides.

New dunes (usually of Holocene age) form a coastal fringe, and are more continuous, with bolder outlines, and accretion of quartzose sand often still continuing. They are fixed by grasses or scrub, except where blowouts and parabolic dunes are developing, or where large transgressive masses of mobile sand are advancing inland. The sand is fresh and yellow or brown in colour, and has not yet been leached of its small shell content or of the iron oxides that stain the sand grains.

The junction between old and new dunes is often well marked, particularly where the new dunes are transgressive, advancing across the more subdued old dune topography. Coastal barriers of quartzose sand show a similar contrast, inner barriers (generally of Pleistocene age) having old dune topography while outer barriers (usually added after the Holocene marine transgression) bear new dunes. In Oregon, the inner, older gently undulating and now vegetated transgressive dunes were formed before the Holocene marine transgression brought the sea to its present level, when younger Holocene foredunes were added along the coast (Cooper, 1958).

A contrast between old and new dunes is also seen on calcareous dunes, where the old dunes are preserved in dune calcarenite, and new dunes remain relatively unconsolidated, but the contrast in vegetation is less marked, the old dunes having only a superficial layer of sand decalcified by percolating rainwater.

While old quartzose dunes are usually of Pleistocene age there are occasional examples of grey dunes of Holocene age with podzolic soils bearing heath vegetation. On the shores of Wilson's Promontory, Australia, beaches derived from weathered granite consist almost entirely of grey quartzose sand without an iron oxide staining, and in Leonard Bay Holocene dunes derived from these are already grey, and have been podzolised beneath heath vegetation. On South Haven Peninsula, Dorset, heath vegetation has spread on to quartzose Holocene dunes where leaching of sparse carbonates and iron oxide coatings has proceeded rapidly.

The sequence of old (Pleistocene) dunes inland or on inner barriers, and new (Holocene) dunes on the coastal fringe or on outer barriers, is widespread in Australia, but not in Britain or NW Europe. Pleistocene coastal dunes are found in the Naples region in Italy and on the Landes coast in SW France. Further north it is possible that they were largely removed by glacial or periglacial processes during the Last Glacial phase, but in SW England old dune sands are found overlying emerged Late Pleistocene shingle beaches in coastal outcrops at Godrevy Rocks, Trebetherick Point and Fistral Beach in Cornwall and Saunton in Devon. In each case they are capped by periglacial Head deposits and yellow Holocene dune sand. The Pleistocene dune deposits are here beneath, rather than landward of, Holocene dunes.

9.14 Dune sandrock

Podzolic soil profiles form on dunes where iron oxides and organic matter leached from the surface are washed down to the subsoil and deposited as cemented layers. This lightly cemented sand is known as humate, and is commonly found in Pleistocene coastal barriers of quartzose sand. Humate is a form of sandrock that can be exposed as cliffs or backshore ledges of slightly more resistant material where the dunes have been dissected by wind action or trimmed back by the sea. There are humate cliffs where an inner barrier is truncated by the coast at Evans Head in New South Wales, and on the east coast of Chatham Island, New Zealand. Similar ledges may be formed of sandrock that originated where sandy swamps or peaty sand in low lying seasonally or permanently waterlogged sites have been overrun and compressed beneath advancing dunes, as at Rainbow Beach in SE Queensland. This type of sandrock often contains compressed plant remains, which are not found in humate.

9.15 Dune lakes

Hollows in dune topography that pass beneath the level of the water table are occupied by dune lakes, some of which may be intermittent, forming only when the water table rises after heavy rains, and drying out subsequently. They are also known as window lakes. Dune lakes are usually round or oval in shape as the result of wave and current action generated on them by wind action, the resulting configuration being related to the wind regime. Some lakes have been impounded where dunes built across the mouth of a valley have ponded back the stream, while others occupy hollows excavated by deflation during dry weather, particularly between the trailing arms of parabolic dunes, as on the Queensland coast north of Cooktown, where they provide a sand source for dunes downwind from the lake basin. Alternatively, dune lakes may form along the inner edge of a transgressive dune where it abuts a rising hinterland, as at Bridgewater Lakes near Portland, in SE Australia.

There are good examples of perched dune lakes within the high dune topography of Fraser Island in Australia, where there are more than 40 of them, standing at various levels in depressions where the presence of underlying impermeable humate (sandrock) prevents the water draining down through the sand. Several have been impounded in hollows in the old dune topography enclosed by the advance of transgressive newer dunes. Where a dune coast has been cut back by marine erosion, peat deposits that formed in dune lakes may be exposed overlying, or interbedded with, dune sands in cliff sections, as at Ocean Beach, near Strahan in Tasmania.

Infilling of dune lakes by blown or inwashed sand or by the formation of peat deposits may produce flat floored enclaves within a stabilised dune topography. Similar flat floored enclaves may be formed by erosion rather than deposition, where the wind has blown away dry sand to expose flat wet sand (or a humate outcrop) at the level of the water table. If the water table falls the lake will dry out, leaving a flat floored basin.

9.16 Summary

Coastal dunes are generally supplied with sand blown from adjacent beaches. They include foredunes, where wind-blown sand has been trapped by vegetation behind a beach. These may be cliffed by storm wave erosion, and subsequent formation of more foredunes may result in the evolution of parallel dunes. Breaching of the vegetation cover leads to the formation of blowouts, which may grow into larger parabolic dunes with advancing noses of bare sand and trailing arms retained by vegetation.

Where backshore dunes are not retained by vegetation (or where the vegetation cover is destroyed) they drift downwind as transgressive dunes. Cliff-top dunes originate in various ways, and dunes behind shingle beaches may have been cut off from their sand source.

Dune calcarenite is formed where calcareous dunes become lithified by carbonate precipitation. There are often superimposed calcarenite formations containing calcrete layers and palaeosols marking former dune surfaces, and dune calcarenites may be durable enough to form cliffs and shore platforms on eroded coasts.

Machair is a landscape of calcareous dunes (and associated vegetation) well known in Scotland and Ireland.

Dunes are poorly developed on humid tropical coasts, largely because of generally light winds and meagre sand supplies. A distinction is often clear between older and newer dunes, particularly on quartzose sands where podzolic soils develop.

Dune sandrock (humate) forms where the sand is bound by organic deposition, and may develop beneath lakes that occupy hollows in dune topography.

10

Intertidal wetlands

10.1 Introduction

The intertidal zone, exposed as the tide falls, may be rocky, gravelly, sandy or muddy. It includes vegetated areas such as seagrass beds, salt marshes, mangrove swamps and reedswamps, which are termed wetlands, subject to frequent submergence by the sea. They are extensive on macrotidal and megatidal coasts, and particularly within relatively sheltered bays and inlets, and in estuaries and lagoons (Chapter 11). Surveys of these areas are inevitably complicated by the fact that their features are partly or wholly submerged as the tide rises. Theodolite and staff are then put aside, and use made of soundings (including echo-soundings) to survey the submerged morphology in relation to a selected datum. In recent years use has been made of remote sensing with high resolution satellite imagery and light sensing and ranging (LIDAR) techniques for intertidal and nearshore surveys (Finkl, 2005).

10.2 Intertidal morphology

Examples of coasts with extensive intertidal zones include Westernport Bay (Figure 10.1) in SE Australia, a landlocked embayment linked to the sea (Bass Strait) by wide entrances east and west of Phillip Island. The tide is augmented as it passes into Westernport Bay, so that tide range increases from about 2 m in the entrances to more than 3 m north of French Island. As a result, intertidal morphology becomes increasingly tide dominated (Section 2.3.4) in the upper part of the bay.

As the tide falls in Westernport Bay the waters subside into diverging steep sided intertidal creek systems on either side of a tidal divide, and extensive intertidal mudflats are exposed (Figure 10.2). The intertidal creeks on the NW side show incision with waterfalls at heads of rejuvenation and are extending headward, whereas those on the SE side are becoming broader and shallower as the result of infilling and are being truncated. The tidal divide is thus migrating SE. The coasts bordering this tidal divide have high tide shorelines that are partly cliffed, partly sandy and extensively mangrove fringed.

Westernport Bay is essentially marine, but shows estuarine features at the mouths of Bass River and a number of smaller streams. Somewhat similar features are seen in Poole Harbour, Dorset, where the intertidal mudflats front salt marshes. Southampton Water, Portsmouth Harbour, Langstone Harbour and Chichester

Coastal Geomorphology: An Introduction, Second Edition.   Eric Bird.
© 2008 by John Wiley & Sons, Ltd

Figure 10.1 The intertidal zone in Westernport Bay, Australia

Harbour on the south coast of England were formed by Holocene marine submergence of low lying parts of the Hampshire and West Sussex coastal plain. They have extensive intertidal mudflats and backing salt marshes, and have persisted because the small inflowing rivers have not supplied much sediment and because this is a tectonically subsiding coast.

Intertidal mudflats and salt marshes are extensive in Bridgwater Bay, the coast of the

Figure 10.2 The tidal divide in Westernport Bay, Australia (location indicated by arrows in Figure 10.1). The shaded area consists of mudflats, apart from a sand bar (S) washed up from the SE

Bristol Channel, the Bay of Mont Saint Michel in France and the Bay of Fundy in eastern Canada. In the tropics, where salt marshes give place to mangrove swamps, intertidal mudflats occupy macrotidal bays, as on the north coast of Australia in Cairns Bay, where they are backed by a sandy beach ridge plain.

Some features of the intertidal zone have already been discussed: rocky and bouldery shores and shore platforms in Chapter 5 and sandflats, bars, troughs and ripples in Chapter 8. Details of the morphology and sediment distribution in intertidal areas are often poorly displayed on coastal maps and charts. In Britain, for example, Ordnance Survey maps show the high and low water lines, with shaded sand areas (which may include dunes) and marsh symbols (but not muddy areas), but supplementary surveys are needed to document the pattern of intertidal sediment types.

Intertidal areas consisting largely or entirely of muddy deposits are often called mudflats, even though they usually have a gentle transverse gradient seaward (e.g. 1:1000), often with a smooth, convex profile, and include relatively steep slopes, especially where they decline into tidal channels. They are generally contiguous with the coastline, extending from the high tide shoreline down to the low tide shoreline, but they also occur as shoals separated by deeper channels or straits. In places mudflats may have undulating surfaces with bars, troughs and shallow tidal channels, much like those seen on sandflats.

10.3 Sources of intertidal sediments

There are several possible sources of sediment (mainly silt, clay and organic matter, sometimes with a little fine sand) deposited in the intertidal zone. Some is delivered to the shore by rivers and distributed along the coast by wave and current action, some originates from the erosion of bordering cliffs (including alluvial coastal plains and deltas) and some drifts in from the sea floor. Off the West African coast submerged deltas are being reworked by waves to produce mud that makes the sea turbid and is washed onshore to build mudflats.

Fluvially supplied sediment is extensive on intertidal shores bordering deltas, off river mouths and in estuaries and lagoons. The type of sediment depends on the rock formations that outcrop within a river catchment. Sandy deposits are derived mainly from weathered granites and sandstones and silt and clay from outcrops of fine grained sediments, notably mudrocks. Gravels may be carried downstream from older gravels or from intricately fissured rock formations. In the humid tropics rivers carry vast quantities of mud derived from deeply weathered hinterlands to the coast, to be deposited in intertidal areas.

Cliffs cut into soft sand (including coastal dunes) and clay formations (including boulder clay) are sources of intertidal sand and mud around the North Sea. Muddy sediment derived from the clay cliffs of Holderness and NE Norfolk has been dispersed by waves and currents, and deposited in The Wash and the North Sea estuaries.

Sand has been swept in from the sea floor to beaches and sandy shores (Section 6.4.3), and muddy intertidal deposits can also be supplied in this way, particularly where there are offshore outcrops of weathered mudrock or fine grained bottom deposits such as glacial boulder clay. Muddy sediment from glacial drift on the floor of the North Sea has been delivered to intertidal areas in the Wadden Sea on the coasts of Denmark, Germany and the Netherlands, and has contributed to mudflats in eastern England.

10.4 Mudflats

Muddy sediment is widely dispersed by waves and currents. Strong currents keep silt, clay and organic matter in suspension, but as they slacken deposition begins, and mudflats are formed in the intertidal and nearshore zones. Where mudflats remain unvegetated they consist of an almost flat or gently sloping terrace bordered by steeper slopes descending seaward or into tidal channels. This represent a moulding of the intertidal morphology in relation to the ebb and flow of tides, the steeper slope being in the zone where the tide rises or falls quickly and the terrace in the upper intertidal zone where the rise and fall is more gradual.

Mud is deposited particularly at high and low tide slack water in the upper and lower parts of the intertidal zone, the intervening mid-tide zone having more sustained current flow as the tide rises and falls, inhibiting mud deposition. This can result in a steeper lower slope, which may be sandy, as in The Wash in eastern England. Wave action erodes and sorts intertidal sediment as the tide rises and falls, and mud taken into suspension is deposited in calm sheltered water as the waves weaken. While the flow and ebb of tides produce net shoreward drifting and a shoreward reduction in grain size from sand to silt and clay, mudflats are sometimes backed by a beach of coarser material (sand and shells) emplaced by wave action that is often stronger at high tide, and may even cut low backshore cliffs. On many coasts they rise to a landward fringe of salt marsh or mangrove swamp.

Deposition of muddy sediment in the intertidal zone is impeded by frequent tidal oscillations and disturbance by wave action, but can occur as the result of flocculation, due to the electrolytic effect of mainly sodium and chloride ions in brackish estuarine or seawater (Nichols and Biggs, 1985). This results in the clustering and coagulation of clay particles into flocculated silt-sized particles that settle more readily during brief episodes of diminished current flow and wave turbulence to form muddy shoals and intertidal mudflats. Mud deposition is also assisted by the effects of marine organisms, which ingest clay and organic matter and excrete it as pellets large enough to be precipitated on mudflats, and also generate sticky mucus that aggregates fine particles. Often the fine sediment is initially fluffy (fluid mud), but becomes consolidated as pore water escapes. Once deposited, mudflats are more cohesive than sandflats because of their high water content, electrolytic binding and associated organic stickiness. They are not as readily mobilised by waves and currents as sand grains, and stronger currents are required to disrupt and disintegrate them into particles that can be rolled or taken into suspension.

In estuaries, mudflats curve down to gentle slopes exposed at low tide on either side of a deeper channel, steepening as they decline to low tide level. They are moulded largely by current action as the tide rises and falls, but wave action can also influence their morphology. As deposition continues, vertical accretion may be accompanied by progradation of the seaward margins of mudflats, as on the shores of the Gulf of Bo Hai, served with muddy sediment from the Hwang Ho River, and in Turnagain Arm, Alaska, where progradation of up to 12 m/yr results from the inflow of sediment from the melting Portage glacier (Klein, 1985).

The surfaces of intertidal mudflats have a variable topography subject to frequent, rapid changes in response to strong wave and current action (Pethick, 1996). Deposition gives place to erosion as current flow exceeds about 10 cm/sec (when shear stress exceeds shear strength), and there are alternations between slack water deposition and mid-tide ebb or flood current scour. Strong tidal currents produce irregularities, such as mounds and banks separated by hollows and troughs. Mudflats are thus reshaped as the tide rises and falls, particularly during spring tides, and modified in stormy periods when wave action is strong. Their surfaces are generally smooth: the ripples found on sandflats fade out as sandflats pass laterally through muddy sandflats to sandy mudflats. However, mudflat surfaces are often disturbed by marine organisms, including burrowing worms and shells. In the tropics, mudflats sustain a rich variety of highly productive marine organisms, including cockles. On the west coast of Malaysia intertidal mudflats are often hummocky with clumps of worm-secreted tubes and patches of shelly gravel sorted from the mud by wave action (Section 6.4.3).

Mudflats are generally diversified by channels or creek systems, often beginning in embayments or on the flanks of tidal divides and converging in dendritic patterns, but sometimes running almost parallel to each other down to the low tide shoreline. Such creeks are better defined on mudflats, as in the southern part of San Francisco Bay, than on sandflats, as in the Dutch Wadden Sea.

Where the tide range is large, as on the west coast of South Korea, tidal currents dominate the evolution of well defined creek systems in mudflats (Wells et al., 1990), but in microtidal areas wave action may modify them to more subdued outlines. Verger (1968) illustrated frequent changes (notably lateral migration) in the estuarine tidal creeks on mudflats off the Sée and Sélune rivers in the Bay of Mont Saint Michel in NW France. In high latitudes intertidal mudflats are scraped and grooved when waves push disintegrating sea ice and boulders shoreward, as on the shores of Gotland in Sweden (Philip, 1990).

Some intertidal mudflats are actually platforms of marine erosion cut into soft clay, often with a veneer of muddy deposits. Examples of this are seen on the NE coast of Westernport Bay in Australia, where the mudflats are backed by a metre-high receding cliff cut in dark Holocene peaty clay. On the SW coast of Peninsular Malaysia the tidal mudflats are backed by eroding microcliffs at the seaward edge of the mangroves. On the isostatically emerging eastern shore of Hudson Bay in Canada a wide tidal mudflat has been interpreted as a wave-eroded platform, with a veneer of sediment coarsening seaward (Ruz *et al.*, 1998).

Some mudflats have formed in areas that were formerly reclaimed from the sea and protected by embankments, and have recently been abandoned. An example is Butterwick Low at Freiston on the western coast of The Wash, where a gap was cut in the coastal embankment to allow the sea to return to reclaimed meadows. These became mudflats, dissected by creeks where former drains are becoming incised, with waterfalls working headward. Similar features are seen in Porlock Bay, Somerset, where a shingle barrier has been breached by storms, and the backing alluvial plain was excavated as a retreating mud cliff, cut by waterfalls.

10.5 Intertidal vegetation

Vegetation can play an important part in the evolution and stability of intertidal landforms. In marine areas salt-tolerant plants such as seagrasses can grow in intertidal and subtidal areas, while other halophytic plants colonise the upper intertidal zone to form salt marshes or mangrove swamps. There has been much discussion of the extent to which vegetation promotes sedimentation and contributes to the shaping of intertidal morphology, rather than merely occupying areas regularly submerged by the sea. Seagrass beds, salt marshes and mangrove swamps can certainly be eroded by waves and currents, but there is also evidence that loss of vegetation (die-back) can be followed by the dispersal of sediment that had previously accreted in areas occupied by seagrasses, salt marshes or mangroves.

10.6 Seagrass beds

Many intertidal areas carry hydrophytic vegetation in the form of carpets of algae, such as *Enteromorpha*, and seagrasses such as the marine wrack or eel grass (*Zostera* spp.) and Poseidon's seagrass (*Posidonia*), which grows in the Mediterranean and southern Australia. These partially stabilise the surface, and may trap muddy sediment to form low depositional banks or terraces (Figure 10.3).

Seagrasses extend into nearshore waters to depths of about 5 m. They trap sediment because the plants are erect when the tide is high, and can diminish wave heights by up to 40 per cent and wave energy by two-thirds (Fonseca, 1996). They also form a sediment-binding root network beneath the surface. Zhuang and Chappell (1991) observed the building of a seagrass terrace of silt and clay upon a generally sandy intertidal substrate in Corner Inlet, Australia, as the result of wave attenuation by the plant cover. At low tide the seagrasses form a slumped carpet, but as the tide rises the stems stand erect as a sediment-filtering meadow. A seagrass bank or terrace may have vegetation spreading along its borders, or it may be sharp-edged as the result of wave or current scour. As a seagrass terrace expands intervening channels may form, becoming deeper and narrower as currents are confined through them. Seagrass terraces commonly show marginal cliffing as the result of wave action, and may be trimmed back by current scour alongside tidal channels (Figure 10.3). Potholes that form in seagrass terraces, where stones are circulated by waves and currents, may

Figure 10.3 Seagrass terrace exposed at a low spring tide at Quiet Corner, Port Phillip Bay, Australia

grow into round ponds as the result of swirling currents associated with wave action. Seagrasses are not regarded as salt marshes, but terraces built by accretion in seagrass areas may in due course evolve into salt marshes.

Seagrasses occupied broad mudflats across the tidal divide in NE Westernport Bay, Australia (Figure 10.3), but after they died in the 1980s the interfluvial areas were lowered by wave erosion and the intertidal creeks became wider and shallower, so that the morphology became rounded and more subdued. In Florida Bay seagrasses, mainly turtle grass (*Thalassia testudinarium*, common in subtropical and tropical seas), trapped shelly deposits as well as mud, but these were released and mobilised by wave action in areas where the seagrass died (Prager and Halley, 1999).

Rocky outcrops are habitats for other forms of marine vegetation, including seaweeds and kelp, which create relatively sheltered environments and may diminish wave energy through nearshore areas to the coast. They are less effective in trapping sediment than seagrasses.

10.7 Salt marshes

Salt marshes are vegetated areas in the upper part of the intertidal zone on the shores of inlets, estuaries and embayments sheltered from strong wave action (Allen and Pye, 1991). They can exist where wave action is sufficient to wash sediment into them, but occasional storm waves cliff their seaward margins. They can extend down to about mid-tide level, and their substrate is usually muddy, but sometimes there is an admixture of sandy sediment.

Halophytic (salt-tolerant) grasses, herbs and shrubs grow between high spring tide and mid-tide level on shores exempt from strong wave action: their ecology has been described by Ranwell

(1972) and Adam (1990). They are extensive on macrotidal and megatidal coasts, forming wide zones on the shores of the Severn estuary and the Bristol Channel (as in Bridgwater Bay), the Bay of Mont Saint Michel in NW France and the Bay of Fundy in Canada (Davidson-Arnott *et al.*, 2002). They are also found between landward recurves on spits: Hurst Castle spit protects the Keyhaven salt marsh on the Hampshire coast, and salt marshes are developing in the lee of The Bar, the barrier island that began to form a century ago on the Scottish coast at Culbin.

Salt marshes are less frequently inundated by the sea than sandflats and mudflats at lower levels, and currents produced as the tide rises and falls are weaker in the salt marsh. The vegetation diminishes wave action, for swards of salt marsh grass can reduce wave heights by 70 per cent and wave energy by over 90 per cent: the water velocity profile is modified in much the same way that dune grasses modify wind velocity profiles (Section 9.1). Salt marsh species can tolerate varying depths and durations of tidal submergence, and so spread forward to the limits thus set. As a result there is often a well defined zonation of species parallel to the coastline, with plants such as *Salicornia* dominating the outermost zone that is most frequently submerged by the tide, and the shore rush, *Juncus maritimus*, and other salt marsh plants occupying higher zones that are less frequently submerged. These zones could simply represent the occupation by each species of a suitable habitat that moves seaward as accretion continues, but where the vegetation is trapping muddy sediment and adding organic matter (falling leaves, decaying plants) to build up and prograde the substrate it is preparing the way for the seaward advance of the plant zones (a vegetation succession).

A distinction can be made between open marshes spreading seaward on the shores of embayments and estuaries, and vertically accreting closed marshes, formed behind sheltering spits and barriers, as between the shingle recurves on

Blakeney Point and Scolt Head Island, which become lagoons at high tide, and then drain out through a system of converging tidal creeks (Steers, 1960).

Salt marshes have similar features throughout the temperate zone, but in cold regions they can be modified by snow and ice processes. Dionne (1972) described the effects of winter snow cover and ice action on salt marshes in the St Lawrence estuary. Ice-rafted debris, including rocks, beach material, dislodged marsh fragments and driftwood, is left behind on the salt marshes after the spring melt, and when the air trapped under winter ice on mudflats finally escapes with the thaw the surface may develop small mamillated mud domes, termed monroes, after the attributes of the famous Marilyn.

10.7.1 Evolution of a salt marsh

Salt marshes begin to form when vegetation (perhaps initially seagrasses) spreads from the high tide shoreline on to an accreting mudflat (Frey and Basan, 1985). Often the pioneer plants (e.g. *Salicornia* spp.) form spreading clones, within which low mounds of muddy sediment are retained. As these coalesce a depositional terrace is built up by the trapping of mud washed into the vegetated area by waves and currents as the tide rises, and retained by the filtering network of stems and leaves as it falls. Terrace formation is also aided by the development of a subsurface root network, which binds the accreting sediment. Although often characterised as tide-dominated morphology, most salt marshes are influenced by wave action as the tide rises and falls. Accretion of sediment results in the formation of a persistent depositional terrace that slopes gently from the high spring tide line to the high neap tide line, then more steeply to the mid-tide line. In the absence of salt marsh vegetation the substrate remains a mobile intertidal slope, and if the plant cover dies, or is cleared away, the depositional terrace is dissected and degraded by erosion.

Sedimentation in salt marshes is aided by the presence of adhesive algal mats on the muddy surface, and by flocculation and precipitation of clay by the salt exuded from marsh plants (Pethick, 1984). When mud that adheres to leaves and stems dries off it falls to the substrate. The spread and upward growth of salt marshes is aided by an abundant supply of sediment, but salt marshes can still aggrade by the accumulation of peat derived from the decaying vegetation. Vertical accretion of up to 15 mm/yr has been measured on salt marshes in Essex and the Netherlands (Ranwell, 1972), and the progressive burial of artificial markers inserted in a salt marsh on Scolt Head Island in Norfolk showed vertical accretion of up to 8 mm/yr, with variations related to marsh elevation and inundation frequency and the retention of sediment from turbid water overflowing from tidal channel margins. Norfolk salt marshes grow vertically by accretion during ordinary high tides, but the higher parts receive sediment only in storm events (French and Spencer, 1993). Examples of rapid vertical accretion in salt marshes have been documented from tidal lagoons in Argentina and on the Red River delta in Vietnam. Such accretion may be stimulated by a slow relative rise in sea level, but a rapid rise will lead to drowning and die-back of salt marsh plants.

Providing there is a supply of fine grained sediment, and wave action is gentle, salt marshes can spread rapidly. Jacobson (1988) showed that since the 18th century a salt marsh had advanced on to an accreting mudflat in Maine to cover an area of 8 km². The supply of mud to a salt marsh increases where fluvial sediment yields are augmented by catchment soil erosion, where the dredging of channels increases muddy sediment in suspension or where dredged material is dumped on or near marshes, accelerating vertical accretion and progradation. An excessive rate

of mud deposition may however blanket and kill salt marsh vegetation.

Salt marshes have been forming in sheltered sites around the coasts of Britain since the Holocene marine transgression brought the sea up close to its present level about 6000 years ago (Pethick, 1981). At St Osyth Marsh in Essex salt marsh initiation was dated by radiocarbon assay at about 4200 years B.P. Early stages in the development of a salt marsh can be studied in accreting intertidal areas, especially where shelter from strong wave action is increasing because of the growth of protective spits, barriers or shoals. Evolution of salt marshes has been documented on the North Norfolk coast from initial colonisation of muddy or sandy areas in the upper intertidal zone by individual halophytic plants, many of which expand vegetatively as circular clones that eventually coalesce in a broad sward (Steers, 1960; Pethick, 1980). Pioneer plants such as *Salicornia* spp. and *Spartina* spp. begin to trap muddy sediment and organic material (peat and shells) and thereby build up their substrates. Other species then colonise, and in due course the salt marsh terrace attains high spring tide level, where it is submerged only rarely by the sea. Sedimentation then proceeds very slowly, but with peat accretion and the accumulation of strand debris the surface may be raised to a level where rain and runoff leach out the superficial salt, and the marshland can be invaded by reedswamp communities dominated by such species as the common reed, *Phragmites communis*. The vegetation sequence then continues with the invasion of the landward edge of the salt marsh by less halophytic species, reeds and rushes, and then swamp scrub and eventually land vegetation.

A transition from salt marsh to willow and alder scrub occurs locally around Poole Harbour in Dorset (Bird and Ranwell, 1964), but these later stages in vegetation succession are rarely seen because of hinterland drainage and land reclamation. Succession from salt marsh to freshwater swamp and land vegetation is accelerated by coastal emergence, as on the shores of the Gulf of Bothnia in the Scandinavian region of isostatic uplift following deglaciation. On the other hand the poor development of a freshwater swamp transition behind salt marshes on the Atlantic coast of the United States may be a response to a rising sea level, the salt marshes having failed to aggrade at sufficient rate to attain the level where other vegetation can colonise (Kearney and Stevenson, 1991).

On the west coast of Britain salt marshes are generally firmer than those on the east coast because of higher proportions of sand in the muddy sediment. *Salicornia* spp. are again the pioneers, but later stages are dominated by grasses such as *Puccinellia* that form a sward on a depositional terrace of sandy mud, dissected by winding tidal creeks.

Similar stages in evolution can be traced in salt marshes on other coasts in the temperate zone. Salt marsh genera are very widespread, but there are some variations in species around the world. For example, the woody shrub *Arthrocnemum halocnemoides*, extensive in salt marshes in Australia, does not occur in Europe or North America.

10.7.2 Salt marsh terraces

Some salt marshes form a single terrace that extends outward from the high spring tide shoreline to the high neap tide shoreline, then descends a slope that passes seaward into a muddy surface continuing below the mid-tide line. Salt marsh vegetation spreading on to this slope indicates that the terrace is prograding. Others form a terrace ending seaward in a short, steeper slope, or in a microcliff that may be up to a metre high. Where the tide range is large there is sometimes an upper (mature) salt marsh of firm clay ending seaward in a microcliff, below which is a lower (pioneer) salt marsh terrace on soft accreting mud (Pethick, 1992). A double terrace of this kind borders Solway Firth, where the Ordnance Survey maps show an upper salt marsh landward of the high water line and a lower salt marsh to seaward, as in the Nith estuary near Dumfries. Similar features are seen on the northern shores of Walney Island in Cumbria and in Loch Gruinart on Islay in Scotland. It is possible that the upper terrace has been cliffed and cut back during a stormy phase, and that the lower terrace represents a stage in rebuilding. In the Severn estuary there are at least three salt marsh terraces, representing cycles of marsh erosion and accretion. Accretion is most rapid (12.1 mm/yr) on the lower terrace, submerged by every high tide, slower (6.4 mm/yr) on the middle terrace and slowest (2.3 mm/yr) on the higher terrace, which is inundated only by high spring tides (French, 1996).

Sections through salt marsh terraces (exposed in the sides of tidal creeks or in cliffs at the seaward edge) generally show stratified deposits, with layers of fine sand or organic material within the mud. These variations are related to wave conditions, storm waves washing fine sand into the salt marsh and mud accretion continuing as the tides rise and fall in calm weather. In the Severn estuary Allen (1996) found that the grain size of salt marsh sediments diminished from fine sand to silt and clay landward from the edge of the marsh as the result of sorting of sediment washed in from the seaward side, and that there was a similar diminution vertically through the aggraded salt marsh terrace because of progradation. However, there are often storm-carried sediments on the upper salt marsh, including sand and organic litter (Stumpf, 1983).

Upward and outward growth of salt marshes can be accelerated by an increase in the rate of sedimentation of the kind that occurred in Cornish estuaries in the 18th and 19th centuries when river sediment yields were augmented by mining waste (Section 6.4.1). On the south coast of England rates of accretion have been very slow in areas where excavations made in salt marshes (e.g. for salt manufacture) have persisted for many decades, as at Budleigh Salterton in Devon. In the Medway estuary in Kent large quantities of clay have been cut for brick-making and cement production, leaving numerous pits and access canals that have not yet been infilled, although clay extraction ceased at least 30 years ago (French, 1997).

Rates of accretion of sediment in salt marshes can be measured by laying down identifiable layers of coloured sand, coal dust, of similar material on the marsh surface, and returning to put down borings and measure the thickness of sediment added subsequently. Measurements made on salt marshes in Britain have shown that vertical accretion has been relatively slow at the upper and lower limits of a marsh, and more rapid in the intervening zone, where salt marsh vegetation forms a relatively dense sediment-trapping cover and is regularly invaded by sediment-laden tidal water (Steers, 1960). Møller (1963) measured changes in marshland topography on the Danish coast, based on maps prepared in 1941 and 1959–62 on a scale of 1:10 000 with marshland surfaces contoured at 5 cm intervals. Patterns of vertical and lateral erosion and accretion were thus located. On the north Norfolk coast Pethick (1981) found the upper limit of accretion on salt marshes to be about 2.4 m above OD. Accretion is also slower away from tidal creeks, especially where they turn parallel to the coastline. In salt marshes bordering the Severn estuary, French (1996) used evidence from heavy metal profiles and lead (^{210}Pb) dating to define distinct sedimentary units (between planes dating from 1840–50, 1936 ± 7, 1971 ± 4 and 1958 ± 4), and showed that vertical accretion (3–4 mm/yr) has been proceeding at about the same rate as sea level rise in the area. Very rapid silt accretion has been taking place on marshland in the Yangtze delta, where Yang (1999) measured accretion rates of up to 43 cm/yr, and found that accretion in the middle marsh was 1.5 times that

in the lower marsh and twice that in the upper marsh. Remote sensing techniques are difficult to apply to monitoring of salt marshes because of the vegetation cover and the small dimensions of vertical and lateral changes.

10.7.3 Seaward margins of salt marshes

The outer (seaward) edge of many eroding salt marsh terraces is a muddy microcliff up to a metre high, bordered by a sloping erosion ramp. Examples of this are seen on the Burry Inlet marshes in South Wales, where the microcliff drops sharply to Llanridian Sands, and on the Dengie Peninsula in Essex, where the microcliff has been retreating at up to 10 m/yr, but vertical accretion has continued on the marshland. Salt marshes on the southern shores of the megatidal Bay of Mont Saint Michel are undergoing dissection by ebb runoff and rapid erosion along their seaward margins (Figure 10.4). In some sites cliffing is accompanied by continuing vertical accretion of muddy sediment in the backing salt marsh vegetation, building up the terrace even though seaward advance has come to an end.

Cliffing may result from the lateral migration of a tidal channel, undercutting the edge of a salt marsh, but as it is very widespread (there are now only a few sites where salt marshes are spreading seaward) some more general explanation is required. It may be that, as on the sides of developing tidal creeks, seaward margins become oversteepened and cliffed, particularly during occasional storm wave episodes. Cliffing of this kind can be repaired if there is an abundant supply of sediment to restore the profile, permitting vegetation to spread again, but if there is a sediment deficit a microcliff will persist (Figure 12.8). Guilcher (1981) recognised cyclic patterns of marginal erosion and accretion in the salt marshes bordering estuaries in Brittany.

Figure 10.4   Dissected salt marsh on the shores of the Bay of Mont Saint Michel, NW France

Alternatively, the cliffing of seaward margins of salt marsh terraces could be a response to a rising sea level, deepening the adjacent water and allowing larger waves to attack the shore. This would also explain the widening and shallowing of the tidal creeks that is occurring in the salt marshes of southern England. Salt marshes (*barene*) in the Lagoon of Venice, where mean sea level rose 22 cm between 1908 and 1980, also show marginal die-back and receding cliffed edges, as well as dissection by widening tidal creeks. Wave energy and tide range have increased with the deepening of the lagoon, but the salt marshes are aggrading at an average of 1.54 cm/yr with sediment washed in from the eroding marsh margins and the lagoon floor on high tides (Day *et al.*, 1998).

Allen (1989) found that salt marsh microcliffs were bolder, often vertical, on sandy mud, as in Solway Firth, and more subdued on soft mud in the Severn estuary. Where the top of the micro-cliff was bound by plant roots, recession was by way of calving, toppling and rotational slides.

10.7.4 Salt marsh creeks

As salt marsh terraces build up in the form of a sedimentary wedge the ebb and flow of the tide becomes confined to a system of tidal creeks, the dimensions of which are related to the volume of water flowing up and down them as the tide rises and falls (Pethick, 1992). Typically dendritic and intricately meandering, they are channels within which the tide rises until the water floods the marsh surface, then drainage channels into which some of the ebbing water flows from the salt marsh. They are thus like minor estuaries, particularly where they receive fresh water from hinterland runoff or seepage from bordering beaches and dunes. However, it should be noted that some of the alternating submergence

Figure 10.5 The evolution of a salt marsh creek. As the salt marsh forms it builds up a bordering terrace, the edges of which become oversteepened and cliffed

and drainage of a salt marsh results from inflow and outflow across the seaward fringe rather than through creek systems. In the early stages tidal creeks are relatively wide and shallow in cross-section (like the broad channels on intertidal mudflats), but as salt marsh terraces rise and expand they become narrower and deeper, and their bordering slopes higher and steeper (Figure 10.5).

As the creek banks steepen there is frequent local slumping (Figure 10.6). Blocks of compact mud, often with clumps of salt marsh vegetation, collapse into the creek, especially where the bordering slopes are burrowed by crabs. Some tidal creeks are fringed by natural levees formed by deposition of sediment as the rising tide overflows, especially where such plants as *Halimione*

spp. have colonised the bordering slopes. This pattern is more often found on the lower (and younger) seaward fringes of salt marshes, the inter-creek areas becoming flatter as sedimentation proceeds.

Dendritic tidal creek systems give place to straighter, parallel tidal creeks on some salt marshes, as on the shores of Loch Gruinart in Islay and at Morrich More, south of Dornoch Firth in eastern Scotland. Straight parallel creeks across salt marshes are more often found where the tide range is large, the transverse gradient small, or where the rate of seaward spread of salt marsh has been rapid, as on Dengie Marsh in Essex. In Bridgwater Bay on the shores of the Bristol Channel, where the tide range is about 10 m, salt marsh creeks run parallel and orthogonal to the coastline, whereas in Poole Harbour, Dorset, a microtidal estuarine embayment, creek patterns in bordering salt marshes are mainly dendritic.

The morphology of tidal creeks is related to sediment type, plant cover and tide range. There are clearly defined steep edged tidal creeks in salt marsh terraces built largely of cohesive clay, as in Poole Harbour, but they become shallower and wider where the salt marsh is sandier, as in the Welsh estuaries opening into Cardigan Bay. Salt marsh creek patterns are modified and become trellised where linear cheniers of shelly sand have been deposited by storm surges, and channels have been cut through these. In cross-section, tidal creeks tend to be rounded furrows where accretion is in progress, to show asymmetry on meanders where erosion balances accretion and to be rectilinear where erosion is dominant (Figure 10.7).

Studies of creek systems on the salt marshes of Scolt Head Island in Norfolk showed that the exchange of water and sediment with the bordering marshes varied with current velocity as the creek water rose to flood the adjacent levels. Vertical growth of the salt marsh led to increasingly intermittent sediment transport in the creeks,

Figure 10.6   Perranarworthal Creek, Cornwall, bordered by a salt marsh terrace and slumping banks formed as shown in Figure 10.5

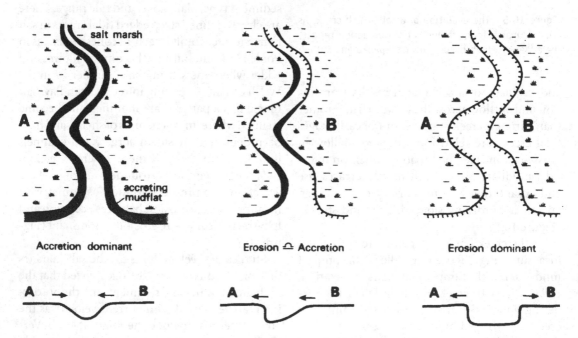

Figure 10.7   Tidal creek morphology in a salt marsh under conditions where accretion is dominant, where erosion balances accretion and where erosion is dominant

fewer tides reaching the velocity required for the entrainment of channel sediments (French and Stoddart, 1992).

On an emerging coast tidal creeks are likely to become confined and incised. There is thus a transition from salt marsh creeks that formed as residual channels within an accreting system to erosional creeks that are excavated in prior land surfaces.

10.7.5 Salt pans

Salt pans are small shallow flat-floored depressions, some of which form as residual unvegetated areas within a developing salt marsh while others (often long and narrow) are the result of the blocking of part of a tidal creek by slumping. A third category results from local die-back of salt marsh vegetation, particularly in *Spartina* marshes (below).

Salt pans are flooded at high tide, and remain bare of plants because evaporation makes the trapped water hypersaline. Many become well rounded and sharp-edged as the result of scour by small waves and circulating currents generated by winds blowing across them, much like dune lakes. In dry periods the water evaporates, leaving flat-floored basins. Numerous salt pans occur on salt marshes between shingle recurves on Blakeney Point in Norfolk, and they are also extensive on salt marsh terraces, such as those bordering the Cree estuary in SW Scotland.

Salt pans may be enlarged to rounded pond holes where pebbles or cobbles lodged within them are rotated by wave action, much in the manner of pothole evolution on rocky shores (Figure 5.2).

10.7.6 The Spartina story

Salt marsh morphology can be modified by the arrival of a new and aggressive species, such as the hybrid *Spartina anglica*, which originated in Southampton Water in about 1870 as a cross between native *Spartina maritima* and American *Spartina alterniflora* (Carey and Oliver, 1918). It spread (or was introduced) to many other British estuaries in the ensuing few decades, advancing across intertidal mudflats and rapidly building up marshland. It has been used as a method of stabilising and land-claiming tidal flats in estuaries in various parts of the world, including the Netherlands, Denmark, New Zealand and the Tamar estuary in Tasmania. The effects of introduced *Spartina* on salt marshes in China were discussed by Chung Chun-Hsin (1985).

Early stages of *Spartina* invasion can be seen where clones are spreading on sandy intertidal areas on the shores of Lindisfarne Lagoon, in the lee of Holy Island and on the Humber mudflats behind the spit at Spurn Head. In Poole Harbour the arrival of *Spartina anglica* in 1899 was followed by the rapid expansion of salt marshes into broader and higher terraces covered entirely by this plant (Figure 10.8). At the same time, intervening creeks and channels became narrower and deeper, indicating that there had been a transference of muddy sediment from these into the areas of spreading *Spartina*. On the North Norfolk coast experimental introduction of *Spartina anglica* modified natural salt marshes and led to the evolution of broad depositional marsh terraces in the intertidal zone. In the Dee estuary, Marker (1967) recorded the rapid spread of *Spartina* grass introduced in 1922, noting that it had become the pioneer colonist on accreting mudflats.

Introduction of *Spartina anglica* to the Tamar estuary in Tasmania, where previously there had been little salt marsh, led to transformation of intertidal mudflats into wide marsh terraces with deep creeks. It has spread rapidly to build depositional terraces on tidal mudflats in Andersons Inlet, Victoria, where it is invaded by mangroves. There can be no doubt that *Spartina* has acted as a sediment-trapping agent, and

Figure 10.8   A depositional terrace formed by *Spartina* in Poole Harbour, Dorset

produced intertidal landforms that would not otherwise have developed.

In Britain some of the older *Spartina* marshes show evidence of die-back, especially around pans and along their seaward margins. The ecological reasons for this are not fully understood, but die-back is often associated with oxygen deficiency, sulphide accumulation and waterlogging. At the seaward margins where the sward dies, and sediment previously trapped is released, there is a receding microcliff. Die-back of *Spartina* along creek margins has led to erosion of marsh edges and resulted in the widening and shallowing of tidal creeks and channels. The process may be cyclic in the sense that released mud is deposited in new or reviving *Spartina* marshes elsewhere. Within Poole Harbour there are sectors where *Spartina* is still advancing, mainly in the upper estuary, as well as sectors of die-back and erosion of the marshland, notably in Brands Bay near the marine entrance, where the *Spartina* sward is disintegrating.

10.8 Mangroves

Mangroves are halophytic shrubs and trees that grow in the upper part of the intertidal area on the shores of estuaries and lagoons and on coasts sheltered from strong wave action, as in inlets or embayments or in the lee of headlands, islands or reefs (Guilcher, 1979). They show their greatest extent and diversity on tropical coasts, where they occupy a similar niche to temperate salt marshes. On some tropical coasts salt marshes or saline flats with little or no vegetation occupy a high tide zone landward of the mangrove fringe.

The ecology of mangroves was reviewed by Chapman (1976) and their global status by Saenger, Hegerl and Davie (1983). A few mangrove species grow outside the tropics in the SE United States, Brazil, South Africa, the Red Sea, Australia and New Zealand. Their poleward limit is in Corner Inlet, SE Australia, where the white mangrove, *Avicennia marina*, is growing at latitude 38° 54′ 35″ S. In New Zealand they extend as far south as Ohiwa Lagoon at latitude 30° 02′ 30″ S. A limiting factor is low temperature, for mangroves are killed by frost.

There are numerous species of mangroves, particularly in the humid tropics, where there are more than 70 in the Indo-Pacific region, but diminishing towards the latitudinal limits northward and southward. Thus in Australia a maximum of 27 mangrove species is found beside the Daintree River estuary in NE Queensland, but the number diminishes rapidly southward to six at Sydney, and in southern and SW Australia there is only one, *Avicennia marina*.

Mangroves grow sparsely on rocky shores and coral reefs (their roots penetrating fractures in the rock) and on sandy substrates, but they are more luxuriant, forming dense scrub and woodland communities, on muddy substrates and shoals exposed at low tide. Where wave energy is low they spread forward to the mid-tide line, but as wave action increases along the coastline the mangrove fringe thins out and disappears. Occasionally the seaward edge of the mangroves has been cliffed by strong waves. On the other hand the longshore growth of sand bars, spits or barriers can make a coastal area more sheltered for mangrove colonisation.

The width of a mangrove fringe generally increases with tide range, and on macrotidal coasts can attain several kilometres, as on the tide-dominated shores of gulfs and estuaries in northern Australia. These have wide mangrove areas backed by sparse salt marshes and saline flats flooded during exceptionally high tides and summer rains. Within the humid tropics mangroves grow to forests with trees 30–40 m high, as on the west coasts of Malaysia and Thailand, in Indonesia, Madagascar and Ecuador. On drier and cooler coasts within, and just outside, the tropics mangroves generally form extensive scrub communities.

Mangrove species or associations of species are sometimes arranged in zones parallel to the coast or to the shores of estuaries or lagoons (Snedaker, 1982). The zonation is accompanied by variations in substrate level related to depth and duration of tidal submergence and exposure to atmospheric conditions, notably rainfall, at low tide. It appears that each mangrove species or association of species grows best at a particular intertidal level, and that competition restricts them to their zone of optimal growth, which may have a vertical range of only a few centimetres in a horizontal zone of several metres. *Avicennia* spp. are often the pioneers, forming a seaward fringe, backed by *Rhizophora* spp. and other mangrove zones dominated by species of several genera, including *Bruguiera*, *Ceriops*, *Laguncularia*, *Lumnitzera* and *Xylocarpus*. *Avicennia* is often accompanied by *Sonneratia* and *Aegiceras* on estuary shores. As accretion proceeds, raising the substrate, the mangrove zones migrate seaward, each species zone displacing its predecessor in a vegetation succession.

In Westernport Bay, Australia, the mangrove fringe (*Avicennia marina*) is backed by upper intertidal zones of salt marsh and swamp scrub vegetation, a zonation that has spread forward as muddy sediment accreted in the form of a depositional terrace (Figure 10.9). In arid regions the upper intertidal zone may become a wide, frequently desiccated and hypersaline mudflat, as in the Townsville district in NE Australia or the bare tannes behind mangroves in West Africa. Above high spring tide level there is further accretion from river floods and downwash from backing slopes, and as the land level is raised hinterland vegetation moves in.

The seaward spread of mangroves is indicated by an abundance of seedlings and young shrubs on the adjacent mudflats and a smooth canopy rising landward as the trees increase in age and size (Figure 10.10). Where the seaward margin of mangroves is abrupt, with trunks and stems of mangroves visible from the sea, and any seedlings fail to survive, the mangroves are no longer advancing. A receding mangrove

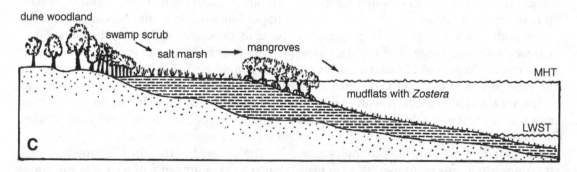

Figure 10.9 Evolution of a mangrove-fringed salt marsh terrace on the shores of Westernport Bay, Australia. (a) Sandy shore at the end of the Holocene marine transgression (about 6000 years ago). (b) Muddy deposition on the shore (with *Zostera* seagrass) is followed by the formation of a mangrove fringe, spreading seaward. (c) As the mangrove fringe advances a depositional terrace is formed and occupied by salt marsh, which is invaded by swamp scrub vegetation on the landward margin. MHT – mean high tide, LWST – low water at spring tide

Figure 10.10 A declining canopy at the seaward edge of advancing mangroves in Westernport Bay, Australia

shoreline is indicated by exposed trunks and stems being undercut, and falling, or where the vegetation has died and there is a receding microcliff in the substrate (Figure 10.11).

Mangroves are structurally and physiologically adapted to survive in a marine tidal environment. Some mangrove species (e.g. *Avicennia marina, Sonneratia alba*) have networks of pneumatophores that rise from sub-surface root systems to project vertically out of the muddy substrate and allow the plant to respire in a waterlogged environment. Pneumatophores occupy a roughly circular zone around the stem of each plant, and can be very numerous and closely spaced, with densities of up to 300/m^2. With the aid of these snorkel-like breathing tubes the mangroves can grow in areas that are submerged by the sea at each high tide, but the necessity for several hours subaerial exposure between each submergence sets a seaward limit, usually close to mid-tide level. Other mangroves, such as *Rhizophora, Bruguiera, Ceriops* and *Lum-*

nitzera species, have subaerial prop or stilt roots that branch downward to the mud and support the stems, while others, like *Xylocarpus*, have no such structures.

Where evidence from the Holocene stratigraphy beneath mangrove areas is available, it generally indicates that they have spread to their present extent during the sea level still-stand of the past 6000 years. Before this, during the Holocene marine transgression, they were confined to sheltered inlets and estuarine sites where they could migrate landward, and to sectors where they could persist on vertically accreting muddy substrates as submergence proceeded. Stratigraphic studies in northern Australia and SE Asia have shown that mangroves were growing in accreting estuaries with a rising sea level until between 7000 and 5500 years ago, when the Holocene marine transgression came to an end. They then spread into other embayments and more exposed sites of muddy accretion along the outer coast. In the estuary of the South

Figure 10.11   A sharp mangrove edge with exposed trunks indicates that the seaward margin has retreated. Pneumatophores protrude from the bordering mudflats

Alligator River mangroves spread rapidly during the Holocene still-stand, when vertical accretion proceeded with sediment washed in from sea as well as down the river, which developed intricately sinuous meanders (Woodroffe, 1993). In general, mangrove swamps spread seaward on coasts where relative sea level is stable or falling, and where there is a sufficient sediment supply.

10.8.1 Mangroves and intertidal morphology

Sediment carried into mangroves by the rising tide is retained as the tide ebbs, gradually building up a depositional terrace between high spring and high neap tides, with a seaward slope (usually about 1:50) descending to the outer edge of the mangroves and continuing across the lower intertidal zone, which is either unvegetated, or has patches of seagrass (Figure 10.12, A).

Mangroves colonise mudflats that are slowly accreting, but once established they promote accelerated accretion within the network of stems and pneumatophores that diminishes current

Figure 10.12   (a) Mangrove terrace formation by vertical accretion as mangroves spread seaward. (b) As the terrace is built up intervening creeks become deeper. If the mangroves are removed or disappear the terrace reverts to the earlier surface (pecked line). If they recolonise, the terrace and creek profiles are restored

flow and wave action (Augustinus, 1995). A mangrove fringe shelters nearshore waters from winds blowing off the land and thus reduces seaward drifting of mud from the shore, whereas shoreward drifting by waves produced by onshore winds and rising tides continues. The depositional terrace thus formed is much the same as that formed beneath salt marshes. There is also peat accumulation, but most mangrove substrates are dominated by muddy sediment. After deposition in the *Avicennia* fringe has raised the mud surface, *Rhizophora* and other mangroves colonise, sedimentation proceeds more slowly, but eventually the terrace is built up to high tide level, whereupon the mangroves are displaced by the backing salt marsh or unvegetated mudflats mentioned previously. These are submerged only by infrequent high spring tides, occasional storm surges or river floods, and sedimentation is thus very slow. Accretion of peat and drift litter may then be required to raise the substrate to levels where it can be colonised by freshwater and land vegetation, a process that is aided by emergence on coasts where sea level is falling relative to the land.

It is possible to measure rates of sedimentation in mangrove areas by the same methods as used in salt marshes (Section 10.7.2). Marker layers can be placed on the substrate, and the depth of subsequent accretion measured by coring, but there are difficulties where burrowing crabs churn up the substrate. Monitoring of changes is more effective when measurements are made on implanted stakes. Measurements in Westernport Bay, Australia, showed that there was sustained accretion of up to 4.5 cm/yr in the *Avicennia* fringe, in contrast with vertical fluctuations on adjacent mudflats. The pattern of mud accretion was strongly correlated with the density of pneumatophore networks, which certainly trap and retain muddy sediment. Low accretion mounds form above the general level of the muddy shore within the pneumatophore network around isolated *Avicennia* trees.

Mangroves with prop roots, such as *Rhizophora* spp., may be less effective in trapping mud, but once it has been retained it tends to become a more compact, firmer clay. This is coherent enough to sustain cliffs where the mangrove terraces bearing *Rhizophora* are eroded by current scour along the sides of tidal creeks, or by wave action on seaward margins. The mangrove community also generates substantial quantities of organic matter from decaying leaves, stems and roots, and from the various organisms that inhabit mangrove areas, and in due course these can form peat deposits, raising the substrate level. In Florida, *Rhizophora* is found growing on vertically accreting peat deposits.

The depositional terrace formed beneath mangroves is submerged at high spring tides, and occasional storm surges may wash sand and shelly deposits up into the mangroves to form cheniers (Figure 10.13). As the tide ebbs, runoff becomes confined to definite channels, similar to salt marsh creeks (Figure 10.12(b)). These may form a dendritic pattern, as in many Australian mangroves, but where cheniers are present the creek network may become reticulate, as on the Niger delta (Allen, 1965). Changes also occur within the mangroves as tidal creeks meander or migrate laterally so that mangrove trees are undermined on one side and mangrove seedlings colonise sediment deposited on the other. As on salt marshes, the spread of mangroves into an estuary slows down as tidal waters become confined to a central channel, in which ebb and flow currents are relatively strong. The channel borders may then oscillate, advancing during phases of local accretion, and receding during episodes of tidal scour, in relation to volumes of tidal ebb and flow.

The widening of a mangrove terrace also depends on a continuing sediment supply, as on deltaic shores close to the mouths of rivers or within embayments where intertidal muddy areas are extensive. Changes within river catchments, such as upstream deforestation, increase

Figure 10.13 A chenier of sand and gravel deposited in mangroves near Karembé, New Caledonia, during the storm surge that accompanied Cyclone Gyan in 1981. Mangroves have been killed in the zone where the chenier was deposited, but have survived, or are reviving, on either side

soil erosion and augment the sediment yield to the coast, thereby accelerating sediment accretion in intertidal areas and promoting the spread of mangroves. This has occurred in the Segara Anakan lagoon in southern Java (Figure 11.14), where a greatly increased sediment yield resulted in siltation and the rapid advance of the mangrove fringe (Bird and Ongkosongo, 1980).

Cyclones (hurricanes, typhoons) may flatten a mangrove forest, but there is usually rapid regeneration among the windthrown trees and little long term change in intertidal morphology. When the mangrove fringe is cut down and removed the substrate is usually soon lowered by erosion. This has occurred where mangroves have been cleared at Oyster Point, near Cardwell in North Queensland, Australia. In Westernport Bay, Australia, *Avicennia marina* was extensively cleared in the 1840s to obtain wood that was burnt to make barilla for soap

manufacture. The depositional terraces that had formed beneath it were dissected, tidal creeks becoming wider and shallower, and degraded to a steeper transverse slope and dissected, so that the roots of former mangroves were laid bare. When *Avicennia* recolonised the dissected areas mud accretion resumed within pneumatophore networks, rebuilding the depositional terrace (Figure 10.12(b)) (Bird, 1986).

These changes support the view that mangroves are land builders, but this has not been universally accepted (Vaughan, 1910; Davis, 1940; Carlton, 1974). Some have suggested that mangroves merely occupy intertidal areas that become ecologically suitable as they are raised by accretion, independently of any effects of vegetation, so that the depositional terraces are landforms that would have developed even if mangroves had not been present (Watson, 1928; Scholl, 1968), while others have envisaged an interaction between colonising mangroves and intertidal deposition (Thom, 1967). It is possible that *Avicennia* and other mangroves with pneumatophores promote accretion and coastline progradation as they spread forward on to the intertidal zone, whereas *Rhizophora* and other mangroves without pneumatophores simply occupy suitable intertidal habitats.

10.8.2 Seaward margins of mangroves

Where mangroves are advancing seaward and mangrove terraces are widening there are usually seedlings and young shrubs on the adjacent mudflats and a smooth canopy rises landward (Figure 10.10). Some mangrove fringes have seaward spurs of shrubs and trees advancing alongside tidal creeks that flow out on to mudflats.

Where the seaward margin of mangroves is abrupt, with trunks and stems of mangroves visible from the sea, and any seedlings fail to survive, the mangroves are no longer advancing. A receding mangrove shoreline is indicated by exposed trunks and stems being undercut, and falling, or where the vegetation has died and there is a receding microcliff in the substrate (Figure 10.11).

Mangrove terraces are being eroded on shores now receiving little or no sediment. Low receding cliffs have been cut into their seaward margins, particularly on deltaic coasts where the sediment supply has been reduced because of dam construction or the natural or artificial diversion of a river outlet. The seaward edge of many mangrove terraces is undercut by a muddy microcliff up to a metre high, similar to that described from salt marshes. Microcliff recession may be accompanied by continuing vertical accretion of muddy sediment in the mangroves, building up the terrace even though seaward advance has come to an end. In some places the cliffing results from lateral movement of a tidal channel, undercutting the outer edge of the

mangroves, but generally it is due to larger waves reaching the mangroves as the result of deepening of the lower intertidal zone, either because of progressive entrapment of nearshore sediment drifting into the upper vegetated area, or because of continuing submergence of the coast (Guilcher, 1979).

It may be that, as on the sides of developing tidal creeks, seaward margins become oversteepened and cliffed, particularly during occasional storm wave episodes. Cliffing of this kind is repaired if there is an abundant supply of sediment to restore the profile, permitting mangroves to spread again, but if there is a sediment deficit a mangrove cliff will persist and recede.

10.9 Freshwater swamps

Freshwater swamps (Section 10.7.1) are found on the landward margins of salt marshes, where they may represent a stage in vegetation succession to land vegetation, but on microtidal or tideless coasts where salinity is low (as in the Baltic Sea) freshwater swamps fringe the shore, particularly in inlets and sheltered embayments. They are dominated by reeds (*Phragmites communis*), often with rushes (e.g. *Typha* spp.) and sedges (e.g. *Scirpus* and *Juncus* spp.) and can spread out into water about a metre deep. Freshwater swamps of this kind also reduce wave action and current flow, and promote accretion of sediment, particularly silt and clay, in such a way as to build up a depositional terrace on the shores of lakes or coastal lagoons. Seasonal decay of freshwater swamp vegetation produces organic matter, which is deposited with the trapped sediment, and where the sediment supply is meagre this organic matter may accumulate as a depositional terrace of fibrous peat deposits. In due course the terrace is built up to high water level, and land vegetation (scrub and forest) then moves in. The outcome is progradation of the coastline by swamp land encroachment, a process that is also important on the shores of coastal lagoons where salinity is relatively low (Section 11.8.5) (Figure 11.17).

10.10 Summary

Intertidal areas may be partly occupied by vegetation, particularly where their substrates are muddy. Their morphology is shaped by tidal and other currents, particularly where they are sheltered from strong wave action. Their sediments are derived from rivers, cliff and shore erosion and sea floor sources, and usually contain proportions of organic matter. Mudflats are dissected by tidal creeks that descend to larger channels, often with shoaly topography.

Seagrass vegetation occupies areas in the intertidal and subtidal zones, and traps fine grained sediment to form terraces, often marginally cliffed. Salt marshes have halophytic plants, often arranged in zones related to depth and duration of tidal submergence. They trap sediment and build depositional terraces where tidal currents are confined to creek systems. They aggrade the intertidal zone until land vegetation can spread on to it above high tide level. Their seaward margins advance as accretion proceeds, but may be cliffed by wave erosion, particularly if sea level is rising. Introduced *Spartina* grass has spread across tidal mudflats and resulted in rapid vertical accretion, but where it has died back erosion has ensued.

Mangroves, mainly on tropical coasts, develop in a similar way to salt marshes by trapping sediment and shaping accretional terraces.

Freshwater vegetation may also trap sediment and organic matter to produce swamp encroachment on the shores of lakes and lagoons, and around the Baltic Sea.

11

Estuaries and lagoons

11.1 Introduction

During the later stages of the Holocene marine transgression, the sea invaded valley mouths and coastal lowlands to produce inlets and embayments of various kinds. These have been given different terms, some of which overlap. In this chapter rias, fiords and fiards, calanques, sharms and sebkhas are considered first, although many of them would fit within a broad classification of estuaries.

11.2 Rias

Inlets formed by partial submergence of unglaciated river valleys are termed rias. In 1886 Von Richthofen defined a ria as a drowned valley cut transverse to the geological strike, between ridge promontories, but the Rias of Galicia, in NW Spain, where the term originated, do not meet this strict definition (Cotton, 1956). The term ria has come to be used as a synonym for a drowned valley mouth (usually with a branching dendritic or treelike outline) remaining open to the sea, as in Carrick Roads (Figure 11.1) and the Tamar estuary in SW England, Chesapeake Bay in the United States and Port Jackson (Sydney Harbour) in Australia. The straight valley-mouth gulfs on the SW coast of Ireland, such as Bantry Bay, are also rias. The term overlaps with the broad concept of an estuary, for many rias have inflowing rivers providing fresh water than meets and mixes with seawater moved in and out by the tide.

Kidson (1971) noted that the rias of SW England are underlain by alluvial deposits in buried channels that were cut when the rivers flowed down to lower sea levels in glacial phases of the Pleistocene. Several phases of valley incision occurred, with removal of at least some of the sediments deposited during interglacial phases. Downcutting was completed in the Last Glacial low sea level phase, and infilling has occurred during and since the Holocene marine transgression.

Rias show varying degrees of sedimentary infilling. On the north coast of Cornwall sand has been washed in from the Atlantic Ocean to fill rias, as at Padstow, where at low tide the Camel River is narrow, flowing between broad exposed sandbanks that are submerged when the tide rises. On the south coast of Cornwall rias generally remain as relatively deep inlets, such as Carrick Roads, some with a threshold sand bar at the entrance, as at Salcombe. Similar rias in Brittany, such as the Aber Benoît, have salt marshes in their upper reaches, and the Rade de Brest

Coastal Geomorphology: An Introduction, Second Edition. Eric Bird.
© 2008 by John Wiley & Sons, Ltd

Figure 11.1 The river systems of south Cornwall about 18 000 years ago (left) and the formation of the ria at Carrick Roads by their partial submergence during the Holocene marine transgression (right)

is noteworthy for its various sand and gravel spits (Guilcher *et al.*, 1957). The Galician rias are generally deep marine inlets, with sandy, gravelly and rocky embayed shores, some inwashed spits, and muddy sediment and salt marshes at their heads (Castaing and Guilcher, 1995). Tectonic movements may have influenced the morphology of these rias, and also of Cork Harbour, a large ria in southern Ireland with converging drowned river valleys, where bordering archaeological sites extend below present sea level.

The rias and estuaries of western Ireland show sandy deposition of sand washed in by Atlantic waves as threshold banks and flanking spits, passing upstream to muddy areas, often with salt marsh at the head, as at Pollatomish in County Mayo. The bordering land is higher, and the coast steeper and cliffed, towards their seaward ends, diminishing to more subdued slopes upstream.

11.3 Fiords

Fiords differ from rias in that they are inlets at the mouths of valleys that were formerly glaciated, as on the steep mountainous coasts of British Columbia and Alaska, Greenland, Norway and Siberia. The sea lochs of western Scotland are essentially fiords. In the southern hemisphere there are fiords on the south coast of Chile, on the west coast of South Island, New Zealand (where Milford Sound is a fine example: Figure 11.2), and on the margins of Antarctica and the Falkland Islands.

The chief characteristics of fiords are inherited from prior glaciation, for they are glacial

Figure 11.2 Milford Sound (the term Sound is incorrect, for this is a fiord) is the product of several episodes of glaciation. It was finally shaped as the ice melted about 14 000 years ago, having scoured a basin up to 290 m deep with a shallower (120 m) rock sill near the entrance. It is bordered by very steep slopes that plunge far below present sea level (as at Mitre Peak) and has hanging valleys (with waterfalls such as Stirling Falls, 146 m), bordering overhanging cliffs, avalanche chutes, striated rock surfaces, perched residual glaciers and snowfields. The tide range is about 2.0 m, and the head of the sound has a marshy intertidal mudflat

troughs scoured out by ice action well below the depths that rivers cut valleys during Pleistocene low sea level glacial phases, and submerged by the rising sea as the ice melted (Syvitski and Shaw, 1995). Depths of more than 1300 m have been recorded in Scoresby Sound, a fiord on the East Greenland coast, and Sogne Fiord in Norway is up to 1244 m deep, but most fiords have depths of about 300–400 m. Fiords are steep sided, narrow and relatively straight compared with most rias, and they have the classic U-shaped cross-profile of glaciated troughs, together with hanging tributary valleys, usually with waterfalls. The shores are generally narrow and rocky, with some talus slopes and morainic ridges. Near the seaward end there is often a shallower sill that may be rocky (as at the entrance to Milford Sound in New Zealand), possibly formed because thinner ice towards the mouth of the glaciated trough scoured less deeply, leaving a sill on the seaward side of a more deeply excavated basin. Some sills may be drowned morainic banks, but most are rocky with a sedimentary capping. Sills are generally less than 100 m deep, and were probably land isthmuses separating a lake in the glaciated trough from the sea, until they were submerged by the Holocene marine transgression.

At the heads of some fiords there are ice cliffs fringing a residual glacier, but many have deltas built by glacifluvial outwash. Although their upper reaches have some of the characteristics of estuaries, with rivers discharging fresh water, most fiords remain as deep marine inlets. Some have been filled with sediment from rivers and melting glaciers. On the west coast of South Island, New Zealand, the glaciated valleys south of Awarua Point are deep fiords whereas to the north similar valleys have been filled with glacifluvial sediment from the Southern Alps. The contrast may be due to the northern valleys having steeper catchments, or there may have been transverse tilting by tectonic movements, the fiords persisting in the downwarped southern sector. Alternatively the contrast between the infilled valleys and the open fiords may be due to lithology, the former on sedimentary and volcanic formations while the fiords are in the older and harder rocks, including gneiss, gabbro, granite, schist and marble, to the south of the Alpine Fault.

11.4 Fiards

Inlets formed by Holocene marine submergence of formerly glaciated valleys and depressions in low lying rocky terrain are known as fiards rather than rias. There are examples on the Swedish coast near Stockholm (e.g. Broviken) and in Denmark (e.g. Hjortsholm), and similar inlets in coastal plains of glacial deposition are termed förden on the German Baltic coast. The Firth of Forth and other elongated firths in eastern Scotland are a closely related category of drowned valley in formerly glaciated lowlands that are partly rocky and partly morainic. Like rias, they are also estuarine.

Beyond the limits of past glaciation many drowned valley mouths have marginal features related to periglaciation, as in SW England, where the rias are bordered by slopes covered with solifluction (head) deposits exposed in fringing cliffs (Section 4.3.3). The drowned mouths of deeply incised river valleys beyond the limits of past glaciation may resemble fiords. Bathurst Channel in Tasmania is a steep sided inlet that was formerly thought to be a fiord, but as

there is no evidence that it was glaciated down to or below present sea level, and as it has none of the features of glacial sculpture that distinguish true fiords, it should really be classified as a ria.

11.5 Calanques

The mouths of steep sided valleys that were deeply incised during low sea level stages into the limestone plateau east of Marseilles in southern France have been submerged by the Holocene marine transgression to form cliff-edged inlets known as calanques. Port Miou, near Cassis, has almost right-angle bends related to the cutting of valleys along joint planes while Port-pins and Port d'en Vau are straighter, with beaches of inwashed sand and gravel at their heads. Calanques are also found on the limestone coasts of the Maltese islands (Figure 11.3), and on the Dalmatian coast. They are essentially steep sided marine inlets on karstic coasts with few of the features of estuaries. Small inlets known as calas are found on limestone coasts in the Balearic Islands in Spain.

There are similar deep (50–100 m) steep sided inlets on other rocks, such as the volcanic formations on the south coast of the Izu Peninsula, Japan. At Cape Nelson in New Guinea the sea has risen to submerge numerous valley mouths around Mount Trafalgar, a volcanic upland dissected by a radial pattern of incised valleys to form deep inlets such as MacLaren Harbour.

11.6 Sharms and sebkhas

On the Red Sea coast, long, narrow marine inlets termed sharms (sherms) have formed where wadis or valleys cut by streams in wetter Pleistocene low sea level episodes were invaded by the sea during Holocene marine transgression. Their shores are lined and their entrances constricted, and in some cases blocked, by fringing

Figure 11.3 Wied il Ghazri, a calanque on the limestone coast near Marsala on the north coast of Gozo, Malta

coral reefs, and they often end abruptly up-stream, not necessarily in a wadi or stream in-flow. A well known example is Abhur Creek, just north of Jeddah in Saudi Arabia, and similar features are seen in Mombasa Harbour in Kenya and on the east coast of Vanuatu (Castaing and Guilcher, 1995).

Sebkhas are broader, often branched embayments found on arid coasts, such as those of the Red Sea and the Arabian Gulf. They are often partly enclosed by sandy spits and barriers, and penetrated by the sea to an extent that varies with tides and weather. Above the normal level of marine submergence there are salt-encrusted areas, invaded by the sea only during storm surges, as at Abu Dhabi. Similar embayments are seen, branching into inter-dune swales bordering King Sound in NW Australia (Figure 11.4): they differ from the Red Sea and Arabian Gulf sebkhas in being subject to large and variable tidal submergence. Sebkhas usually show parallel zones of different evaporite deposits (chlo-rides, sulphates and carbonates in a landward sequence), formed by precipitation from seawater and groundwater (Evans and Bush, 1969).

11.7 Estuaries

Estuaries have been defined in various ways. According to Pritchard (1967), 'an estuary is a semi-enclosed body of water which has a free connection with the open sea and within which seawater is measurably diluted with fresh water derived from land drainage', but this definition does not take account of tides or morphology, and is wide enough to include the Baltic Sea or Port Phillip Bay in Australia. Alternatively, an estuary may be defined as the seaward part of a drowned valley system, subject to tidal fluctuations and the meeting and mixing of fresh river water with salt water from the sea, and receiving sediment from its catchment and from marine sources. Channels shaped by unidirectional river

Figure 11.4   Chattur Bay is a sebkha north of Derby on the west coast of King Sound, NW Australia

flow widen downstream to estuaries subject to alternating inflow (flood currents) and outflow (ebb currents), with rapid variations in current velocity through the tidal cycle. The morphology of an estuary represents an adjustment between the capacity of its channels and creeks and the volume of water moved in and out by tidal oscillations (the tidal prism). General reviews of estuaries have been provided by Emery and Stevenson (1957), Lauff (1967), Nichols and Biggs (1985) and Perillo (1995).

Most estuaries are found on coastal plains. There is considerable overlap between the concept of an estuary and the various kinds of inlet described above. Some rias and fiords are valleys that have been almost completely drowned by marine submergence, receiving so little drainage from the land that they are essentially arms of the sea, but most are fed by rivers, the mouths of which can be described as estuarine. Where valleys have been incised into coastal plateaux, as

in SW England and Brittany, or into hilly coastal country, as in China and the North Island of New Zealand, marine submergence has produced rias that are estuaries with steep bordering slopes. Tectonic movements (folding and faulting) have influenced the shaping of estuarine inlets such as San Francisco Bay in California and Westernport Bay in Australia, and land subsidence in SE England has resulted in wide, short estuaries such as those of the Essex coast and the broad embayments known as Chichester, Langstone and Portsmouth Harbours on the Hampshire coast and Poole Harbour in Dorset. The Wash is a broad embayment, estuarine around the mouths of the inflowing Welland, Nene and Great Ouse Rivers. Distributary river channels on deltas are usually also estuarine.

As infilling proceeds, estuaries that were originally deep and branching become shallower and simpler in configuration. Funnel-shaped estuaries, widening seaward (Figure 11.5), are best

| ZONE OF RIVER DEPOSITION AND REED-SWAMP | ZONE OF TIDAL SHORES WITH SALT MARSH, SANDFLAT AND MUDFLAT | SEA SHORES |

Figure 11.5 Estuary zones related to tidal conditions and salinity. Fresh water outflow from the river meets salt-water inflow from the sea as it rises and falls with the tide. The intertidal zone is indicated by fine shading

developed on coasts with a large tide range. In Britain the Thames, the Humber and the Severn show this form, as do the Rhine and the other rivers of northern Germany, the Seine, the Loire and the Gironde in France and the Tagus in Portugal. On the east coast of the United States Delaware Bay and Chesapeake Bay are much branched coastal plain estuaries, formed where the sea has risen to invade tributary valleys. In Australia, rivers draining into the megatidal northern gulfs have large estuaries, the Fitzroy flowing into King Sound and the Ord and the Victoria into Joseph Bonaparte Gulf.

The morphology of an estuary can be related to hydrodynamic processes, such as river flow, tidal currents associated with the incoming tide wave, wave action, chemical processes such as the flocculation of fine grained sediment, and biological processes such as the growth of salt marsh, mangroves or sea-grass vegetation and the generation of shelly deposits. These processes are complex and variable over time and within an estuary. They have been discussed in

the previous chapter in relation to the intertidal zone generally, and will now be considered in terms of the shaping estuaries and lagoons.

Salt-water penetration from the sea characterises estuaries. The upstream limit of an estuary can be defined as the point where salinity falls below 0.1 ppt or where the dissolved ions (notably carbonates) become radically different from those (sodium, chlorides) found in seawater (Pritchard, 1967). The seaward limit of an estuary can be taken as the point where seawater (salinity about 35 ppt) is undiluted by fresh water from rivers, but this is highly variable and may be well offshore in large estuaries or during river flooding. Alternatively, the upstream limit of an estuary can be taken as the point where tidal oscillations fade out, or where marine and estuarine sediments give place to fluvial sediments. The downstream limit may be definable in terms of the bordering coastline, or where fluvial bottom sediments come to an end. An arbitrary boundary is often used: the seaward boundary of the Severn estuary in the Bristol

Channel, for example, is sometimes taken as the line between Lavernock Point and Sand Point, near Weston super Mare.

Sea water is denser than river water, and as it enters an estuary it forms an underlying salt wedge that moves upstream, while the fresh river water spreads downstream over it, the junction between upper fresh and lower salt water being sharply defined. Fine grained sediment carried out to sea by fresh water can be brought back up the estuary by a bottom current formed by the incoming salt wedge, and coarser sediment may accumulate upstream from the salt wedge limit. Such a pattern of sediment coarsening upstream has been found in Chesapeake Bay, and along the Lower Mississippi. Where the tide range is large and river input relatively small estuarine water is less sharply stratified and there is an inflow of sediment from the sea, diminishing upstream in grain size as the current slackens, as in the Thames estuary. In wide estuaries with strong tidal currents and wave action, and only weak river discharge, vertical stratification disappears and the estuarine water becomes homogeneous. In funnel-shaped estuaries such as the Firth of Forth the inflowing tidal current tends to move towards the left and the outflowing river current to the right in response to the Coriolis force generated by the earth's rotation (the pattern is reversed in the southern hemisphere).

Estuaries can also be classified by tide range, which strongly influences estuarine processes. A microtidal estuary is dominated by wind-generated waves and discharging river currents, and has a relatively narrow intertidal zone, a mouth often encumbered by spits and shoals or a sandy threshold and a salt wedge that moves in and out with the tides. There are often bordering beaches and spits, and estuary-head river deltas. A mesotidal estuary has stronger tidal currents, a wider intertidal zone with mudflats and salt marshes or mangrove swamps, and multiple meandering channels and creeks. The tidal meanders may be a response to upstream movement

and deposition of bottom sediments that impede the outflow of ebb currents in the channels. A macrotidal (and megatidal) estuary is dominated by tides and tidal currents, and is usually funnel shaped, with a broad mouth and linear banks. The funnel shape is of exponentially diminishing width upstream, adjusted to accommodate the tidal prism, particularly where the length of the tide wave is equal to, or is a simple multiple of, estuary length, as in the Ord estuary, Western Australia, where the ratio is 1:4 (Wright, Coleman and Thom, 1973). Intertidal mudflats and marshland are extensive at low tide, and there are elongated bars between sub-parallel deeper channels. Relatively deep round or oval depressions (colks) may be scoured by strong tidal currents (circulating abrasive sand or gravel) into sediments or bedrock on the estuary floor. Wave action is of minor importance on the high tide shoreline, but may contribute to the shaping of intertidal morphology.

Dalrymple, Zaitlin and Boyd (1992) divided estuaries into three zones: an outer, marine-dominated zone with net sediment flow landward, a central zone of relatively low wave energy and convergence of marine and fluvial sediment and an inner river-dominated zone where net sediment flow is seaward. They distinguished between wave-dominated and tide-dominated estuaries. Wave-dominated estuaries have prominent depositional sandy thresholds at the mouth, with bordering beaches and cliffs, a central zone of muddy deposition and a river delta upstream. Tide-dominated estuaries have rivers flowing into tidal channels that wind through extensive mudflats and marshlands to shoaly mouths that may have tidal deltas (the term delta is misleading, for these are produced by ebb and flood tidal currents and not by fluvial deposition) (Wells, 1995).

Tides flow into estuaries in a manner determined by the configuration, the velocity of tidal currents (C) diminishing with depth ($C = \sqrt{gD}$, where g is the gravitational constant). As the tide wave (length $L = T\sqrt{gD}$, where T is the tidal period) enters the mouth of an estuary, maximum velocity theoretically occurs at high tide (wave crest) and low tide (wave trough), with slack water as the current reverses at mid-tide. This changes towards the head of the estuary, where maximum inflow and outflow occur theoretically at mid tide and slack water at high and low tides. In practice, the pattern is complicated by partial reflection of the tide wave, which delays the progress of the wave crest (high tide) upstream.

Currents through the mouth of an estuary depend on the tidal prism, the volume of water that enters and leaves as the tide rises and falls. The mean cross-sectional area of the entrance (A) is related to the volume of the tidal prism (P) according to O'Brien's formula, $A = CPn$, where C ranges between 0.003 45 and 0.235 and n between 0.84 and 1.05 (Pethick, 1994).

In flood-dominated estuaries the narrowing and shallowing of the channel and the outflow of fresh water from the river cause the tide wave to steepen, increasing the tide range, as it passes upstream. Inflowing currents strengthen during the briefer tidal rise and outflowing currents weaken during the longer ebb, producing tidal asymmetry and a net upstream movement of sediment. The inflow of sediment carried by currents produces rapid accretion on sandflats, mudflats and marshlands in the upper intertidal zone, especially during the long slack water period at high tide. In Britain the Severn and Dee have flood-dominated estuaries. By contrast, there are ebb-dominated estuaries with stronger outflow during the briefer ebb tide and a long slack water period at low tide, resulting in a net seaward movement of sediment. The rapidly falling ebb tide causes erosion in the mid-tide zone, steepening the intertidal profile and forming relatively narrow low tide channels. The Ems and Weser Rivers, draining into the Wadden Sea, are examples.

The tidal prism, or total volume of water that enters and leaves an estuary in a single tidal cycle, determines the discharge and therefore the velocities of flow across any cross section (O'Brien and Dean, 1972). In microtidal and mesotidal estuaries narrowing and shallowing cause frictional drag on the incoming tide, and the tidal prism diminishes upstream. Mutual adjustment between morphology and the tidal prism results in a dynamic equilibrium that can be disturbed by storm or flood events, or by changes in sea level or the dimensions of the estuary mouth resulting from spit growth or shoal migration (Pethick, 1996).

Where tide range increases upstream on macrotidal coasts the front of the tide wave may steepen into a wave (known as a tidal bore), propelled upstream, as in the Severn estuary (Section 2.3.3). Estuaries with mouths constricted by promontories or spits have impeded tidal ventilation and a diminished tidal amplitude, but increased ebb and flow current velocity at their mouths, as in the Mersey in Lancashire. Tidal currents are there strongest at high and low water (slack water at mid-tide), and sediments are carried upstream as the tide rises and deposited there as current velocity diminishes in the intertidal zone.

The meeting and mixing of fresh water from the river and brackish water from the sea determine salinity regimes within an estuary. In calm weather, inflowing tides bring in a wedge of brackish marine water, which moves sediment upstream along the estuary floor. As the tide turns, freshwater outflow becomes dominant, initially in surface waters, where it has little effect on bottom sediments, but can discharge fine grained sediment in suspension, forming plumes that extend off river mouths. Areas of turbid relatively fresh water extend far out to sea off the Amazon and many other rivers, especially in the tropics. The salinity distribution within an estuary can be modified by turbulence and by the effects of strong wind action and fluvial discharge, especially during phases of river flooding.

Tidal currents dominate estuarine morphology, but wind action moves water and influences estuary circulation, and also produces waves up, down and across an estuary. Wave action may become significant in shaping shore, nearshore and shoal features in large, wide estuaries. Many estuaries on high wave energy coasts, as in Australia and South Africa, are partly blocked by spits and barrier islands built up by strong wave action from the sea. Such estuaries are in fact estuarine lagoons. Some are sealed off completely in the dry season, when river flow is unable to maintain an outlet, but after heavy rain river flooding builds up the water level until it spills over, reopening the entrance.

11.7.1 Evolution of an estuary

During the Last Glacial phase, late in Pleistocene times, rivers drained valleys that descended to a lower sea level. The land surface at that time can often be traced as a subaerially oxidised horizon that was submerged by the sea, then buried beneath Holocene estuarine sediments. This pre-transgression land surface may have been dissected by the more deeply scoured estuary channels. It surfaces at the landward limits of the Holocene marine transgression, which in the upper estuary may be a valley-side slope showing evidence of basal cliffing by wave action and beaches formed before the deposition of Holocene fluvial sediments by the inflowing river as an alluvial flood-plain. Sometimes there is a transition from intricate river meanders downstream to long, sweeping meanders on the aggraded (essentially deltaic) valley floor in the zone that was formerly submerged by the sea, but this contrast may indicate the extent of modification of river discharge by tidal oscillations or occasional wind-driven upstream flow (Bird, 1978a). Further downstream the estuary

may have been narrowed by shore progradation, marked by beach ridges, salt marsh or mangrove terraces, formed after marine submergence came to an end. In some places the coast has been cut back beyond the limits of the Holocene marine transgression by later marine erosion.

Between the pre-transgression land surface and the existing estuarine channels and bordering intertidal morphology is a body of sediment that has been deposited in the estuary in Holocene times. Some of this sediment has been brought down by the river, some has come from bordering slopes and some has been washed in from the sea: the proportions vary from estuary to estuary (Guilcher, 1967). Woodroffe (1996) gave examples of the stratigraphy of these infills in some European and Australian estuaries. The Holocene stratigraphy commonly includes interdigitating wedges of fluvial and marine sediment of varying grain size (mainly sand, silt and clay), with shell beds, peat horizons formed in salt marshes, mangroves or sea-grass beds that have been submerged and buried by younger sediment.

The sedimentary sequence usually shows evidence of phases of erosion resulting from the lateral migration, enlargement or reduction of tidal channels, resulting in discontinuities and lateral transitions in the thickness, texture and composition of sedimentary horizons, and embedded deposits that originated as shoals, cheniers, levees or vegetated zones. There is usually a transition from fluvially supplied sediment in the upper reaches to inwashed marine sediment around the entrance from the sea, and from bordering beaches, dunes, cliffs and marshland down through intertidal sandflats and mudflats to central channels that carry river discharge at low tide. As infilling proceeds, the zone of fluvial (deltaic) deposition extends downstream, a threshold of inwashed marine sandy sediment (and sand blown from coastal dunes bordering the lower estuary) grows upstream and encroaching marshland, prograding beaches and tributary deltas narrow the estuary, suppressing tidal ventilation and reducing wave energy. The rate of infilling depends partly on fluvial sediment yields and partly on the size and shape of the estuary. Eventually, the estuary fills, and sediment delivered to the coast may be added to beaches or marshes along the shore. Further deposition may then form a protruding delta (Chapter 12).

Stages in the filling of a tropical estuary were documented by Woodroffe (1996) with reference to the Adelaide, Mary, South Alligator and East Alligator Rivers that flow across a broad confluent delta plain to Van Diemen Gulf in northern Australia. He showed that the Holocene sea level rose to flood a wide shallow embayment, submerging a land surface of oxidised sands, muddy sands and gravels. Fluvial sands were then deposited, and overlain by muddy sediment in spreading mangrove swamps, through which the river channels meander intricately.

Before considering the origins of sediments deposited in estuaries it is necessary to consider the form and dynamics of the channels that convey water and sediment through them.

11.7.2 Estuarine channels

At low tide in estuaries there are often a number of channels leading from the river to the sea, between banks and intertidal slopes of sand, silt and clay. These channels and slopes are shaped largely by tidal currents, and subject to rapid changes in configuration. The banks often bear superficial ripple patterns, produced by the action of waves and currents on the estuary floor as the tide rises and falls.

As tidal currents ebb and flow, shoals move and channels migrate, and many estuaries have independent, mutually evasive ebb and flood channels. Ebb channels have a residual outflow of river water and sediment and become wider and shallower seaward, while flood channels have an inflow of seawater into the estuary and become shallower upstream. Where the tidal flood begins before ebb has ended, outflow may continue in ebb channels for up to an hour after low tide, when water has begun to move up the nearby flood channels. Where the tide range is large these channels are straight and parallel, but where it is smaller they curve and interdigitate.

Currents change the position and dimensions of estuarine channels and shoals, cutting away sediment in one place and building it up in another, which is why estuarine ports often require dredging to maintain the depth and alignment of a navigable approach. A knowledge of the characteristic patterns of erosion and deposition is necessary to ensure that dredged material is not dumped where it can be washed straight back into the channel required for navigation, and in recent years use has been made of radioactive tracers to determine paths of sediment flow in the Thames estuary to guide dredging and ensure access to the port of London.

Meanders that develop in estuarine channels, as in the Humber estuary in Britain or Chesapeake Bay in the United States, are related to patterns of alternating ebb and flood currents, notably the brief phase of very high discharge that occurs during the ebb, when the channels carry water volumes comparable with those in major rivers. Intricately meandering fluvial channels give place downstream to elongated pools with alternating cuspate bank projections and deposited bars in the intervening calmer zones (Figure 11.6). Usually there is an accompanying

Figure 11.6 Estuarine meanders

flaring of the estuarine channel seaward, related to the rapidly increasing volume of ebb-augmented discharge towards the river mouth. Examples of estuarine meanders are found in the rivers that drain into Joseph Bonaparte Gulf, in northern Australia.

Many ports are situated beside estuaries, and there have been modifications of morphology and tidal regimes by dredging to create or maintain navigable channels and by excavation and landfill in port areas. Insertion of gates or barriers across estuary channels to prevent storm surge flooding upstream, as in the Thames estuary, has modified tidal regimes downstream, leading to higher high tide levels on bordering salt marshes and increased wave and current scour on mudflats, salt marshes and shore embankments.

11.7.3 Estuarine sediments

Estuaries are typically areas of active sedimentation, the area drowned by the Holocene marine submergence being progressively filled, and contracting in volume, depth and surface area until the river winds to the sea through a depositional plain. It is possible to find stages between a deep, branching estuary, little modified by sedimentation (e.g. Sydney Harbour, also known as Port Jackson), through to estuaries that have been completely filled in (e.g. Snowy River in SE Australia) and deltas growing out beyond the former coastline. The rate of filling depends on the original area and volume of the drowned valley mouth and the sediment yield from rivers, downwash from bordering slopes, shore erosion and inflow from the sea. Guilcher (1967) cited the Loire estuary as an example of infilling mainly by fluvial sediment from a river carrying a large load of suspended silt and clay from the river, the Rance estuary in Brittany as one that has received substantial silt and clay from bordering slopes of periglacial head deposits and the

Kapatchez estuary in Guinea as one filled with sediment washed in from coastal mudflats. Enclosure by spits and barriers accelerates the natural reclamation of an estuary by diminishing wave and tidal scour and impeding outflow of sediment, but increased river discharge and/or channel narrowing can intensify current velocity and increase scour along channels. The pattern of sediments on the floor of an estuary depends on source locations, wave action as limited by fetch along or across the estuary and the rise and fall of the tide, the effects of tidal currents and configuration, including the presence of enclosing spits and the distribution of tidal channels.

Continuing coastal submergence tends to postpone infilling by deepening the estuary and widening its mouth. On the other hand, estuaries on emerging coasts are shallowed and infilled more rapidly. On the south coast of Norway, where there is land uplift due to isostatic recovery following Holocene deglaciation, rapids develop across emerging rocky areas in the shallowing mouths of estuaries, which become elongated lakes as they are cut off from the sea.

Patterns of sedimentation also vary in relation to the relative dominance of inflowing (flood) and outflowing (ebb) tides: where flood tides are stronger (flood-dominated estuaries) sediment tends to move upriver, while in ebb-dominated estuaries sediment is scoured and carried downstream.

Sediment is moved in estuaries by wave and current action, which rolls and bounces sand grains along the floor of the estuary and lifts finer sediment into suspension. Winnowing of fine sediment by waves and tidal currents leaves zones of coarser sediment, mainly sand, gravel and shell deposits, on bordering shores, in shoals and along channel floors. Sand is readily deposited on the floors of estuaries when the current flow slackens, and sandflats form on either side of tidal channels. Mud deposition, assisted by flocculation, forms intertidal mudflats, as described in the previous chapter.

Fluvial sediment yields depend on catchment geology, and rivers draining catchments where argillaceous formations outcrop, or where superficial weathering has yielded fine grained material, are sources of estuarine mud. Many of the rivers of southern and eastern England have derived muddy sediment from extensive clay lowlands for deposition in their estuaries: the London Clay has been a major source of Thames estuary mud, and the Mesozoic clay formations of the English Midlands have yielded muddy sediment to the estuaries of the Severn, the Humber and rivers draining into the Wash. The mud found in the rias of Devon and Cornwall has been washed in by rivers draining granite areas (where there has been extensive hydrothermal alteration of feldspars to clays, mainly kaolin) and Palaeozoic metasediment outcrops, which also weather to yield clay minerals. Much of the valley-floor alluvium and estuarine mud has come from fine grained material washed out of the Pleistocene periglacial head deposits (derived from weathered granites and metasediments) that mantle hillsides and are eroded from low bordering cliffs. The same deposits mantle coastal slopes, but the fine grained sediment eroded from these is quickly dispersed by the open sea, except in low wave energy sectors, as on the muddy shores of Mounts Bay, in the lee of the Lands End peninsula in Cornwall.

The muddy sediments of the Wadden Sea are derived, at least in part, from the north German rivers laden with silt and clay derived from loess and argillaceous outcrops, but there may also have been fine grained sediment washed in by wave action from glacial drift deposits on the sea floor. In South America the River Plate drains a Pampas catchment with extensive silt and clay deposits, and delivers mainly muddy sediment to its estuary, which has only a few beaches of fine sand. In the Fly River estuary in Papua New Guinea suspended silt and clay from tropically weathered material in the catchment are floccu-

lated and deposited on meeting brackish water (Wolanski and Gibbs, 1995). .

In Australia the mud that occupies the wide intertidal zone in Westernport Bay (Figure 10.1) has come partly from erosion of peaty clays and silts that had accumulated in extensive freshwater swamps to the north, and partly from the weathered mantle of clays on basalt outcrops around Phillip Island to the south. Shallow Inlet, in Victoria, is a sandy estuarine area in the lee of a dune-capped barrier spit, with a very small catchment and little fluvial sediment inflow. It contrasts with nearby Andersons Inlet, which is of similar configuration, but has been receiving muddy sediment brought down by the Tarwin River. Both have mouths constricted by spits that have sometimes grown to deflect and constrict their entrances, thereby reducing tidal ventilation. The breaching and truncation of these spits during occasional phases of floodwater outflow or storm surges has been followed by increased tidal ventilation. This diminishes as the spit is rebuilt.

The oscillatory motion of waves moving across estuarine mudflats at high tide can throw silt and clay into suspension, and ebb tide currents can remove this sediment downstream or out to sea. On the other hand, sediment thrown into suspension by waves on a rising tide can be carried forward by these waves, and by flood tide currents, for deposition in the upper intertidal zone or on backing salt marshes or mangrove swamps. Some of the muddy sediment thus mobilised is lost seaward during ebb tides and discharging river floods, and some is carried upstream for eventual deposition on aggrading flood plains. Muddy sediment that is retained in mudflats and marshlands may be recycled by waves and currents within the estuary as tides rise and fall, the cyclic exchange of sediments being indicated by changes in cross-profile.

In the estuaries of eastern England, Pethick (1996) found that wave action from the sea into the outer parts of estuaries was erosive during stormy periods, when the upper mudflats were eroded and sediment moved to the lower inter-tidal zone, widening and flattening the intertidal profile, whereas in calmer weather sediment was washed back up on to the upper mudflats, restor-ing the earlier profile. Waves from the open sea did not reach the upper estuary, but winds generated waves, especially at high tide, which shaped the mudflats into steeper, narrower and higher profiles upstream. There was also a longer term alternation (over decades to centuries) whereby high convex mudflats shaped by depo-sition during flood-dominant phases produce ebb-dominant conditions, resulting in erosion of the upper mudflats to lower concave profiles until flood-dominant conditions return, and the convex profiles are cyclically restored.

The rate of sediment yield from a river catch-ment is a function of lithology, weathering and runoff, and much influenced by climate. Sed-imentation can be accelerated either naturally by uplift, volcanic activity or a climatic change in the hinterland, or artificially by man's im-pact, through mining and quarrying, defor-estation, overgrazing or unwise cultivation of erodible land within the catchment. Stronger runoff then augments the sediment supply to estuaries. Many tropical estuaries have been rapidly filled with abundant fluvial sediment from deeply weathered hinterlands, forming deltas and coastal plains. Extensive clouds of tur-bid water extend seaward off humid tropical es-tuaries, as in northern Java.

Estuary morphology and sedimentation pat-terns have often been modified by human activ-ities in river catchments. Soil erosion in deeply weathered catchments in the humid tropics has supplied vast masses of mud to estuaries and coastal embayments, accelerating the prograda-tion of mudflats and the spread of mangroves, as in Jakarta Bay, Indonesia. Infilling of estuaries with fluvial sediment has also been augmented by soil erosion in the river catchment in Chesa-peake Bay on the Atlantic coast and several Ore-gon estuaries on the Pacific coast of the United States.

On Mediterranean coasts the shrinkage of estuaries during the past 2 000 years results from river sediment yields being augmented by widespread soil erosion following the impover-ishment of vegetation by overgrazing and un-wise land use, notably in Greece and Turkey. In California, Rooney and Smith (1999) used suc-cessive charts to measure volumes of sediment accretion in Tomales Bay at intervals between 1861 and 1994, and found that changes in land use in the catchment resulted in variations in rates of soil erosion and infilling of the bay from 94 tonnes/km^2/yr in 1861–1931 to 357 tonnes/km^2/yr in 1931–1957. They noted that additio-nal sediment generated by deforestation or cul-tivation took decades to reach the river mouth as riverbed deposits moved slowly downstream.

Mining operations have also accelerated sed-iment yields to rivers, as in Cornwall, where the wastes from tin, copper and kaolin mining have been washed into the rivers and carried down to fill estuaries at Par and Pentewan (Section 6.4.1), and build the Fal delta into an arm of Car-rick Roads (Section 12.6). On the coasts of New Caledonia estuaries are being rapidly infilled with sediment generated by open-cast mining of nickel in the steep hinterland, and in Tasmania sandy waste from tin dredging in a hilly head-water region has been carried downstream to be deposited in the estuary of Ringarooma River (Bird, 2000).

11.7.4 Tidal deltas and thresholds

In open, funnel-shaped estuaries sandy sedi-ment washed in from the sea may be deposited in the form of banks or thresholds, sometimes shaped into paired tidal deltas submerged at least at high tide, with channels diverging upstream and downstream. Tidal deltas of this kind are found between barrier lands on the Gulf coast of

Figure 11.7 The inwashed sandy threshold at Wagonga Inlet, New South Wales, Australia

the United States, for example at Yass Abel, one of the entrances to Barataria Bay in Louisiana. On the Atlantic coast, where wave energy is stronger, there are flood tide deltas on the landward side of such entrances, as Moriches Inlet on Long Island. By contrast, ebb tidal deltas have formed off inlets on the New Zealand coast, varying in form with coastline configuration, tide regime and quantity of sand available (Hicks and Hume, 1996).

Where the tide range is small and wave action strong the tidal delta becomes a wider threshold, usually with a single channel winding over a broad sandy flat exposed at low tide, a smooth and gentle seaward slope and a steep inner slope, where sand is advancing into the estuary from the sea. Thresholds of this kind are found in the entrances to estuarine inlets on the New South Wales coast (Bird, 1967a), notably at Narooma (Figure 11.7). At Bremer Bay in Western Australia a broad inwashed sandy threshold is crossed by ebb and flood channels similar to those in estuaries (Section 11.7.2).

11.7.5 Estuary shores

Estuary shores extend from the low spring tide line across the intertidal zone to the high spring tide line, which may be marked by a beach or the transition to land vegetation behind a salt marsh or mangrove swamp, and on to the limit of flooding, whether by rivers, very high tides or storm surges when waves drive water on to bordering land. In passing from an open coast to the shores bordering an estuary there is a reduction in wave energy. Cliffs become less bold and beaches more irregular, and the generally broader intertidal zone is diversified by sandflats, mudflats, shoals and bars, salt marshes and mangrove swamps. As the tide rises and falls in these shallow intertidal zones wave energy is dispersed across broad shore environments.

Estuarine beaches may include sediment brought down by rivers, but generally they consist of sand, gravel or shelly material washed in from the sea, eroded from bordering shores or derived from the nearby foreshore (Nordstrom,

1992). These are moderate to low wave energy shores, with wave processes effective only at and about high tide and wind action on exposed sectors. There is often inward longshore drift, forming beach lobes that migrate alongshore and spits that grow into the estuary. The Thames estuary is bordered by beaches with flint and chalk pebbles derived from the chalk outcrop downstream from London and gravels from the overlying Tertiary formations and Quaternary terraces. Like many estuaries that are major ports, gravel ballast is a prominent constituent of these beaches.

The configuration of an estuary can be changed by wave and current action, particularly at high tide when shores exposed to strong wind and wave action are eroded and the material deposited on more sheltered parts of the shore, or on the estuary floor (Nordstrom and Roman, 1996). Depositional features on estuary shores sheltered from strong wave action and current scour include the terraces built in salt marshes, mangrove swamps and sea-grass beds, described in the preceding chapter. Poole Harbour, a large estuarine inlet in southern England, has been modified in this way, with cliffed promontories and embayments occupied by salt marsh. Deltaic deposition may occur as rivers flow into estuaries, and fluvial sediment carried into an estuary may be moved to and fro before eventually being deposited on the shore near high tide level.

Changes on the shores of the Dee estuary in NW England during the past two centuries have been documented by Pye (1996), who traced the effects of dredging and stabilisation of tidal channels and of land reclamation on estuarine morphology. There has been erosion of salt marshes and intertidal sandflats and mudflats along the SW (Welsh) shore of the estuary, and rapid accretion on the NE (Wirral) shore, with lateral and vertical growth of salt marshes.

Some estuary shores have been modified by artificial structures. Shore walls confine the ebb and flow of tides, and force storm surges further upstream. The embanking of estuaries on the Hampshire and Sussex coast has increased high tides upstream, as at Bosham. Structures built to prevent storm surges travelling upstream, such as the Thames Barrier and the Dartford Gate have raised high tide levels downstream.

11.7.6 Estuarine marshes and swamps

The upper part of the intertidal zone in estuaries is usually vegetated, with salt marshes, freshwater swamps or mangroves. As has been shown in Chapter 8, these vegetation communities have colonised and stabilised the shore, diminishing the effects of waves and currents, and so promoting the accretion of sediment in such a way as to build terraces between mid-tide and high spring tide that would not otherwise have developed. Typically salt marshes and mangroves are fronted by intertidal mudflats and sandflats and backed by a transition through freshwater swamps to land vegetation. Their seaward margins are gentle slopes where the vegetation is spreading seaward, or small cliffs cut back by wave action, or meandering tidal channels. They are dissected by tidal creeks that continue seaward across mudflats and sandflats to deeper low tide channels. The existing morphology of salt marshes or mangrove swamps, the mudflats and sandflats that front them and the systems of channels and creeks that receive tidal ebb and flow and river discharge has been moulded largely by currents, and partly by wave action.

As infilling proceeds, bordering salt marshes and mangrove swamps encroach on the estuary, and as the intertidal sandflats and mudflats are built up into depositional terraces the central channel becomes narrower and often deeper. Thereafter, river flooding brings fluvial sediments to aggrade the marshlands and form an alluvial valley floor. In terms of the division by Dalrymple, Zaitlin and Boyd (1992), the inner river-dominated zone thus displaces the central low wave energy zone and the outer marine-dominated zone progressively seaward. This sequence can be reversed by a relative rise of sea level if fluvial discharge and sediment yield remain unchanged.

11.8 Coastal lagoons

Coastal lagoons are areas of relatively shallow water that have been partly or wholly sealed off from the sea by the deposition of spits or barriers, usually of sand or shingle, built up above high tide level by wave action (Emery and Stevenson, 1957; Colombo, 1977; Phleger, 1981; Kjerfve, 1993; Cooper, 1994). This definition excludes lagoons enclosed by coral reefs, either within atolls or between fringing or barrier reefs and the mainland, because these are essentially marginal marine environments linked with the open sea at high tide. However, there are examples of coastal lagoons formed by localised tectonic subsidence, such as the Sissano Lagoon on the north coast of New Guinea (Section 11.8.3).

Most lagoons are to some extent estuarine, being subject to tidal oscillations and the interaction of salt water from the sea with fresh water from inflowing rivers, and much that has been said about estuaries also applies to coastal lagoons. Indeed, some classifications regard coastal lagoons as bar-built or back-barrier estuaries (Fairbridge, 1980).

Coastal lagoons range in size from over 10 000 km^2 (Lagoa dos Patos, Brazil) down to less than a hectare. There are very many small lagoons at river mouths (often termed barred, blocked or blind estuaries), some of which are of interest despite their small size. Thus Oyster Pond, on Cape Cod in the United States is a small lagoon that was studied intensively over a long period by Emery (1969), and research by Cambridge scientists has made famous the little lagoon at Swanpool, near Falmouth in SW England (Barnes, 1980).

The simplest lagoons are found where the mouth of a river has been enclosed by a wave-built barrier. Such a barrier may be breached from time to time by storm waves, or when river floods pour out over it after heavy rain, but usually it is soon rebuilt by wave action when fine weather returns. Lagoons of this type are common on oceanic coasts, where barriers have been built across drowned valley mouths by the action of strong swell.

Some lagoons are long and narrow, parallel to the coast and separated from the sea by barriers built up in front of the former coastline. The Coorong in SE Australia is an example. Others show a branched configuration, elongated at an angle to the coastline, formed where river valleys have been submerged, then enclosed by a depositional barrier built up across their mouths. Drake's Estero in California is a lagoon formed where a branching inlet (ria) incised into the hilly country of the Point Reyes peninsula has been partly enclosed by bordering sand spits (Figure 11.8).

At Orbetello, on the west coast of Italy, a lagoon has been enclosed between twin barriers that form a double tombolo, attaching a former island to the mainland, and on the Chatham Islands, east of New Zealand, barriers of dune calcarenite have formed between three high islands to frame Te Whanga Lagoon, which has a small outlet on the east coast. The largest and most complicated lagoon systems are found where broad embayments have been sealed off from the sea by successive depositional barriers. The Gippsland Lakes in SE Australia (Figure 8.13) are a chain of coastal lagoons enclosed by a Holocene outer barrier (behind the Ninety Mile Beach) and separated by remnants of Pleistocene inner and prior barriers.

Coastal lagoons usually have one or more entrances from the sea, which are permanent or intermittent gaps through the enclosing barriers (Bird, 1993b). The definition of a coastal lagoon implies that their entrances are narrow

Figure 11.8 Drake's Estero, a lagoon of branched outline impounded by paired spits on the Californian coast. The curved outline of Drake's Bay passes from a barrier beach into eroding cliffs

compared with the coastwise extent of the enclosing barrier or the lagoon. The European Wadden Sea, behind the Frisian barrier islands, is not usually considered a coastal lagoon because the combined width of the openings to the sea (including the broad gap east of Wangeroog) is more than one-third of its coastwise extent, but it can be classified as an open lagoon. The same is true of the area behind the sandy barrier islands (Sea Islands) of the coast of Georgia, in the United States. In general, the term coastal lagoon can be applied where the width of marine entrances at high tide is less than one-fifth (20 per cent) of the total length of the enclosing barrier. A problem arises where the lagoon is elongated at an angle, rather than parallel to the coastline, so that a substantial water area is enclosed by quite short barriers or spits across a narrow seaward entrance. Features of a sheltered lagoon environment may then occur even though the marine entrance is substantially more than one-fifth of the length of enclosing barriers. Indeed there are some inlets of intricate configuration with rocky entrances that have no enclosing depositional features, and yet show lagoonal characteristics: the Knysna Lagoon in South Africa, Lake Maracaibo in Venezuela and the large landlocked bay on Sumbawa, Indonesia are examples. On a larger scale the Baltic and Mediterranean Seas also have some of the characteristics of large coastal lagoons, with relatively narrow entrances from the Atlantic Ocean.

Coastal lagoons occur on about 12 per cent of the length of the world's coastline. An inventory prepared by UNESCO listed about 450 coastal lagoons (minimum surface area at high tide 1 km²) on the world's coastline. There are good examples on the Gulf and Atlantic coasts of the United States, where they and their enclosing barriers have transgressed landward in response to a continuing sea level rise, a process that can also be seen on the submerging shores of the Caspian Sea (Section 8.6). Coastal lagoons occur in Brazil, where they formed as the sea attained a slightly higher level, and their enclosing barriers prograded after sea level fell. They have formed behind storm-built barriers in SW Iceland and on the southern shores of the Baltic Sea. They are found around the Mediterranean, notably between Perpignan and Marseilles in France, in eastern Corsica, the northern Adriatic, Egypt and the NW Black Sea, including the Sea of Azov. An intermittent chain of barrier-enclosed lagoons extends from the Ivory Coast to the Cameroons in West Africa, and lagoons are also present in Natal, on the Indian and Sri Lankan coasts, on the north coast of Hokkaido and in eastern Sakhalin. They are also well developed on the arctic coasts of Alaska and NE Siberia.

On the south coast of Britain there are a number of small lagoons impounded by shingle barriers, notably Loe Pool in Cornwall, Slapton Ley in Devon and The Fleet, behind Chesil Beach in Dorset. In the Orkney and Shetland Islands there are many small shingle barriers, known as ayres, that have been built up across the mouths of embayments to enclose, or partly enclose lagoons, known as oyces, which rise and fall with the tide. On the Australian coast, lagoons are best developed behind sandy barriers formed during Pleistocene phases of high sea level and

during and since the Holocene marine transgression, on sectors of the coast of Western Australia south of Perth, and intermittently along the coast of SE Australia from the mouth of the Murray around to southern Queensland (Bird, 1967b). New Zealand has coastal lagoons on the shores of the Bay of Plenty in the North Island, and Lake Ellesmere, the Southland lakes east of Bluff and Okarito Lagoon on the west coast of the South Island.

Coastal lagoons are generally about 6000 years old, having formed where valley mouths or lowlands have been submerged by the sea during the later stages of the world-wide Holocene marine transgression. Some also existed during Pleistocene phases of relatively high sea level, enclosed by Pleistocene barriers, parts of which survive as inner barriers, as in the Gippsland Lakes. Lagoons that were enclosed by Pleistocene barriers drained out during the Last Glacial low sea level phase, leaving subaerial basins that were flooded when the sea rose again. In the Gippsland Lakes surveys have detected drowned river levees and submerged red gum trees, marking the course of the extended Latrobe River along the floor of Lake King during the low sea level phase.

In South Australia, the Coorong is the latest in a series of long, narrow lagoons that were enclosed by successive barriers on a coast that has been uplifted during Quaternary times. Its predecessors are marked by tracts of lagoon and swamp (now largely drained) lying between successive emerged sand barriers in the country behind Encounter Bay (Figure 3.4). The majority of coastal lagoons are simply the product of barrier and spit deposition across inlets and embayments that formed in Holocene times.

Coastal lagoons have a variety of shapes and sizes, related to the configuration of the pre-existing coastline and the enclosing spits and barriers, as modified by internal erosion and deposition around their shores and on their floors. They are best developed on low lying coasts

Figure 11.9 Lagoon zonation related to tidal levels and seasonal salinity variations

behind shallow coastal seas. They are poorly developed on coasts dominated by high retreating cliffs, as in the Great Australian Bight, on the steep and rocky coasts of Norway, British Columbia, Chile, and southern New Zealand, on the ice-girt Antarctic and Greenland coasts and on the rapidly emerging coasts of northern Canada and the Gulf of Bothnia. They are also rare on megatidal coasts, such as the Bay of Fundy in Canada or the Bay of Mont Saint Michel in France, because strong tidal currents have prevented the formation of enclosing spits and barriers.

Lagoon morphology (water depth) can be mapped using conventional survey techniques (traverses with echo sounders), or from satellite multispectral band imagery (checked by on-site sample surveys). Typically depositional lagoon floors are subhorizontal, with deeper channels that may be inherited from prior submerged topography, or scoured by existing currents. Deep scour holes (colks) may be found in narrow straits, especially where there is sharp curvature in the lagoon causing current eddies, as at Metung in the Gippsland Lakes.

Coastal lagoons show a wide variety of geomorphological and ecological features, but their essential characteristics are summarised in Figure 11.9. There are often three zones: a freshwater zone close to the mouths of rivers, a saltwater tidal zone close to the entrance and an intervening transitional zone of brackish (moderately saline) but relatively tideless water. The three zones may occur in a single lagoon system, as in the Gippsland Lakes. The proportions of each zone vary from one system to another: the Myall Lakes, on the New South Wales coast, consist largely of the freshwater zone; Lake Illawarra consists largely of the intermediate zone and Wagonga Inlet largely of the salt-water tidal zone. The extent of each zone depends on the relative proportions of freshwater and seawater inflow to the lagoon system and on climatic conditions, lagoons tending to be more brackish in arid regions. Lagoons completely cut off from the sea, like those on the Landes coast north and south of Arcachon, in France, are essentially freshwater lakes. In Dorset the Fleet is a typical estuarine lagoon with salinity augmented by percolation of seawater in through the shingle barrier (Chesil Beach) at high tide and during storms, and Slapton Ley is an almost fresh coastal lake (Figure 6.7).

11.8.1 Lagoon entrances

Some lagoons have been completely cut off from the sea by barriers, but most have at least one marine entrance (or tidal inlet). The entrance to a coastal lagoon may be bordered by a spit or paired spits, as in the estuarine lagoons at Poole, Christchurch and Pagham on the south coast of England, or there may be several entrances separating barrier islands, as in the Dutch, German and Danish Wadden Sea.

Some lagoon entrances are residual gaps that persisted between spits or barrier islands where the lagoon was never completely sealed off from the sea. Others are the outcome of breaching, either by storm waves or floodwater spilling out of the lagoon. Many lagoon entrances are artificial, having been excavated and stabilised, usually between bordering breakwaters, to facilitate navigation or hasten the drainage of floodwaters to the sea. In Italy the three entrances to the Lagoon of Venice have been artificially stabilised by breakwaters up to 2 km long, the intervening

barrier islands having been armoured by large sea walls (Figure 11.10).

Lagoon entrances are larger, more numerous, and more persistent on barrier coastlines where relatively large tide ranges generate strong currents, as on the German North Sea coast, than where tidal action is weak, as on the long sandy barriers that fringe the southern shores of the Baltic Sea (Section 8.6). Lagoon entrances are often located on parts of the coastline where wave action is relatively weak, and the action of inflowing and outflowing currents therefore more effective. On wave refraction diagrams, such sections are indicated where there is a marked divergence of wave orthogonals, as at the head of the embayment in Figure 2.2(b). Persistent lagoon entrances are found alongside rocky headlands, as at the entrance to the Tauranga lagoon at Mount Maunganui on the north coast of New Zealand. Others are in the lee of islands or reefs where wave action is weakened and the ebb and outflow currents are sufficient to maintain a gap. The entrance to Lake Illawarra in New South Wales is protected by Windang Island immediately offshore, while the entrances to several other lagoons on the New South Wales coast are situated at the southern end of sandy bays, close to rocky headlands, where the dominant SE ocean swell is much refracted, and therefore weakened.

The configuration of a natural lagoon entrance (like that of an estuary mouth) is the outcome of a contest between the currents that flow in and out and the effects of onshore and longshore drift of sand or shingle that tend to seal them off. As in estuaries the current velocity depends on the tidal prism, the volume of water that enters and leaves a lagoon as the tide rises and falls, and the mean cross-sectional area of the entrance increases with the volume of the tidal prism.

Currents are generated through entrances in several ways. There are tidal currents produced by tides entering and leaving the lagoon, their

Figure 11.10   The Lagoon of Venice, showing the three entrances, the network of tidal channels and the site of the city of Venice

strength increasing with tide range. There are currents due to outflow from rivers, particularly after heavy rain, when floods build up the level of the lagoon so that water pours out through the entrance, and there are currents generated by wind action, onshore winds driving seawater into the lagoon and offshore winds driving lagoon water out to sea. Strong currents tend to maintain the dimensions of a lagoon entrance, the cross-sectional area varying in relation to the volume of water passing through, being widened or deepened during episodes of floodwater discharge. When the outflow is weak the entrance may be modified by wave action. Waves that arrive parallel to the coast move sand from the sea floor shoreward on to beaches and into a lagoon entrance, and waves that reach the shore at an angle cause longshore drift, supplying sand to build spits that deflect a lagoon entrance. The growing spit impacts transverse ebb and flow of tidal currents against the farther shore, which is cut back by current scour. Some lagoons have

entrances maintained between the end of a longshore spit and a rocky headland, as at Wingan Inlet in Australia. Lough Hyne in SW Ireland is a tidal lagoon in a glacially scoured basin that has an outlet through a rocky channel where rapids and waterfalls develop as the tide rises and falls.

The position and dimensions of lagoon entrances thus change frequently in response to variations in the processes at work on them. In SE Iceland lagoon entrances are deflected further by longshore spit growth in winter, when wave energy is high and outflow from melting glaciers low, and more direct outlets are cut in summer when outflow increases and wave energy is reduced. Migrating entrances punctuate the sandy outer barrier on the Atlantic coast of the United States: the Outer Banks of North Carolina have had a long and complicated history of breaching, enlargement, reduction, migration and closure of tidal inlets to Pamlico Sound as sand moved to and fro along the coast. Captain Sams Inlet in South Carolina has been migrating southward at 60–70 m/yr. Many such tidal inlets on the Atlantic seaboard have been stabilised by the building of bordering breakwaters.

Lagoon entrances may show seasonal variations, shallowing or becoming sealed in dry seasons when outflow is weak, and reopening, widened or deepened, when the wet season brings greater volumes of water outflow from the lagoon. This sequence is well known on the South African coast, where most lagoons are sealed off by sand deposition in the relatively dry winter season (May to August) and entrances are reopened in the summer when rains in the hinterland increase fluvial discharge into the lagoons until the water spills over into the sea. In SE Australia, lagoon entrances are usually reduced in size or sealed off altogether in dry summer periods, then reopened or enlarged in the wetter winters. When a barred entrance has persisted for several seasons, local people may reopen it by digging an outflow channel, which widens and deepens as the head of water disperses. After several dry years, discharge of water from the River Murray through the Murray-mouth lagoons (Figure 11.12) into Encounter Bay has been insufficient to maintain a channel through the outer barrier near Goolwa. In 2002 about 300 000 m³ of sand was dredged from the mouth of the River Murray to maintain tidal flushing in the adjacent Coorong lagoon, but the channel thus formed was soon narrowed and shallowed by the inwashing of sand by waves. Repeated dredging will be necessary to maintain this outlet.

Lagoon entrances are generally backed by partly or wholly submerged shoals or fans of inwashed sediment, similar to the thresholds seen in estuaries. There may also be tidal deltas, lobate or triangular shoals deposited on the inner or outer sides of a lagoon entrance by inflowing or outflowing tidal currents. Examples of thresholds have been documented from the South Coast of New South Wales, as at Burrill Lake (Jennings and Bird, 1967) and at the entrance to the Murray-mouth lagoons (Bourman and Harvey, 1983; Walker and Jessup, 1992). Such features are less common on the seaward side, where stronger wave action disperses outwashed sediment offshore or alongshore. At Lakes Entrance, Victoria, an outlet from the Gippsland Lakes, cut in 1889, was initially deep, but a sandy threshold has formed on the landward side and is growing into Lake King, evidently nourished by sand inflow caused by the dredging of the looped sand bar off the lagoon entrance.

Migration of lagoon entrances has occurred on many barriers, and is well illustrated on the Danish North Sea coast, where the barrier enclosing the lagoon known as Ringköbing Fiord (the term fiord includes lagoons in Danish terminology) has had an entrance in various positions since 1650 (Figure 11.11). This variable entrance became sealed off altogether when it was replaced by an artificial canal at Hvide Sande, maintained between stone jetties, and a sluice was added in 1931 to reduce seawater inflow. Sand accumulation on the northern side of the breakwaters is indicative of the southward drifting that formerly diverted the lagoon entrance.

In New Jersey curving channels lead from lagoons into the rear of the sandy barrier towards entrances that have been deflected along the coastline by the longshore drift of sand, and behind the Ninety Mile Beach in Australia similar curved channels lead towards the sites of tidal entrances to the Gippsland Lakes that have since been sealed off by sand deposition.

Several lagoon entrances have been maintained by building breakwaters and then transporting sand from the updrift to the downdrift side: a procedure known as by-passing. Examples include Scheveningen in the Netherlands, Figueira da Foz in Portugal and several inlets in Florida.

Lagoon entrances thus show a variety of features. At one extreme is the permanent entrance, natural or man-made, which allows a perennial unhindered exchange of water, sediment,

Figure 11.11 Ringköbing Fiord, a coastal lagoon in Denmark, showing the southward migration of the former entrance as a result of longshore drifting of sand, and (inset) the artificial entrance cut at Hvide Sande

dissolved materials and organisms between the lagoon and the open sea. The lagoon then shows estuarine characteristics, such as a transverse salinity gradient declining towards the mouths of inflowing streams and an inwardly diminishing tidal ventilation, as in the Peel-Harvey Inlet in Western Australia. Tide ranges within a lagoon diminish rapidly away from the entrance(s), the more remote sections of large lagoon systems being unaffected by marine tides: shore width generally decreases with tide range. In the Gippsland Lakes the range of spring tides at Lakes Entrance is slightly less than a metre, but at Metung, 10 km to the west, it is less than 30 cm, and tides are not perceptible further away from the entrance, in Lake Victoria and Lake Wellington. There are, however, irregular changes of level due to heavy rain or river flooding, oscillations (termed seiches) during and after periods of strong wind, fluctuations in atmospheric pressure and the effects of storm surges. Similar variations have been observed in other coastal lagoons, for example the Swan River estuarine lagoon at Perth in Western Australia.

At the other extreme is the completely enclosed lagoon with an impermeable barrier preventing exchanges with the sea. As has been noted, such lagoons tend to become freshwater lakes in humid environments and hypersaline in arid regions. Between these extremes are lagoon entrances that vary in form, dimensions and location, and are sometimes completely sealed off. These variations modify the extent to which tides invade a lagoon (tidal ventilation), river flood levels, marine incursion, salinity regimes and related ecological conditions within a lagoon system, and can influence patterns of sedimentation and geomorphological change.

11.8.2 Tides and salinity in lagoons

The hydrological characteristics of a coastal lagoon are determined partly by its configuration, partly by the dimensions of entrances from the sea and partly by the balance between atmospheric precipitation, freshwater inflow from rain and rivers and evaporation on the one hand, and salt-water inflow from the sea, related to tide range and tidal ventilation of the lagoon, on the other. Winds blowing over a lagoon lower the level at the windward end and build it up to leeward, so that when the wind drops normal level is restored, often by way of seiches of diminishing amplitude.

Entrance dimensions also influence the pattern of salinity in a coastal lagoon. As in estuaries, salinity is determined by the meeting and mixing of fresh water from rain and rivers and salt water from the sea, and generally diminishes from the lagoon entrance towards mouths of the rivers. In regions with seasonal variations of precipitation the salinity regime is also seasonal, for in the dry season seawater flows in through the entrance to compensate for the diminished river flow and the loss of fresh water by evaporation, but in the wet season evaporation losses are reduced and the lagoon is freshened by rain and rivers. In humid regions lagoons typically have estuarine salinity regimes, salt-water inflow from the sea being diluted by rainfall and freshwater runoff into the lagoon system. However, on arid coasts lagoons may lose more water by evaporation than they receive from rainfall and runoff, and if the inflow of seawater is insufficient to prevent the development of high concentrations of salt they become hypersaline, or even dry out altogether as saline flats. This has happened in the Laguna Madre, on the Texas coast, which has shallow hypersaline areas and saline flats with gypsum and algal mats. Similar features are seen at the southern end of the Coorong in South Australia, away from the lagoon entrance, where tidal ventilation is too weak to prevent the development of high salinity in evaporating water. During the summer months hypersaline conditions develop in enclosed lagoons such as Lake Eliza and Lake St Clair on the coast of South

Australia, and Lake Daningdella is a desiccated lagoon (salt flat) behind Israelite Bay in Western Australia where saline evaporite deposits have formed.

Salinity conditions influence modes of sedimentation in lagoons. Clay carried in suspension in fresh water is flocculated and precipitated by the electrolytic effect of sodium chloride in solution when saline water is encountered. Inflowing fresh water is often brown or grey with fine grained sediment in suspension, and becomes clear when salt water is encountered, the junction often being sharply defined, with coagulated sediment raining down. Salinity is also of ecological importance, affecting the development and distribution of shore vegetation around lagoons and thus influencing patterns of sedimentation in encroaching swamps.

Lagoons with more restricted or temporary entrances are less influenced by tidal movements and more protected from the effects of waves from the open sea. Geomorphologically they may resemble inland lakes. In New South Wales, for example, Lake Macquarie is essentially a marine lagoon, with almost tideless shores, a barrier that excludes ocean waves, and water salinity similar to that of the open sea, except in small areas of dilution near the mouths of inflowing streams. Marine influences are also much reduced in lagoons where the entrance is in the form of a long, winding channel through the enclosing barrier, as in the Myall Lakes, also in New South Wales, and the Castillo Lagoon in Uruguay. Usually the water is fresh or slightly brackish, but during droughts salinity increases as seawater spreads in along the connecting channel.

Some coastal lagoons have been sufficiently cut off from the sea by enclosing impermeable barriers to become freshwater lakes, as at Slapton Ley in SW England (Figure 6.7), while the Murray-mouth lagoons in South Australia (Figure 11.12) have become fresh since their entrances were artificially sealed by barrages. In New Zealand, the Waitangitaona River changed its course during a flood in 1967 to flow into Okarito Lagoon, a brackish lagoon that was thus freshened (Soons, 1982).

By contrast, the cutting or enlarging of an entrance to a lagoon that was previously fresh or slightly brackish results in a salinity increase that can have geomorphological as well as ecological consequences. This has happened in the Gippsland Lakes, Australia, and two West African lagoons: Lake Nokoué in Benin, after the opening of a marine entrance at the port of Cotonou, and the Ebrié lagoon on the Ivory Coast, after the cutting of the Vridi Canal through the enclosing barrier in 1950. Reduction of runoff from rivers by reservoir construction or soil conservation schemes in the hinterland can also increase salinity in coastal lagoons, and may have contributed to the recent rise in salinity in the Gippsland Lakes.

11.8.3 Evolution of coastal lagoons

The initial form of a coastal lagoon depends on the shape of the inlet or embayment enclosed and the inner shores of the barriers that enclosed it. Some lagoons were originally broad embayments (e.g. Lake Illawarra in New South Wales), others show the much branched form of submerged valley systems (e.g. Lake Macquarie in New South Wales), and in some the enclosing barriers incorporate high islands (e.g. the Tuggerah Lakes in New South Wales). The evolution of coastal lagoons has been influenced by the geological and geomorphological history of the coastal area, and the sequence of changes in the levels of land and sea that have resulted in coastal submergence, forming the inlets and embayments. The growth of coastal barriers shaped the initial morphology, and determined the position and dimensions of entrances from the sea.

Once enclosed, coastal lagoons are modified by erosion and deposition. The inner shores of

Figure 11.12 The Murray-mouth lagoons, enclosed by coastal barriers on the shores of Encounter Bay. Five barrages were constructed across gaps in the inner barrier in 1940 to prevent tidal incursion and control salinity

the enclosing barriers are usually simple in outline, but there may be protrusions of recurved ridges, which marked stages in the prolongation of a spit that became an enclosing barrier, or promontories formed where sand has blown over as an advancing dune, or been washed over or through low lying sections of the barrier by storm waves or exceptionally high tides. Washover fans are trimmed and re-shaped by lagoon waves and currents, and may evolve into cuspate spits or forelands on a lagoon shore. Deposition is also common in the zones behind barrier islands where tides flowing in from neighbouring entrances meet, as in the Wadden Sea and on the southern side of Scolt Head Island,

where deposition forms a tidal divide with sandflats and mudflats. As accretion proceeds this may become marshy.

Dunes spilling from the outer barrier into the Coorong lagoon in South Australia squeeze out plastic mud on the lagoon shore, and on Sperm Whale Head in the Gippsland Lakes similar loading produces recurrent mud islands off the inner shore of the barrier. These are soft sediments, soon dispersed by wave erosion.

Sediment is carried into a coastal lagoon by rivers, by tidal currents entering from the sea and by winds blowing sand from bordering coastal dunes. Deposits include material of organic origin, such as shells, guano and peat, and in arid regions chemically precipitated salt, calcite and dolomite. In one way or another, most lagoons are being gradually filled in and will be replaced by depositional coastal plains.

Lagoons fed by rivers receive sediments ranging from coarse sand to silt and clay, some of which may be deposited in deltas. In the sheltered waters of the Gippsland Lakes small deltas of silt and clay have been built by the Latrobe and Avon Rivers into Lake Wellington, and the Tambo and Mitchell Rivers into Lake King (Bird, 1978a). The deltas of the Latrobe, Avon and Tambo Rivers are cuspate in form, while the Mitchell delta, built into the more sheltered water in the northern part of Lake King, consists of elongated silt jetties similar to those built by the much larger Mississippi River in the United States. As on the Mississippi delta, the growth of all of these deltas was assisted by the presence of reedswamp on their shores, in which sediment washed down to the river mouths during floods was trapped and retained. Where the reedswamp fringe has disappeared as the result of salinity increase deltaic sediment is no longer retained by shore vegetation, and waves are attacking the unprotected shores of the deltas. The Tambo delta and the Mitchell delta in Lake King show advanced stages in dissection, but the Latrobe and Avon deltas, in the less brackish waters of Lake Wellington, remained reed fringed and were still growing by the addition of intercepted silt and clay until the 1970s. Now only patches of reedswamp remain, and these deltas are no longer growing. Evidently the presence of reedswamp promoted sedimentation so that deltas could be built at these river mouths, but when the reed fringe disappeared the deltas were no longer stable, and erosion began to consume them. Figure 11.13 uses outlines from an 1848–49 survey (when the delta was reed fringed) and air photographs taken in 1940 (when the reed fringe had disappeared) and 1990 to show the shrinkage of the Mitchell River silt jetties.

Sand and gravel deposited as the river enters a lagoon may be added to lagoon beaches and spread around the shore by wave action, while the finer sediment is carried out into the lagoon and deposited on the floor, progressively reducing the depth.

Fluvial sediment yields to lagoons may be modified and accelerated by the reduction of vegetation cover, the onset of soil erosion or mining activities in the river catchment. The rivers that drain into the Gippsland Lakes formerly delivered mainly silt and clay, but deforestation and soil erosion in the catchment has resulted in downstream movement of sand and gravel. So far little of this has reached the lagoons, but successive floods are moving the sand and gravel downstream, and this coarser sediment will eventually be deposited on the lagoon floors and shores. Lake Wellington, which is now bordered by a mainly swampy, eroding shore, will then become a lagoon fringed with sandy beaches.

As has been noted, deforestation and agricultural development in the catchment of the Citanduy River in southern Java greatly increased the flow of muddy sediment from that river into the Segara Anakan lagoon, which is rapidly silting (Figure 11.14).

In NE Tasmania sand generated by tin dredging in has moved down valleys into the George River, which is now building a sandy delta into the coastal lagoon at George Bay. Medea Cove, formerly an arm of the George Bay lagoon, has been largely filled with sandy mining waste washed in by Golden Fleece Creek to form a broad sandflats and mudflats, colonised by rushy salty marsh and riverine scrub (Bird, 2000).

On the other hand, there could be a reduction of sediment yield to a coastal lagoon where a dam constructed on an inflowing river is intercepting sediment in the reservoir, or by successful soil conservation works in the hinterland.

Subsidence of coastal regions, as in the northern Adriatic or along the Gulf and Atlantic

Figure 11.13   Deltas in the Gippsland Lakes, Australia. The Mitchell delta, built into Lake King, is digitate and was still growing by sedimentation in reedswamp (as in the modern Mississippi delta, Figure 12.2) when European explorers arrived in 1830. The outline of the reed-fringed Mitchell delta was shown on an early survey (1948–49). Air photographs taken in 1940 show that the reed fringe had disappeared, and the delta shores were cliffed and eroding. By 1970 the delta had diminished further, but subsequently much of its coastline has been stabilised by walls and boulder ramparts. The Tambo delta, more exposed to waves generated by the prevailing SW winds, is cuspate, as is the Latrobe delta, built into Lake Wellington. The Tambo delta had lost its former reedswamp fringe by 1940 and was eroding, whereas the Latrobe delta was still reed fringed and growing, but by 1980 this reed fringe had become sparse, and delta growth has come to an end

coasts of the United States, may deepen and maintain coastal lagoons, delaying their infilling. Cavazzoni (1983) concluded that subsidence had deepened the water in the Lagoon of Venice and allowed larger waves to erode the lagoon floor. On the north coast of New Guinea the 1907 earthquake in the Torricelli Ranges resulted in local subsidence on the coastal plain and the formation of a lagoon up to 2 m deep in an area that previously carried villages and coconut plantations, behind a sandy coastal barrier at Sissano, near Aitape. There was extensive, though brief, marine submergence here during the 1998 tsunami. Along the northern flank of the Sepik delta in New Guinea subsidence has widened and deepened an old meandering channel to form the chain of lagoons known as the Murik Lakes behind a narrow sandy barrier.

Accumulation of inwashed sediment, organic deposits such as peat or shells, and precipitated salts results in the shallowing and shrinkage of lagoons. The following sections show how changes in configuration are related to the effects of wind-generated waves and the currents produced by rivers, wind action and tides within the lagoon, and how ecological conditions, particularly water salinity and temperature, are

Figure 11.14 The Segara Anakan is an estuarine lagoon in southern Java, which has been shallowing rapidly as the result of inwashing of sediment from the Citanduy and other rivers. In recent decades mangroves have advanced along the lagoon shores

important in the geomorphological evolution of coastal lagoons.

Isla (1995) suggested that the evolution of lagoons in warm and arid coastal regions has been dominated by physical processes, with relatively coarse sediment mobilised by waves, currents and wind action, and chemical processes, notably the formation of evaporites, with biological processes playing a subsidiary role (notably where there are mangroves and corals). By contrast the evolution of lagoons in cooler and wetter coastal regions has been influenced more strongly by biological processes, including extensive wetland vegetation and the production of shelly organisms. However, there is plenty of biological activity in humid tropical lagoons, while physical processes in cold regions include physical weathering, especially in cold environments where frost and ice processes are active.

11.8.4 Rounding and segmentation

As a barrier develops to enclose a lagoon sea waves are excluded and the effects of marine salinity and tide changes of level are reduced. Winds blowing over the lagoon generate waves and currents that are related to the direction and strength of local winds and the lengths of fetch across which these winds are effective. Long, narrow lagoons have the strongest wave action in diagonal directions, along the maximum fetch. The lagoon shore may be cliffed by wave attack, yielding sediments that are carried alongshore by waves arriving obliquely.

Waves coming in at an angle to the lagoon shore move sediment to and fro along beaches, eroding embayments and depositing spits and cusps. These may grow to such an extent that the lagoon becomes divided into a series of

Figure 11.15 Segmentation of a long, narrow lagoon into a chain of smaller, rounded lagoons separated by paired or coalescent spits, as the result of erosion and deposition by wave action. The shape of the segmented lagoons is related to waves generated by local winds, and reflects the wind regime. After Zenkovich, 1967

small round or oval lagoons, linked by narrow straits, or sealed off completely (Figure 11.15). This process was called segmentation by Price (1947), who described it from lagoons on the Texas coast, and the growth of spits into cusps and cuspate forelands has led to segmentation of Koozata Lagoon on St Lawrence Island, Alaska (Fisher, 1955) and the Kosi lagoon in Zululand. Zenkovich (1959) described a similar process at work in the Shagany lagoon on the Ukrainian Black Sea coast, where sediment moved along the shore had been built into spits and bay barriers, rounding and smoothing the initially irregular configuration.

Segmentation is essentially an adjustment of lagoon shape to patterns more closely related to the waves and currents generated within a lagoon. Wind-driven currents play a part in smoothing the curved outlines of the shore in the later stages of segmentation and may also maintain the connecting straits between segmented lagoons. With equivalent winds from all directions, the segmented lagoons produced by wave action would be circular, but they are more often oval, with a long axis parallel to the direction of the prevailing winds.

Segmentation takes place most readily in tideless lagoons or in parts of lagoons where the tide range is small, for tidal currents interfere with the wave processes by deflecting growing spits and forelands so that they trail towards, or

away from, the point of tidal entry and the coalescence of opposing spits is prevented. Tidal changes of level prevent continuous wave action at a particular level, so that a neat adjustment of coastlines to wave resultants is less likely. Instead, an initially long and narrow lagoon with straight parallel shores (between a former coastline and a parallel barrier, or between successively formed barriers) may develop a meandering outline, as in the lagoon at Cananéia in southern Brazil and the Hopetoun Channel in the Gippsland Lakes, Australia. It is possible that the meandering outline results from current flow (perhaps alternating), and develops in a similar way to meandering river and estuary channels.

In the Gippsland Lakes segmentation is illustrated by the growth of a recurved spit on the eastern shore of Lake Wellington, which has been almost isolated from Lake Victoria, except for the link maintained by currents through the intervening McLennan Strait. The oval outline of Lake Wellington reflects the prevalence of westerly winds. In adjacent Lake Victoria, erosion of embayments and growth of intervening cuspate forelands are further signs of segmentation in progress. Also in the Gippsland Lakes, Cunninghame Arm shows a series of cuspate spits which, towards the eastern end, have grown to such an extent that they have almost cut off a chain of shallow pools, the Warm Holes, linked by connecting creeks maintained by wind-driven and tidal currents. On the South Australian coast, the lagoons between Robe and Beachport (Lake Eliza, Lake St Clair, and Lake George) have been formed by the segmentation of a long narrow lagoon, originally similar to the Coorong farther north. Many lagoons are bordered by growing spits and scoured embayments indicating incipient segmentation. The Lagoa de Araruama, near Cabo Frio in Brazil, has cuspate spits (e.g. Punta do Aceira) that have been enlarged as a result of emergence, which shallows lagoons and hastens segmentation.

Another form of segmentation occurs where tidal inflow from more than one marine entrance meets behind barrier islands. Deposition in the meeting zone can produce a tidal divide (Figure 10.2), and eventually a land isthmus segmenting the lagoon, as in the Laguna Guerrero Negro, Mexico (Phleger, 1969).

11.8.5 Swamp encroachment

As shown in the preceding chapter, vegetation has a strong influence on patterns of sedimentation. This is illustrated on coastal lagoon shores by the process of swamp encroachment. Near tidal entrances to lagoons, salinity conditions are similar to those in estuaries, with foreshores and banks of sediment exposed at low tide, which may be colonised by salt marshes or mangroves. These can spread forward and build up depositional terraces. Salt marshes can spread from tidal lagoon shores, and mangrove encroachment has been described from Lagos lagoon in Nigeria and the Segara Anakan in southern Java (Figure 11.14).

Away from the entrance, where the water is brackish and tidal fluctuations diminish, the salt marsh or mangrove fringe is much reduced. Lagoon shores in this intermediate zone may be unvegetated and bordered by beaches of sand or gravel. Towards the mouths of rivers where the water is relatively fresh, reedswamp dominated by species of *Phragmites*, *Scirpus* and *Typha* may colonise the shore, and spread into water up to 1.5 m deep. Reedswamp promotes sedimentation by trapping silt and floating debris, and by contributing organic matter (peat) so that new land is built up. Reedswamp is then invaded by swamp scrub communities, followed by swamp forest as the substrate aggrades (Figure 11.16). In Slapton Ley, Devon, reedswamp that has spread across a freshwater lagoon is being replaced by a natural vegetation succession to willow scrub and woodland.

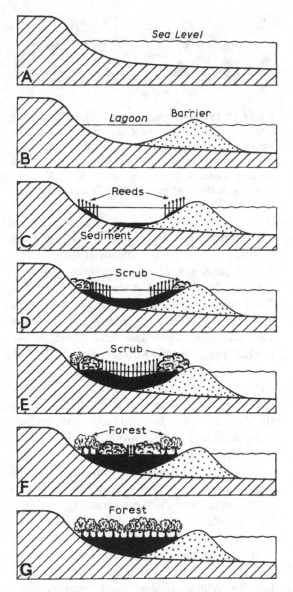

Figure 11.16 The formation of a coastal lagoon by barrier formation (a), (b) is followed by reedswamp encroachment (c) and a vegetation succession to scrub (d), (e) and forest (f), (g) as the lagoon fills with sediment and peat deposits

Swamp encroachment is only possible where ecological conditions are favourable. If the reeds and rushes are not present, sedimentation occurs across the lagoon floor, and is not

Figure 11.17   Reedswamp encroachment on the shores of the Gippsland Lakes (see Figure 11.16(c))

concentrated near the shore, so the lagoon becomes shallower instead of shrinking. Strong waves or current scour impede reedswamp encroachment, the reed fringe that is broad on sheltered parts of a lagoon shore thinning out where wave exposure is greater. Reedswamp can be reduced by cutting, or damaged by boat scour or water pollution, but the limiting factor in coastal lagoons is usually salinity. The reedswamp fringe is best developed in fresh water around river mouths, and thins out, disappearing towards marine entrances where the lagoon becomes more saline.

The Gippsland Lakes were formerly relatively fresh, with shores extensively fringed by encroaching reedswamp (Figure 11.17), but in 1889 the intermittent natural outlet was replaced by a permanent artificial entrance cut to improve navigation in and out of the Gippsland Lakes, and this allows unrestricted inflow of seawater in dry seasons. Over subsequent decades salinity has increased in the Gippsland Lakes, and the reedswamp fringe has died back, surviving only in relatively freshwater areas close to river mouths. Without its protection the lagoon shores, including deltaic silt jetties built by inflowing rivers, have been eroded (Figure 11.18), and sediment removed from them deposited offshore. The change to more brackish water has led to the Gippsland Lakes becoming larger and shallower.

This sequence is reversed where lagoons have been sealed off from the sea, either naturally by the development of barriers or artificially by the insertion of weirs and barrages, so that salinity has diminished. Such freshening stimulates the spread of reedswamp around lagoon shores and initiates new patterns of sedimentation, as in the coastal lagoons at the mouth of the Murray River in South Australia (Figure 11.12), which were naturally brackish, estuarine systems with several natural entrances from the

Figure 11.18 Shoreline erosion of previously formed swamp land on the shores of the Mitchell River silt jetties, Gippsland Lakes (see Figure 11.13)

sea until barrages were built to exclude seawater. Salt marshes that formerly bordered their shores have been replaced by reedswamp, which is spreading forward into the lagoons and trapping sediment that would otherwise have been dispersed or swept out to sea. Sealing off natural entrances to the Murray-mouth lakes thus reversed the geomorphological sequence seen in the Gippsland Lakes.

Similar changes are taking place in the Etang de Vaccarès, a lagoon on the Rhône delta in the Camargue region of southern France, where sluices now exclude seawater and freshening has resulted in the rapid spread of reedswamp. Kalametiya Lagoon in southern Sri Lanka showed a similar response following the building of a sluice at the entrance to keep out seawater, and increased inflow of fresh water discharged from irrigated rice-fields in the hinterland has resulted in extensive swamp encroachment (Mahinda Silva, 1986).

11.8.6 Lagoon configuration

Lagoons bordered by encroaching reedswamp contract in area until they are completely occupied by swamp land, whereas those without shore vegetation are reshaped by waves and currents as segmentation proceeds. Eventually, after many changes in configuration, coastal lagoons receiving sediment are filled in, and replaced by coastal plains, across which rivers and residual creeks wind, uniting to pass out to sea through the tidal entrance. Sediment and soil patterns on coastal plains sometimes show the outlines of former lagoons that have been extinguished by deposition, as in the Anzio district in Italy and on the Sussex coastal plain.

Lake Reeve, behind the Ninety Mile Beach in Australia, is an elongated shallow lagoon that often dries out completely. Its shores have advanced by the accretion of numerous low beach ridges of shelly sand, emplaced by the small waves generated across the shallow lagoon, which collect shelly and sandy material (mainly *Coxiella* shells) from the lagoon floor (Jenkin, 1966). Similar features are seen on the shores of the Laguna Madre in Texas.

On parts of the Gulf and Atlantic coasts of the United States lagoons have become narrower, and in places extinguished, by transgressive barriers that have been driven landward (Section 8.6). Napier Lagoon on the east coast of North Island, New Zealand, emerged and drained out during the 1931 earthquake.

Coastal lagoons may be reopened, reviving an embayment, where erosion breaches and removes the enclosing barrier. Guichen Bay, at Robe, on the South Australian coast and Rivoli Bay, at Beachport, farther south, were at one stage enclosed lagoons comparable to Lake Eliza, Lake St Clair and Lake George, which lie behind a barrier of calcarenite on the intervening coast. Reefs and islands of calcarenite in Guichen Bay and Rivoli Bay indicate the former extension of the enclosing barrier northward and southward (Bird, 1967b). The breaching of similar calcarenite barriers on the coast of Western Australia has led to the formation of Cockburn Sound, and lunate embayments have also developed where dune calcarenite barriers have been breached on the coast of Israel near Nahsolim, and on the north coast of Puerto Rico.

In terms of the geological timescale, coastal lagoons are ephemeral features, likely to be replaced by depositional plains or opened as coastal embayments, depending on the subsequent evolution of the coastal region in which they have developed.

11.9 Summary

The Holocene marine transgression invaded valley mouths and parts of coastal lowlands to form inlets and embayments. The generally branched inlets formed by submergence of unglaciated river valleys are known as rias, while the deeper, steep sided troughs cut by glaciers are submerged to form fiords. Fiards (förden or sea lochs) are formed by marine submergence of glaciated lowlands. Calanques are steep sided marine inlets where limestone gorges have been invaded by the sea. On arid coasts narrow marine inlets are termed sharms, and branched embayments with zoned evaporites are sebkhas.

Estuaries are river mouths influenced by tides and salinity from the sea. They are shaped largely by currents, particularly where the tide range is large, but waves may shape the shores of wide estuaries. There are often distinct ebb and flood channels, as well as threshold bars and tidal deltas at the marine entrance. They are subject to interactions between fluvial and marine processes and sediments, and are generally being filled with sediment. Salt marshes and freshwater swamps have formed in many estuaries, and influence patterns of sedimentation.

Coastal lagoons have been produced by marine submergence, then partly or wholly enclosed by longshore spits or barriers. They have one or more marine entrances, generally maintained by tidal ebb and flow, and incursions of seawater through these produce a salinity gradient from marine through brackish to fresh water at river inflows. Tide range in a lagoon generally diminishes from a marine entrance, except on arid coasts where lagoons may become hypersaline or dry out as salt flats. The evolution of a coastal lagoon is related to wave and current processes, the inflow of water and sediment from rivers and the sea and the development of salt marshes, mangroves and freshwater swamps. Most lagoons become smaller and shallower as the result of sedimentation and swamp encroachment. Waves and currents may reshape lagoon shores, eroding bays and building spits, leading eventually to segmentation into smaller, rounded lagoons. Where the enclosing barriers are transgressive they advance into the lagoon, and on submerging coasts the barrier–lagoon system may migrate landward.

Lagoons are strongly influenced by ecology, particularly in response to salinity changes: if they become fresher, freshwater swamp encroachment increases, but if salinity increases there may be die-back and erosion of formerly freshwater swamps.

12
Deltas

12.1 Introduction

Deltas have been built where sediment brought down by rivers has filled the mouths of valleys drowned by the Holocene marine submergence to form a depositional formation that protrudes from the general outline of the coast. They have formed where the rate of sediment accumulation at the river mouth has exceeded the rate at which sediment is eroded and dispersed by waves and currents. The volume of sediment deposited in the world's deltas in Holocene times is enormous, but collectively they occupy only about one per cent of the world's coastline.

The term delta was introduced by the Greek scholar Herodotus in the fifth century BC to describe the large alluvial lowland at the mouth of the River Nile, which resembled the Greek letter Δ. It became a geomorphological term for depositional lowlands formed around river mouths, even if (like the Rhine delta) they do not protrude from the general outline of the coast. The extensive deltas built by the several large rivers that flow to the north coast of Java have coalesced to form a wide deltaic plain (confluent deltas).

12.2 Delta components

Most deltas have subaqueous and subaerial components, above and below the low tide line (Coleman, 1981). The subaqueous component includes a nearshore sea floor plain sloping gently out to a more steeply sloping delta front that declines to a flatter prodelta apron on the sea floor. These features been formed largely by deposition of sediment by outflowing rivers, the calibre of sediment generally decreasing from sand and silt on the nearshore sea floor plain and delta front to clay in the prodelta. These subaqueous features have all advanced seaward as the delta prograded. The nearshore sea floor plain may however include segments that are essentially wave cut, formed where parts of the delta have been cut back by marine erosion.

The subaerial component (above sea level) consists of a lower and an upper delta plain. On some deltas the river divides into distributaries that diverge across a delta. The upper delta plain has been built above high tide level by the deposition of alluvial sediment during episodes of river flooding, storm surges or exceptionally high tides. They may include natural levees

Coastal Geomorphology: An Introduction, Second Edition.   Eric Bird.
© 2008 by John Wiley & Sons, Ltd

alongside river channels, declining into back-swamp depressions that contain lakes or wetland vegetation, accumulating peat. Some deltas have grown to enclose former high islands as bedrock hills, as on the Klang delta in Malaysia.

In the lower delta plain the river channel becomes tidal. Distributaries reach the coast as salients between embayments, which contain salt marshes or mangrove swamps and beaches built by wave action along the shore. In the coastal fringe there may be beach ridges, dunes and cheniers.

On some large deltas there are active zones where vertical accretion and seaward progradation are continuing, and abandoned zones in which there is no longer river deposition, and where the coastal fringe may be submerging and eroding.

12.3 Deltaic processes

Deltas have been built at the mouths of rivers delivering an abundant water and sediment yield to the coast, derived from runoff and erosion in extensive drainage basins and depending on their climate, geology and topography. The Mississippi drains 3.3 million km^2/yr and the Amazon 5.9 million km^2. Deposition of fluvial sediment takes place in and around river mouths as the velocity of river flow diminishes on entering the sea, but most deltas also incorporate sediment drifting alongshore, and marine sediment moved in from the sea floor. In general, deltaic sediments show gradations from coarser material (sand) to finer silt and clay downstream along channels, resulting from a diminution in flow velocity.

Deltas are found on coasts in various climatic zones. Climate within river catchments influences runoff and river regimes, and with geology and topography determines the nature and rate of sediment yield down to the coast. Humid tropical deltas carry luxuriant vegetation including mangrove swamps, except where the vegetation has been cleared to make way for agriculture and aquaculture. On cold and arid coasts delta vegetation is sparse, and physical processes predominate.

Natural levees border river channels on alluvial valley floors and deltas, and are backed by low lying, often swampy or flooded depressions (Russell, 1967). These are the outcome of an unequal building up (aggradation) of the alluvial plain. When the river rises and overflows its banks the flow of water is most rapid along the line of the river channel and much slower on either side so that the coarser load of sand and silt carried by floodwaters is relinquished in the zone immediately adjacent to the river channel, where water velocity diminishes, and only the finer clay particles are carried into the calmer water beyond, to be deposited on the valley-floor plain. On macrotidal deltas such as the Irrawaddy natural levees may be built in a similar way by the overflow of water flowing up-channel during high spring tides. As natural levees are built up along the sides of the river channel they develop gentle outward slopes, passing down lateral depressions, known as levee flank or backswamp depressions, as in the Mississippi valley. These depressions are floored by clay deposits that settle from floodwaters, and are often occupied by freshwater swamp vegetation that can build up peat deposits. In dry regions repeated evaporation of water from enclosed backswamp depressions leads to a concentration of salt, increasing soil salinity so that they are occupied by salt marshes, or even unvegetated saline flats. Deflation from dry depressions may lead to the building of silt or clay dunes, as on the Senegal delta in west Africa, and river channels that dry out during periods of low flow may provide a source of wind-blown sand for dunes built leeward of river channels during periods of low water flow. Baer's Mounds are low hills of wind-blown sand and silt on the Volga delta. Overbank splays of sediment are deposited where the river has flooded over or through the natural levees.

Distributaries may form as the result of breaching of natural levees during floods, particularly after deposition of sediment on the floor of a river channel has lifted the river, so that avulsion occurs as it spills out over its banks and finds a new outlet. Alternatively, a river mouth may be split into two or more channels by the formation of shoals that grow up as islands. During a major flood in 1878 the Clutha River in the South Island of New Zealand established twin courses on a wide deltaic plain on either side of Inchclutha, and deposited gravel to seal off the former Port Molyneux. There may be two or three diverging distributary channels, as on the Rhône delta, or more complex bifurcations, sometimes rejoining or anastomosing, as on the Volga delta: the more complicated patterns are best developed where the offshore gradient is very low, or where the fluvial sediment yield is coarse (sand and gravel). Distributaries wax and wane: some may become major river channels while others silt up. On the Rhône delta the former main outlet is now the relatively unimportant Petit Rhône, the present river discharging through the Grand Rhône channel farther east. The Nile delta downstream from Cairo used to have many distributaries, but all but two of the outflow channels are now defunct (Figure 12.1).

River channels on deltas are influenced by tides. Tidal ebb and flow currents tend maintain river mouths, whereas longshore drifting may divert them or seal them off altogether. Where the tide range is large, delta distributaries widen seaward to funnel-shaped estuarine outlets such as those of the Rhine (Berendsen, 1998).

River-mouth processes depend partly on nearshore water depth, the velocity of discharge diminishing in shallow water because of bottom friction. Fluvial sediment is deposited as the current slackens, forming one or more bars in shapes depending on the type of discharge, the influence of tidal currents and the effects of wave action. Where the nearshore water is relatively deep a turbulent jet flow forms a lunate bar across the river mouth, often of coarse sand or gravel on the inner side, grading to fine sand, silt and clay on the seaward slope. The pattern of deposition then follows the pattern described by Gilbert (1890) from a section through an emerged and dissected Late Pleistocene delta on the shores of the former Lake Bonneville, now reduced to the Great Salt Lake in Utah. He described almost horizontal radially dispersed bottomset beds, overlain by inclined (10–25°) coarser foreset beds deposited on a steep prograding bar front, capped by horizontal or landward dipping topset beds behind the bar crest.

Where the nearshore zone is shallower and bottom friction stronger the river mouth has a triangular middle ground bar between divergent channels, a pattern characteristic of the microtidal Mississippi mouths. Strong tidal currents can divide this into linear shoals diverging slightly off a funnel-shaped river mouth, with intervening mutually evasive channels, some ebb dominant and some flood dominant, as in the macrotidal Ord estuary (Wright, Coleman and Thom, 1973). As deposition proceeds, natural levees bordering the river channel are prolonged seaward, and these various bars and shoals move forward in front of them. However, where wave energy increases these river-mouth deposits are reshaped into smooth arcuate shore-parallel swash bars that are driven shoreward and eventually incorporated in beaches on the delta coast.

On deltas where the main river branches into distributaries, progradation is by means of sedimentation at and around the mouths of these channels, particularly during river floods. Waves and currents can spread the sediments alongshore, sorting them into sandy beaches and spits and separating finer sediment to settle in backing lagoons and swamps or be dispersed seaward. As tide range increases, tides penetrate further upstream, impeding or reversing river discharge and causing overbank flow, crevassing

Figure 12.1 The Nile delta, showing the pattern of past and present distributaries. Erosion is now rapid (>10 m/yr) on the arrowed shore sectors

and splay deposition. Deltaic river channels subject to such tidal oscillations are essentially estuaries, with features similar to those described in the previous chapter.

12.4 The Mississippi delta

So much attention has been given to the large, complex delta built by the Mississippi, which has a catchment of 3.3 million km^2 and delivers about 240 billion (10^9) kg of sediment to the river mouth each year (Coleman, Roberts and Stone, 1998), that it requires separate treatment. The sediment yield to the delta consists of fine sand, silt and clay (the clay fraction being about 70 per cent), and during Holocene times the volume of fluvial sediment deposited to form the 28 500 km^2 delta was about 2800 km^3.

Numerous oil borings on the Mississippi delta have shown that it is underlain by a great thickness of Holocene sediment, occupying a crustal depression, the subsidence of which is partly due to isostatic adjustments of the earth's crust

Figure 12.2 Successively formed Mississippi subdelta lobes, now partly submerged and dissected as the result of subsidence. The most recent delta growth has been SE from New Orleans. D – Iles Dernières

beneath the accumulating sedimentary load (Section 3.3.4). A long history of subsidence can be deduced from this stratigraphic evidence, for the deposits on river terraces bordering the Mississippi valley can be traced down into the wedge of sediment beneath the delta (Russell, 1967).

The delta has formed in several stages during this subsidence, and consists of seven successive subsiding and partly overlapping subdelta lobes (Kolb and Van Lopik, 1966). As Figure 12.2 shows, one of these grew southward between 5500 and 4400 years ago, and others have been added subsequently in different positions to the east and SE. As each lobe grew, channel gradients in the river were reduced, and the outflow of water and sediment diminished. Bordering levees were then breached during floods to form crevasses that became shorter and steeper

outflow channels. Some of these were later sealed by deposition but others persisted to form distributaries, at the mouths of which sedimentation formed a new subdelta lobe, producing eventually the branched pattern shown in Figure 12.2. The most recent outgrowth is to the SE, where elongated sedimentary jetties (digitate or bird's foot delta) are extending seaward beside diverging distributary channels. Meanwhile subsidence has continued, and the older lobes, no longer maintained by sedimentation, are sinking beneath the sea in coastal embayments of irregular configuration. As submergence proceeds wave action has sorted the sandy fraction from the more widely dispersed silt and clay and shaped it into beaches, spits and barrier island chains, such as the Iles Dernières to the east, which commemorate the margins

Figure 12.3 The reduction of Iles Dernières, on the eastern side of the Mississippi delta (Figure 12.2), by subsidence and erosion. After Kolb and Van Lopik (1966)

of the partly submerged fourth subdelta lobe (Figure 12.3).

The delta has grown out into the Gulf of Mexico, a microtidal and generally low wave energy environment, the nearshore area being a broad shallow plain up to 20 km wide. Wave action is weak, so that sediment delivered to river mouths during recurrent Mississippi floods is deposited to form parallel subaqueous ridges alongside channels that continue out across the sea floor. As sedimentation proceeds these ridges are augmented to become emerged jetties of sand and silt, prolonging the natural levees on either side of the river mouths. This growth of a digitate delta has been aided by colonising reedswamp (known as roseau in Louisiana), within which sediment is trapped (Figure 12.4), much in the manner described from coastal lagoons, notably the Mitchell River silt jetties in the Gippsland Lakes (Figure 11.13).

Figure 12.4 Reedswamp growing in shallow water traps sediment and initiates the growth of a digitate delta at South Pass, Mississippi delta

Confinement of the Lower Mississippi river between artificial levees during the past 250 years has intensified its outflow to the Gulf and increased the sediment supply to the river mouths. The modern delta coastline is highly indented and marshy, with only minor sandy beaches and spits, and off the river mouths deposition of shoals has resulted in local upwellings known as mud-lumps (diapirs), which are offshore extrusions of prodelta clay, squeezed up as the result of nearby sediment loading.

Although the Mississippi delta has grown on a low wave energy coast, it is subject to occasional hurricanes. In 1957, for example, Hurricane Audrey submerged the Mississippi delta when it raised sea level by up to 3.6 m. Under these conditions, large waves swept sandy sediment inland from the shore, and deposited it as cheniers, low ridges on the deltaic plain (Section 6.19). Similar changes occurred when New Orleans was flooded during the 2005 hurricane.

The features of the Mississippi delta are of great interest, but they are not representative of the world's deltas, most of which are of simpler configuration.

12.5 Delta outlines

The size and shape of a delta depends partly on the pattern and rate of sediment yield, on the configuration of the coast and on the nearshore bathymetry of the water body (open sea, gulf, lagoon, lake) into which it has grown, and partly on the effects of waves and currents on the accumulating sediments. Small rivers have built deltas at the heads of estuaries or the sheltered shores of coastal lagoons, or on low wave energy shores of tideless seas, but on coasts dominated by strong wave action and tidal scour protruding deltas have been built only by large rivers draining catchments that have yielded abundant sediment, especially during floods. A gravelly intertidal delta has been built by recurrent floods off the mouth of the River Lyn in north Devon, and was enlarged during the Lynmouth flood in 1952 (Figure 4.13). Such rivers, with steep hinterlands, often deliver coarse gravelly sediment to their mouths. The Var on the Mediterranean coast of France is an example of a river with a gravelly delta supplied from a mountainous hinterland. Deltas have been built on arid coasts where intermittent streams flowing from a wadi have deposited cones of sediment extending into the sea as arcuate salients, as at Ghabour, south of Hurghada on the Gulf of Suez coast in Egypt.

12.5.1 Wave-dominated deltas

The shape of a protruding delta is related to the effects of wave action and accompanying currents that tend to disperse sediment and smooth the coastline. Wave action is reduced where the nearshore water is shallow, as on deltas (such as the Mississippi, where the 10 m depth contour is about 17 km offshore) where fluvial deposition has built a broad shallow submarine profile.

It is possible to classify deltas in terms of outlines related to rates of fluvial sediment supply and incident wave energy. The branching digitate outline of the Mississippi delta is found where wave energy is usually low, and fluvial sediment supply abundant (Figure 12.5). Similar long, narrow projections built by rapid deposition of large quantities of fine grained sediment during floods from the Cimanuk and Solo Rivers in Indonesia extend out into the shallow Java Sea, where wave energy is low and protruding silt jetties produced by river-mouth deposition can persist. The Volga delta has numerous distributaries opening to an irregular marshy shore that prograded rapidly as the level of the Caspian Sea fell between 1930 and 1977, and has been partly submerged by the ensuing sea level rise (Figure 3.11). These deltas can be classified as river-dominated deltas, as distinct from wave-dominated and tide-dominated deltas.

Figure 12.5 Delta configuration. The Mississippi delta is digitate (finger-like), the Ebro delta (Spain) cuspate with bordering spits, the Niger delta (Nigeria) lobate with a sandy fringe and the São Francisco delta (Brazil) blunted by high wave energy

Legend labels visible within figure:

Digitate- MISSISSIPPI
Gulf of Mexico
SOUTHEAST PASS
SOUTH PASS
SOUTHWEST PASS
Swamp

Cuspate- EBRO
Mediterranean Sea
Beach ridges
Swamp and salt flats
Abandoned channels

Lobate- NIGER
Atlantic Ocean
Sand ridges
Mangroves
Channel deposits
Alluvium

Blunted- SAO FRANCISCO
Atlantic Ocean
Coral and algal reef
Coastal dunes
Beach ridges
Mangroves
Swamp
Alluvium

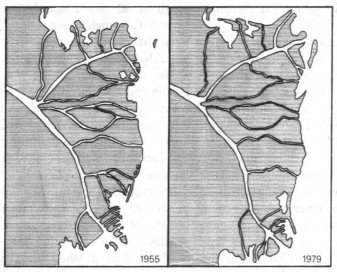

Figure 12.6 Historical stages in the growth of part of the Danube delta

Where wave action is somewhat stronger (and the nearshore sea floor less shallow), deltas develop smoother, cuspate outlines. The Danube delta (Figure 12.6) (10 m depth 5 km offshore) has several distributaries, which have waxed and waned as large quantities of fluvial sediment have been delivered to their mouths, but, although progradation has been rapid, digitate outgrowths have not formed. Instead, wave action has shaped newly deposited material into cuspate salients, sorting sand from silt and clay to form beaches and spits.

The Ebro delta on the Mediterranean coast of Spain (10 m depth 3.5 km offshore) has a cuspate outline and a generally sandy coastline, with bordering beaches prolonged as trailing spits by waves that come in from the east and break obliquely as they pass along the northern and southern coasts (Figure 12.5). The Burdekin river in NE Australia divides into distributaries in its delta region, and a series of spits have been built northward by longshore drifting resulting from the prevalence of waves generated by the SE trade winds, deflecting distributary outlets and culminating in a long recurved spit built northward to Cape Bowling Green. A similar pattern is seen on the Godavari delta in eastern India, shaped by SE waves that have drifted sand from the river mouth northward to the Kakinada spit, and on the Kelantan delta in Malaysia, where the large recurved Tumpat spit has also grown northward from the river mouth. Although it has formed on a coast exposed to high wave energy the Nandi delta in Fiji has a cuspate outline shaped by wave refraction around an outlying coral reef.

With stronger incident wave action, delta outlines become lobate, as on the Niger delta in West Africa (Figure 12.5), a large delta containing 1400 km^3 of Holocene sediment, fringed by a sandy barrier. High energy ocean swell from the South Atlantic is diminished across a shelving sea floor (10 m depth 9 km offshore) to waves that are sufficient to suppress delta outgrows and distribute sandy sediment alongshore. The delta has several channels diverging across a marshy flood plain and a mangrove zone to outlets through the barrier, each of which has an offshore sand bar. The Nile delta (Figure 12.1) is also rounded and lobate, with a fringing sandy beach, because the nearshore sea floor has a concave profile (10 m depth 3 km offshore) that allows moderate Mediterranean wave action to reach the shore.

Still stronger oceanic waves, reaching the coast across a narrow nearshore slope (10 m depth 2 km offshore) have blunted the delta of the São Francisco in Brazil (Figure 12.5). Sediment delivered to the river mouth is quickly dispersed and sorted by waves, the sandy fraction being deposited on bordering beaches so that the delta coast has prograded by the addition of sandy beach ridges shaped by refracted ocean swell. Finally, in a high wave energy envi-ronment on a coast where the nearshore profile is steep (10 m depth 1 km offshore), the Senegal River opens on a sandy Atlantic Ocean coastline that is straight rather than protruding and has a blocked delta, the river mouth having been deflected southward by longshore spit growth.

Deltas are poorly developed on high wave energy coasts because sediment delivered by rivers has been dispersed by strong wave action, the coarse fraction (sand and gravel) being deposited in beach and barrier formations along the coast and the finer material being carried away in suspension by the sea or settling on the sea floor. However, on the stormy west coast of South Island, New Zealand, rivers such as the Haast have brought down sufficient gravelly material from glaciated mountain catchments to build slightly protruding, blunt deltas.

Protruding deltas are rare on the coasts of Britain and NW Europe, partly because of high wave energy and strong tides, and partly because fluvial sediment yields have been generally low. Most of the rivers draining to the Atlantic and Pacific coasts of the United States flow into estuaries and lagoons, and have not yet built deltas that protrude into the sea. In Australia protruding deltas have generally not developed because of high wave energy and meagre fluvial discharge. An exception is the De Grey delta on the NW coast, where wave energy is moderate because the continental shelf is broad and the coastal waters megatidal. Most Australian rivers not yet provided sufficient sediment to fill valley mouths drowned by Holocene marine submergence. The Murray–Darling system has a catchment basin of more than a million square kilometres, but in its lower reaches the Murray flows through a semi-arid region and loses so much water by evaporation that it has not delivered sufficient sediment to fill the lagoons (Lakes Albert and Alexandrina) at its mouth (Figure 11.12). Some rivers such as the Snowy and the Shoalhaven have brought down more sediment and reclaimed their drowned valleys as depositional plains, but as in the Senegal the building of protruding deltas has been prevented by strong ocean swell arriving through relatively deep water. On the Queensland coast the growth of protruding deltas such as that of the Burdekin has been possible because ocean swell is excluded by the Great Barrier Reef.

12.5.2 Tide-dominated deltas

Although deltas are best developed on microtidal coasts (as around the Mediterranean Sea) they also occur where the tide range is large, as on the Irrawaddy delta (tide range 5.5 m) and the Ganges delta where the tides rise and fall 4.5 m. These are on low wave energy coasts where broad shallow areas offshore diminish incoming waves. Deltas on macrotidal coasts show distinctive features, notably funnel-shaped estuarine river mouths produced by the ebb and flow of tidal currents, and associated linear river-mouth bars.

The Mahakan delta in Kalimantan, Indonesia, is an example of a protruding delta with several distributaries opening to wide mouths on a mesotidal coast where wave action is moderate. Following hinterland deforestation large quantities (about 8 million m³/yr) of fine grained sediment are delivered by the river to its distributaries and deposited at and around their mouths, but much of the delta plain is occupied by intertidal mangrove swamps passing landward into swamp forests. Because of tidal currents the delta outline is irregular, rather than cuspate.

On the west coast of Peninsular Malaysia, the Klang delta, near Kuala Lumpur, has been shaped by tidal processes as fluvially supplied sediment was delivered to the coast. The delta has been built into the Strait of Melaka, here about 50 km wide. There is low to moderate wave energy, chiefly from the NW, and mean spring tide range at Port Klang, beside one of the river mouths, is 4.2 m, producing strong currents flowing SE as the tide rises and NW (augmented by a monsoon-generated northward current) as it falls. These currents disperse sediment mainly to the NW, where it has been deposited to form the large trailing intertidal Angsa Bank, with parallel mud-crested ridges and current-scoured sandy troughs. Tidal currents have been more important than waves in shaping the Klang delta, much of which is submerged at the highest tides. There are extensive tidal flats with mangrove islands and inter-connecting tidal channels, and rapid peat accumulation has contributed to aggradation in these areas. The southern shores of the delta are eroding, but the seaward spread of mangroves has prograded parts of the western and northern shores at rates of up to 6 m/yr. Tidal channels between mangrove islands meander and migrate laterally, one bank building up as the other is undercut, and some become sealed off as the mangroves spread.

The Ord River, opening into Cambridge Gulf on the north coast of Australia, has extremely variable seasonal flow, and discharges an average of 22 billion kg of sediment annually (Wright, Coleman and Thom, 1973), but the large delta occupies the head and fringes of the funnel-shaped (estuarine) gulf and a great deal more infilling will be necessary to form a coastal protrusion. Meanwhile, the fluvially supplied sediment is widely dispersed by strong tidal currents (often >3 m/sec), mean spring tide range being of the order of 6 m. Off the mouth of Cambridge Gulf there are sub-parallel tidally scoured channels between submerged linear sand ridges (King Shoals). The gulf-head delta has extensive sandflats and mudflats exposed at low tide, and there are several long inlets that may have been former outlets of the Ord River. Wave action from the Timor Sea is weak, and the coastal morphology is almost entirely tide dominated.

The Red River delta in Vietnam has both wave-dominated and tide-dominated shores in a mesotidal environment where wave energy diminishes northward along the coast (Mathers and Zalasiewicz, 1999). The wave-dominated southern part has beach ridges built of sand reworked from deposits swept into the sea by occasional major floods interspersed with muddy tidal lagoon deposits, while the tide-dominated northern part has numerous inlets and tidal creeks fringed by mangroves.

12.5.3 Effects of nearshore currents

Nearshore currents can move fine grained sediment supplied by rivers along the coast. The Amazon River has filled a former gulf with a large swampy deltaic plain, but growth of a protruding delta has been prevented by the Guyana current, which flows NW past the mouth of the river and sweeps fine grained sediment along the coast past Surinam to Trinidad (Gibbs, 1970). Deposition of this sediment has formed extensive shore mudflats, and sand sorted from the fine grained sediment by waves forms low beaches and cheniers. On the coast of Papua muddy water from the Purari River is swept westward by waves arriving from the SE and deposited in mudflats and mangrove swamps (Thom and Wright, 1983).

12.5.4 Deltas in cold regions

Arctic deltas, such as those of the Ob, Lena and Yenisey in Russia, and the Mackenzie, Yukon and Colville in North America, show features related to cold climate weathering and erosion processes (Walker, 1988). Frost action produces ice wedge polygons and ice-heaved soil-capped conical hills up to 50 m high and 400 m in diameter, known as pingos. Fluvial sediments in cold regions are generally coarse, with much sand and gravel. Sand deposits on river bars are frozen and snow covered in winter and saturated by river flooding in the spring thaw, but in summer they dry out and are built up by the wind into

river-bank dunes (including barchans) that may become sparsely vegetated. There are numerous lakes and peaty swamps on the delta plain, modified by interactions between river floodwaters and coastal and sea ice (Hill et al., 1994). Some of the sediment swept down by spring floods is deposited on the frozen sea and carried away when the sea ice disintegrates into dispersing floes in summer. In Alaska the Point Barrow coast has a number of deltaic salients built by former glacifluvial deposition and now being reshaped by wave action. In Finland the deltas of such rivers as the Siikajoki are several kilometres wide because of extensive flooding when their mouths are blocked by shore ice in spring.

12.6 Delta evolution

Stages in the evolution of a delta may be traced from the time when the sea attained its present level relative to the land, and sedimentation began to fill a valley-mouth inlet, through to the complete filling of this inlet and the formation of a depositional landform protruding into the sea. The Ord delta in northern Australia is still at the stage of a gulf-head delta, but many rivers have filled the former valley-mouth inlets and now protrude in various forms.

The outlines of such a delta change as the result of deposition and erosion along its coasts. Deltas continue to prograde as long as the supply of sediment, mainly from the river, exceeds its removal by wave and current action. Progradation can be spectacular in shallow seas, especially on parts of tropical deltas, where progradation of over 200 m/yr has been recorded on some Javanese deltas (Figure 12.7) (Bird and Ongkosongo, 1980).

There is historical evidence of delta growth, especially around the Mediterranean Sea over the past 2000 years. The port of Ostia, near Rome, became silted as the result of the growth of the Tiber delta, and is now 3 km inland. In

LAND ADDED
- 1947-1966
- 1966-1976

1947 coastline

P. Balok

P. Payang

1966 coastline

Course of Cimanuk

N

Cimanuk

P. Song

Course of Cimanuk until 1947

1947 coastline

0   1 km

Figure 12.7 The Cimanuk delta, Indonesia, showing the extent of growth following river diversion in 1947. A former delta at the mouth of the abandoned Cimanuk course has been removed by erosion, and a new delta built by sedimentation from the diverted outlet, which branched into three distributaries. Two phases of delta growth are shown

Greece the ancient port of Ephesus, built where the River Kayster opened into the Bay of Anatolia, became stranded by delta growth and the spread of marshland, particularly after a breakwater was built across the mouth of the bay in 150 BC, and is now 24 km from the sea (Kraft, Aschenbrenner and Rapp, 1988).

Delta progradation can be accelerated by increased sediment yield due to deforestation, overgrazing or unwise cultivation leading to rapid soil erosion, or mining activities in the river catchment. The Mahakan delta in Kalimantan expanded rapidly following forest clearance in its catchment. The rapid growth of the George River delta into a Tasmanian coastal lagoon was the result of increased fluvial sediment yield caused by tin mining upstream (Bird, 2000) (Section 11.8.3). Sluicing for tin has augmented the sandy loads of several rivers in Malaysia, where the Pahang delta has been enlarged by sand washed down from a mined catchment.

In Cornwall the River Fal has built a small delta into a sheltered arm of the Carrick Roads ria (Figure 11.1). It grew rapidly during the phase when the White River brought down large quantities of kaolin from the china clay quarries in its upper catchment, but growth ceased and delta shore erosion began after this supply of sediment

Figure 12.8 The growth of the Fal delta into Carrick Roads, Cornwall, during a phase when the river was delivering a sediment load augmented by china clay waste from its upper catchment. Control of china clay pollution has been followed by delta shore erosion, so that by 1998 the delta had returned to approximately its 1930 outline

was curtailed by conservation works in the mining area upstream (Figure 12.8). The river is no longer white, but kaolin is exposed in salt marsh cliffs along the delta shore (Figure 12.9).

The Jaba delta on the island of Bougainville began to grow in 1972 after tailings from copper mines in the catchment flowed down the Jaba River to the coast at the rate of 26 million tonnes/yr. By 1977 an arcuate salient had formed on the shores of Empress Augusta Bay, but surveys by Wright, Thom and Higgins (1980) showed that delta growth was then curtailed by moderate SE wave action (and a 1975 tsunami), which spread sandy sediment northward to prograde beaches along the coast, leaving a small wave-dominated lobate delta at the river mouth.

The Colorado delta in Texas prograded rapidly after log jams were removed from the river channel in 1929, forming several distributaries and extending across a lagoon at Matagorda Bay towards a sandy coastal barrier. When an artificial outlet was cut in 1932 through the barrier to the sea, fluvial sediment carried through to the Gulf of Mexico was dispersed by wave action, so that a protruding delta failed to develop.

Delta growth is slower into deep water, or where a submarine canyon exists offshore, as off the mouth of the River Ganges. There has been subsidence and erosion of earlier subdelta lobes on either side of the modern Mississippi delta (Figure 12.2), and similar features are seen on a smaller scale on the Rhône delta in southern France. Subdelta lobes built by the Petit Rhône and the Vieux Rhone have been cut back by marine erosion following the decay of these distributaries and the shifting of the river to a SE outlet. Sediment eroded from these former subdelta lobes has been reworked and sorted by marine erosion, producing sand that has been carried westward by the dominant SE waves to be built into depositional forelands at Pointe de Beauduc and Pointe de l'Espiguette respectively (Suanez and Provansal, 1998).

Erosion of deltaic coastlines may follow diminished sediment yield from the river. The delta of the Argentina River on the Ligurian coast in Italy grew for as long as the river delivered sediment, but extensive headwater soil erosion has laid bare rocky outcrops, and sediment yield

Figure 12.9 Microcliff on the eroded margin of the Fal delta, exposing a wave-cut ramp in white china clay (kaolin) deposits

has diminished, so that the delta is now eroding. More often, fluvial sediment yield has been reduced by dam construction upstream. Erosion on the shores of the Rhône delta accelerated following dam construction that reduced fluvial sediment yield from 11 to less than 5 million tonnes annually. In Ghana the building of the Akosombo Dam on the Volta River greatly reduced discharge and sediment yield to the coast, resulting in an increase in the rate of coastline recession from 2–3 m/yr up to 8–m/yr (Ly, 1980).

Erosion has been prevalent on the shores of the Nile delta since the beginning of the present century, and in recent decades it has been locally rapid, up to 120 m/yr near the Rosetta mouth (Figure 12.1) (Sestini, 1992). This coastline previously prograded as the result of the delivery of Nile sediment to the coast by floods, but the building of dams, especially the Aswan High Dam in 1964, depleted the supply of Nile

sediment to the delta coastline and so accelerated erosion (Stanley and Warne, 1998). Fluorescent tracers were used by Badr and Lotfy (1999) to show net eastward longshore drifting of sand along the eroding coastline of 1.48–3.21 million m^3/yr, and losses offshore of 0.39–0.44 million m^3/yr. Some of the drifting sand has been washed into the mouths of the Rosetta and Damietta distributaries and the Burullus lagoon outlet.

Natural diversion of a river mouth is generally followed by erosion of the former delta and the building of a new one. An example is the Huanghe River in China, where there has been erosion of the delta abandoned after the river changed its outlet during an 1855 flood and growth of another delta at the new outlet northward into the Gulf of Bo Hai. Similar changes followed the diversion of the Hwang Ho River in China in 1852, the Rio Sinu in Colombia in

1942, the Rioni River in the Republic of Georgia in 1939, the Ceyhan in Turkey in 1935 and the Medjerda in Tunisia in 1973.

The modern Po delta has grown since a new outlet was cut in the late 16th century to divert the river away from the Venice region, where it threatened to fill the lagoon and deprive the city of its natural defensive moat. The Po branches into distributaries, and the seaward fringes of intervening marshy deltaic islands are lined by sandy barriers. In recent decades, delta enlargement has ceased because of a diminishing sediment yield due to river impoundments and soil conservation works upstream (Cencini, 1998).

When the fluvial sediment supply to a delta is reduced or cut off, and erosion of the delta coastline follows, there is usually an interval between the halting of the fluvial sediment supply and the onset of erosion because the change from a convex, aggrading sea floor profile to a concave, eroding sea floor profile begins offshore, and is transmitted landward. As the concave profile intersects the deltaic coastline, erosion begins, or is accelerated (Figure 12.10).

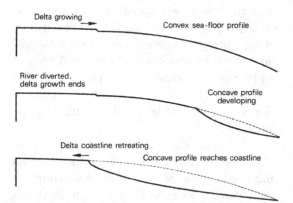

Figure 12.10 A growing delta usually has a convex depositional nearshore sea floor profile. When sea level rises and the delta coastline is submerged a concave erosional profile is initiated offshore. This migrates shoreward until it intersects the coastline, which then recedes more rapidly

12.7 Delta shores

The advance or retreat of deltaic coastlines can be traced from successive maps and air photographs, as in Figure 12.6, which shows stages in the growth of the Danube delta since 1830. Nossin (1965) used cartographic records to reconstruct the evolution of the North Padang delta in Malaya since the early 17th century. Historical changes on the Ganges–Brahmaputra delta front were determined from successive charts and modern Landsat imagery by Allison (1998), who found that progradation had added about 7 km^2/yr to the delta coastline since 1792, while the submerged delta front had expanded eastward and been reduced by erosion on the western side. It is possible that construction of artificial levees along the River Ganges has confined and increased downstream flow, thereby augmenting sediment yield to the coast and accelerating progradation.

Beaches and spits border many deltaic coastlines, supplied with sand or gravel by rivers, derived from erosion and sorting of delta shore sediments by wave action, or washed in (often with shells and other marine organisms) from nearshore shallows. Beaches, spits and barriers built by marine processes on the shores of a delta often enclose lagoons and swamps (as in the Kizilirmak delta in northern Turkey), and are sometimes themselves incorporated as the delta grows larger. In northern Italy the Tagliamento delta has a series of symmetrical parallel sandy beach ridges marking stages in the progradation of a cuspate delta coastline. Barrier islands have formed from beaches fringing subdelta lobes of the Mississippi that have subsided (Figure 12.3), and similar features are seen on the Apalachicola delta in Florida (Donoghue and White, 1995). Some delta-shore beaches are backed by dunes formed by onshore winds, as on the Ebro delta in Spain, but on the Rhône delta dunes are poorly developed because the prevailing winds blow from the land.

Cheniers (Section 6.19) are long, narrow, low lying strips of sand that have been deposited by wave action during occasional high tides or storm surges on deltas and coastal plains, and marked out by contrasts in vegetation. They are found on the Mississippi delta and in Surinam on the coast of South America, where several sand ridges have been emplaced on broad swampy deltaic plains adjacent to the Amazon and the Orinoco.

Where wave energy is low on delta shores, or in bays and lagoons behind spits and barriers along these shores, there are freshwater swamps and salt marshes, and in low latitudes mangrove swamps, fronted by mudflats. The shores of the Mississippi and Mahakan deltas have sectors of salt marsh or mangroves that form irregular, sometimes crenulate, mid-tide shorelines that prograde as fine grained sediment is deposited. If the delta coast becomes exposed to stronger wave action (e.g. if a sheltering spit is submerged or truncated, or if there is a relative sea level rise, perhaps due to delta subsidence) progradation ceases, and these swampy shores may be eroded. Alternatively, higher wave energy may deliver sandy sediment to form new fringing beaches.

Some delta-like protrusions at river mouths are the outcome of erosion rather than deposition. On the west coast of Lough Foyle in NW Ireland there are residual lobes of glacial drift around stream mouths where fluvial sediment deposited in the nearby intertidal zone has shallowed the sea and diminished incident wave action, thereby allowing the lobes of glacial drift to persist. Deposition off river mouths can form intertidal deltas, such as that exposed at low tide off the Lang Lang River in Westernport Bay, Australia, where continued aggradation may eventually build a true delta, if this is not prevented by wave erosion at high tide.

12.8 Summary

Deltas form where rivers have deposited sufficient sediment to build land protruding from the coast. Often the rivers branch into distributaries, which may build subdeltas. The load of deposited sediment causes isostatic subsidence, and if they are not maintained by fluvial sedimentation deltaic areas become submerged and eroded. Large deltas like that of the Mississippi consist of a series of partly superposed subdelta lobes. The outlines of deltas vary with exposure to wave action, and range from digitate deltas built into low energy sea areas to blunt deltas built where wave energy is high. Some deltas are wave dominated; others are shaped partly by tides and currents. In the Arctic deltas are influenced by freeze–thaw processes and ice formation.

Stages in delta evolution may be traced with reference to historical and archaeological evidence, as well as dating procedures. Damming or diversion of rivers can reduce water and sediment flow and result in erosion of deltas, for example the Nile Delta after the building of the Aswan High Dam in 1964. Deltas may be bordered by beaches and spits or marshes and swamps that prograde as long as the delta grows. They are eroded if the sediment supply is reduced.

13
Coral and algal reefs

13.1 Introduction

Reefs built in the sea or on the shore by organisms are termed biogenic or bioconstructional. By far the most extensive of these bioherms are coral reefs, which incorporate algae and other organisms, but there are also algal reefs and reefs built by oysters, bryozoans, serpulids, mussels and tubeworms. Coral reefs are coastal landforms where they fringe the coastline, where they have emerged to form limestone islands (usually also reef fringed), and where they are surmounted by islands of coralline sediment. Barrier reefs and atolls are not strictly coastal landforms, but are included here because they provide evidence of reef-building processes.

13.2 Coral reefs

Coral growth is confined to warm seas, where the mean temperature of the coldest month does not fall below 18°C, and the warmest month does not exceed 34°C. Reefs built by coral and associated organisms thus occur extensively in tropical waters, particularly between latitudes 30°N and 30°S in the western parts of the Pacific, Indian and Atlantic oceans (Davis, 1928; Guilcher, 1988). Their distribution is too intricate to be shown adequately on a textbook map, but is portrayed in the Russian *Atlas Okeanov* (1974, 1977). In the Pacific coral reefs extend north to Japan and Hawaii, in the Atlantic to Florida and the Bermuda Islands and in the Indian Ocean to the Red Sea and the Arabian Gulf. They are well developed in the Caribbean Sea, the Philippines and Indonesia, and around the coasts of Australia, particularly off the east coast of Queensland, where the Great Barrier Reef extends from Torres Strait in the north to the Bunker and Capricorn reefs in the south (latitude 24°S). In the Coral Sea reefs extend farther south, to the Middleton and Elizabeth atolls (29° 30′S) and the west coast of Lord Howe Island (31° 34′S). There are scattered reefs off the northern coast of Australia, and off the west coast of the continent they extend as far south as Houtman Abrolhos, in latitude 28–29°S (Fairbridge, 1967). In the Indian Ocean they extend south to Mozambique and Natal, and in the Atlantic Ocean to the Brazilian Abrolhos.

The global distribution of coral reefs is related to the dispersal of free-floating planktonic coral larvae by ocean currents to areas warm enough for coral growth. The paucity of reefs in the eastern Atlantic and Pacific (there are no reefs on the Galapagos Islands or on Easter Island) results from the westward flow of equatorial

Coastal Geomorphology: An Introduction, Second Edition. Eric Bird.
© 2008 by John Wiley & Sons, Ltd

currents away from these areas (as well as the up-welling of cold water along the coast), whereas the richness of the reefs off NE Australia (including the Great Barrier Reef) is related to an abundant larval supply in currents arriving from the Pacific and the Coral Sea. There is relatively poor development of coral reefs off the west coast of Australia, where reefs running parallel to the mainland coast are submerged ridges of dune calcarenite (Section 9.10), with only a veneer and fringe of living corals and calcareous algae. The flow of ocean currents across the Indian Ocean results in more extensive coral reefs to the west, and in a similar way the poor development of reefs in the eastern Atlantic gives place to more extensive reefs in the Caribbean. Sandstone reefs similar to those off the west coast of Australia are found along the NE coast of Brazil, where the supply of coral larvae has been meagre, both the northward summer current and the southward winter current arriving from ocean areas where there is little or no development of coral reefs.

13.3 Origin of coral reefs

Coral reefs bordering the coast are termed fringing reefs and those that lie offshore and parallel to the coast are barrier reefs. There are also patch reefs, isolated coral reef platforms of various shapes and sizes, sometimes with lateral spurs that have grown to leeward (Hopley, 1994). Atolls are oceanic reefs that encircle a lagoon. The term atoll comes from the Maldives, where *atolu* are government districts, each being a circular reef enclosing a lagoon.

Coral reefs are built by polyps, small coralline organisms that extract calcium carbonate from seawater and grow by accretion into a variety of branching skeletal structures, forming a coral garden (typified by stag-horn coral growth). These relatively fragile structures can form a habitat for calcareous algae, which grow with

the coral, as well as foraminifera, molluscs and other shelly organisms. They have a similar effect to mangroves in that they diminish current flow and promote sedimentation, so that fragments of shells, corals and algae (such as the sand-producing *Halimeda*) are deposited in the spaces between the coral garden structures; with precipitated carbonates they form a solid calcareous reef limestone. The seaward slopes of coral reefs are often very steep (up to 50°), descending to aprons of reef-derived sand and gravel that decline more gradually to the sea floor. While coral is essential for reef building, it generally forms only a small proportion of a solid reef structure, as seen in sections or quarries in emerged coral reefs.

Corals require a firm sea floor substrate (usually rocky) and coral larvae cannot establish on mud or on mobile sand and gravel, or where sea floor sedimentation is proceeding rapidly. An adequate supply of sunlight is essential for algal photosynthesis, and growth of coral is best in clear, warm water. Intensity of sunlight diminishes downwards into the sea, and although live corals have been found in exceptionally clear water at depths as great as 100 m the maximum depth at which reefs are being built is rarely more than 50 m. In coastal waters turbidity due to land-derived sediment (chiefly silt and clay) in suspension reduces penetration by sunlight and impedes growth of reef-building organisms. Off river mouths this cloudiness and the blanketing effects of deposits of inorganic sediment often prevent the growth of reef-building organisms, and break the continuity of fringing reefs. Corals can dispose of small accessions of sediment, but are choked by continued deposition of large quantities of detritus, or by heavy loads of sediment dumped suddenly from discharging river floods. In Indonesia coral reefs are missing from sectors of the coast that receive lava flows (as on Anak Krakatau) or ash deposits from volcanic eruptions (as at Parangtritis, on the south coast of Java, where rivers supply sand erupted from the Merapi volcano). In general, coral reefs are not found on coasts where sandy beaches are extensive.

Corals are marine organisms, found where sea salinity is within the range of 27–38 ppt, and most luxuriant in 34–36 ppt. Dilution by fresh water discharged by rivers impedes coral growth and contributes to the persistence of gaps in bordering reefs. In NE Queensland corals are occasionally killed by the outflow of large quantities of fresh water and sediment during heavy rainfall. On the other hand, excessive salinity may explain why reefs are absent from certain parts of the coast such as Hamelin Pool, in Shark Bay, Western Australia, where salinity is up to 48 ppt in summer. Where ecological conditions are suitable, corals begin to grow with associated algae to form a coral garden and eventually a solid reef.

Oceanic coral reefs have been forming since at least Eocene times, and borings have shown that they extend to great depths: on Bikini Atoll the base of the coral rests upon volcanic rocks 1400 m below sea level, well below the limits at which reef-building corals now grow.

13.4 Rates of growth

Where ecological conditions are favourable, and there is an adequate supply of mineral nutrients in the seawater, corals grow up towards the sea surface. There are differences between the growth rates of individual organisms, which can be quite rapid, the branches of some staghorn corals (*Acropora* spp.) extending by up to 20 cm/yr, and the reef formation as a whole. Measurements of mean upward growth rates of coral reefs are generally in the range 0.4–0.7 mm/yr (Hopley and Kinsey, 1988), with up to 1 cm/yr in favourable conditions (Buddemeier and Smith, 1988). Studies of reef growth during the Holocene marine transgression, when the sea rose at an average rate of about a metre per century, indicate average rates of upward growth of up to 8 mm/yr (Davies, 1983). This enabled a variety of growing corals in the reef framework to be within a few metres of sea level when the transgression slackened about 6000 years ago, and then to extend upward and outward to form existing reefs.

Corals grow upward until they reach a level where they are briefly exposed at low tide (Figure 13.1). As they cannot survive prolonged exposure to the atmosphere they then die, but associated algae, other organisms and sediments may continue to build the solid reef platform up to this level. Algae (chiefly *Porolithon* or *Lithophyllum*) may build a slightly higher algal reef, awash at high tide, and on ocean coasts where there is strong wave action a seaward rise (an algal rampart or rim typically several metres wide) may be formed at the outer edge of a reef. Networks of algal ridges may grow to enclose shallow pools or basins on coral reef flats. Reefs built in more sheltered waters, such as the Java Sea in Indonesia, are flatter and without algal ramparts. It is possible that some reef platforms have been built up to a slightly higher Holocene sea level and then planed off by marine erosion following subsequent emergence (Hopley, 1982).

Coral growth continues on the steep bordering slopes of reefs declining below low tide level. Optimum coral growth occurs where seawater is being circulated sufficiently to prevent clogging by silt deposition, to maintain a uniformly high temperature, to renew the supply of plankton and other nutrients on which the coral polyps feed and to maintain the supply of oxygen, particularly at night when it is no longer replenished by algal photosynthesis and the concentration of dissolved carbon dioxide tends to rise in reef waters. A high concentration of dissolved carbon dioxide impedes coral growth, and could become corrosive.

Where the slopes of reefs are exposed to strong wave action, as on the seaward flank of a barrier reef or the outward slopes of an atoll, and

Figure 13.1 Fringing coral reef on the SE coast of Bali

there is adequate circulation of seawater for vigorous coral growth, waves may break off, or inhibit the formation of, the more intricate skeletal corals, so that the seaward slopes consist of more compact coral growth. On the leeward side, in more sheltered waters, a greater variety of growth forms coral gardens. The breaking of skeletal corals by storm waves generates large quantities of coralline sand and gravel, much of which is banked up as sedimentary aprons on the lower slopes of reefs or laid down on lagoon floors, while some is thrown up by wave action on to reef platforms.

Coral growth is also limited by predatory organisms, such as the crown-of-thorns sea-star (*Acanthaster planci*), which in the past few decades has grown in plague proportions on Indian and Pacific Ocean reefs, notably on parts of the Great Barrier Reef. The sea-star feeds on, and thus kills, corals. The Great Barrier Reef outbreak began near Cairns in the 1960s, possibly as the result of excessive collecting of triton shells

diminishing predation on the sea-star, which has subsequently spread northward and southward. The impact has been severe on branching *Acropora* corals, broken sticks of which form abundant gravelly debris on and around the affected reefs, locally augmenting gravel deposits on reef platforms. Although many consider the outbreaks to be a result of human impacts, stratigraphic evidence of sea-star remains in coral lagoon sediments shows that there have been previous outbreaks of this kind.

Coral reef ecosystems have certainly been modified by human activities, especially during recent decades. Marine pollution has occurred in many coral reef areas, and some reefs have been quarried for sand, gravel and building stone, or damaged by the cutting of boat access channels, or by boat anchors. Coral reefs have also been damaged by human activities such as weapons testing and fishing with explosives. Such activities generate sediment turbidity, which impedes coral growth in neighbouring areas. Around Sulawesi, in Indonesia, coral reefs have been impoverished by increased sedimentation resulting from soil erosion, the greater turbidity having made them less vigorous, and reduced the number of species, especially near coastal towns. Excessive nutrients from eroding soils, agricultural fertilisers and sewage pollution have caused eutrophication, which is detrimental for coral growth, and can lead to the killing of corals by the growth of other organisms. There have been reports of corals damaged by bleaching, possibly as a consequence of higher sea temperatures associated with the El Niño Southern Oscillation in the Pacific Ocean, or as a result of increasing ultra-violet radiation due to atmospheric ozone depletion (Brown, 1990). Bleached corals become brittle, and break up into coralline gravel.

13.5 Fringing reefs

Fringing reefs have been built upwards and outwards in the shallow seas that border continent or island shores (Figure 13.2). They consist of a reef platform at low tide level, similar in many ways to the low tide shore platforms found on cliffed limestone coasts (Section 5.2.4), except that fringing reefs have been built up, rather than cut down, to this level.

Fringing reefs are usually thin, resting on a rocky foundation, and are generally widening by the growth of coral along their seaward margins. As coral growth is inhibited by fresh water and active sedimentation, fringing reefs are not found near river mouths, but are well developed around offshore islands and on coastal salients, extending round headlands. They originated on shores that had ecologically suitable conditions for colonisation by reef-building coral communities when the Holocene marine transgression brought the sea close to its present level, establishing the general outlines of the present coast, and they have grown seaward during the ensuing still-stand of sea level. Some fringing reefs may embody the relics of Pleistocene fringing reefs that developed at an earlier stage close to the present shore, and were dissected during low sea level phases. Others are really shore platforms cut into emerged older coral reefs, but with a veneer of modern coral, as on the shores of Mbudya Island, in Tanzania.

Figure 13.2 The structure of a fringing reef. Some fringing reefs have raised seaward rims backed by a moat (known as the boat channel), which remains flooded at low tide

Where conditions have become adverse for the growth of coral reefs (e.g. because of the diversion of a river mouth towards them) they are dead and disintegrating. Incipient fringing reefs are rare, presumably because ecological transitions from adverse to suitable conditions for coral establishment have been exceptional during the past few centuries. Scattered corals occupy bouldery shores in NE Queensland, as along the coast of the Macalister Range, and coral gardens have formed on shores around the volcanic islands of Krakatau, Indonesia, since the explosive eruption of 1883 (Bird and Ongkosongo, 1980).

Some fringing reefs, especially in East Africa, have a broad higher outer segment backed landward by a shallow depression known as a boat channel, typically 100 to 200 m wide and up to 3 m deep. At low tide the outer segment is exposed and the boat channel becomes a lagoon, while at high tide the whole area is submerged, and waves wash over and through to the coastline. There is often a beach (or low beach ridges) consisting largely or wholly of coralline sand and gravel derived from the reef and swept onshore by occasional storm waves. There is greater vigour of coral and algal growth at the outer margins of such a reef, but the boat channel may indicate a relative rise of sea level since the fringing reef was initiated (Bird and Guilcher, 1982).

At Yule Point on the NE coast of Queensland the reef platform adjoining the coast is strewn with sand, partly of terrigenous origin, and there are patchy mangroves bordering the shore near the high tide line. Live coral is confined to the outer edge, which is briefly exposed at low tide. This fringing reef originated as a nearshore reef that became attached to the mainland as the result of progradation of the adjacent coast by sand deposition.

Fringing reefs are well developed around high islands in the Pacific, as in Tahiti, New Guinea and Indonesia, and off the NE and northern coasts of Australia, as on Lizard Island, Hayman Island and Snapper Island. They also border promontories on the north coast of Australia, notably on the shores of Arnhem Land and between Port Hedland and North West Cape. They are missing from the swampy shores of the Gulf of Carpentaria and other northern gulfs that are receiving deposits of land-derived sediment, where they give place to intertidal sandflats.

13.6 Barrier reefs

Barrier reefs have been built offshore and roughly parallel to the coastline. Geological studies have shown that many barrier reefs were initiated as fringing reefs in Eocene times, and have grown upward during the Tertiary and Quaternary. In 1835 Charles Darwin observed barrier reefs and atolls during his voyage in the *Beagle*, and proposed the subsidence theory, which suggested that barrier reefs (and atolls) were the outcome of upward growth from ancient fringing reefs bordering continental margins and islands that had subsided beneath the oceans. There is no doubt that oceanic reefs have grown upward as their foundations subsided, but there have also been major oscillations of sea level, especially in the Quaternary, so barrier reefs have periodically emerged and been dissected by erosion, then submerged again to be rebuilt as coral growth revived (Figure 13.3).

Reefs that subsided or were submerged too rapidly for corals to maintain their upward growth are now submerged below the depth at which reef-building organisms could revive them. Submerged barrier reefs have been detected by soundings in the Pacific Ocean, particularly off the SE coast of New Guinea, in the Fiji archipelago and the Coral Sea. The patchy development of reefs off northern Australia may be due to excessive subsidence in the Timor Sea, where a submerged barrier reef off the Sahul Shelf indicates that sea floor subsidence has

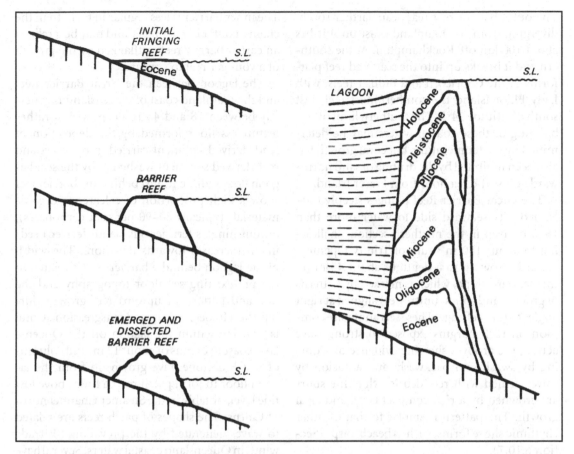

Figure 13.3 Evolution of a barrier reef from a fringing reef that grew upward as sea level rose, relative to the land. Barrier reef stratigraphy shows that there were successive phases of upward growth, interrupted by emergence and dissection in phases of low sea level followed by the revival of corals to widen and raise the reef structure. The modern barrier reef has formed in Holocene times (during and since the Holocene marine transgression), and encloses the dissected remnants of Pleistocene and earlier reefs

been too rapid for reef growth to be maintained (Davis, 1928).

Existing barrier reefs attained their present form during the later stages of the Holocene marine transgression and the ensuing sea level stillstand. The largest and most famous is the Great Barrier Reef, which extends for about 2000 km from north to south off the Queensland coast; the following description is based on the detailed study by Hopley (1982). It commemorates the alignment of a Queensland coastline that

has subsided tectonically. In the north it consists of a chain of elongated or crescentic reefs with intervening gaps, generally about a kilometre wide, lying about 130 km off the tip of Cape York Peninsula, 50 km off Port Stewart, less than 15 km off Cape Melville, and about 56 km seaward from Cooktown. Off Cairns there is a broad transverse passage known as Trinity Opening, which allows ocean waves to penetrate to the mainland coast (swell reaches the shore between Buchan Point and Yule Point).

The outer barrier reef reappears farther south, diverging from the mainland coast until it lies about 100 km off Rockhampton. At the southern end it breaks up into the scattered reef platforms of the Capricorn and Bunker reefs, with Lady Elliot Island the southernmost reef, just south of latitude 24°S. The southern limit of reef building in these coastal waters may be determined by sea temperature, but coral growth has also been inhibited by the drifting of sand northward across the sea floor from Fraser Island.

The Great Barrier Reef plunges steeply (40–50°) on its seaward side to water more than 1800 m deep in the northern section, shallowing to about 180 m south of Trinity Opening. Coral is growing to a depth of about 45 m on this seaward slope, which is marked by patterns of grooves and spurs (buttresses) aligned at right angles to the reef edge. These serrations are common on reef margins exposed to strong wave action, the grooves showing evidence of scouring by swash and backwash, and abrasion by waves armed with reef debris, while the spurs are crowned by a rich compact coral and algal growth. The pattern is similar to that of other rhythmic shore forms, such as beach cusps (Section 6.10.7).

At low tide the upper parts of the Great Barrier Reef emerge as a chain of elongated platforms up to 25 km long and up to 1 km wide, typically with an outer algal rim (rampart) on which the surf breaks, and a backing slope declining gently (5–10°) into the calmer waters of the lagoon. Yonge Reef, north of Cooktown, is a typical outer barrier segment, crescentic in form and recurved at the northern and southern ends bordering gaps in the reef. These gaps may commemorate the sites of river outlets at low sea level stages, but they persist because the rising and falling tides produce strong scouring currents that prevent them from being sealed off by coral growth and sedimentation. In one section, SE of Cape Melville, there is a double barrier, with parallel reefs separated by a lagoon about 8 km wide.

The inner barrier is less regular in form than the chain of outer barrier reefs, and may be a relic of an earlier barrier reef outflanked by the growth of a younger reef to seaward.

The lagoon between the Great Barrier Reef and the mainland coast of Queensland is generally between 18 and 45 m deep, with a rather featureless floor formed by the deposition of land-derived sediment carried in by rivers and reef-derived sediment washed in by the sea. Lagoon floor sediments just behind the barrier reef show a high proportion of calcareous organic material (typically 80–90 per cent carbonates), diminishing as terrigenous (land-derived) sediment increases towards the shore. The width of the lagoon behind a barrier reef is related to the pre-existing sea floor topography and the rate and pattern of upward reef growth during the Holocene marine transgression. Coral is growing within the lagoon off the Queensland coast to depths of about 12 m, and columns of reef limestone have grown up from the lagoon floor to form patch reefs a little above low tide level, notably in the Steamer Channel north of Cairns. The shapes of patch reefs are related to waves generated by the prevailing SE trade winds in Queensland coastal waters. Several have a horseshoe form, with arms trailing NW. Cairns Reef has this form, and a more advanced stage is represented by Pickersgill Reef, the arms of which curve round and almost enclose a shallow lagoon.

Other major barrier reefs include those that run parallel to the coasts of New Caledonia for over 600 km, enclosing lagoons up to 12 km wide and up to 100 m deep, the Great Sunda Reef south east of Kalimantan and the Great Sea Reef, 260 km long, to the north of Fiji. In the Caribbean the Belize barrier reef, extending 220 km along the coast, has three offshore atolls and over a thousand sand cays and mangrove islands. The Belize barrier reef differs from Pacific barrier reefs in having extensive coral gardens, with delicate corals growing upward, and fewer solid coral reef platforms with dead corals on the surface exposed at low tide.

Coasts bordered by barrier reefs show evidence of Holocene submergence by the sea in the form of inlets, embayments and drowned valley mouths. The general absence of cliffing on such coasts is evidence that the barrier reefs grew up as the continental shelf subsided, so the coast was consistently protected from the action of strong ocean waves. There is thus a contrast between the reef-protected Queensland coast, which has numerous promontories and high islands offshore with only limited cliffing and the New South Wales coast, farther south and beyond the protection of the barrier reef, where headlands are more strongly cliffed.

Where coastlines bordered by barrier reefs have fringing reefs as well (as in NE Queensland) the latter are secondary forms, initiated along the coast only after the Holocene marine transgression, which allowed the barrier reef to build to its present level, came to an end.

13.7 Atolls

Atolls are reefs built up to sea level, typically circular or ovoid in plan, more or less continuous and surrounding a lagoon. They range from less than 1 to more than 100 km in diameter, the lagoon being typically several kilometres wide and 30 to 100 m deep. Atolls originated as fringing reefs around islands that were lowered by crustal subsidence during Tertiary and Quaternary times, the fringing reef growing up to become a barrier reef enclosing a lagoon (an almost-atoll retains a central high island, as at Bora Bora), and eventually an atoll, with the central island lost from view as submergence continued. Like barrier reefs, atolls were exposed to subaerial karstic weathering during low sea level phases of the Pleistocene, and rebuilt when sea level rose again (McLean and Woodroffe, 1994).

Atolls are of three kinds. There are oceanic atolls, which have localised (generally volcanic) foundations at ocean depths exceeding 550 m, shelf atolls, which rise from the continental shelf and have foundations at depths of less than 550 m, and compound atolls, where the ring-shaped reef surrounds or encloses relics of earlier atolls.

Oceanic atolls are common in the west Pacific and are present in the Coral Sea and the northern Tasman Sea. Typical features of an oceanic atoll are shown in Figure 13.4. The reef platform on the windward side is often wider, with an algal rampart in the breaker zone, and sometimes one or more low islands of coral debris, while outlets to the ocean are generally on the leeward side. The lagoon floor is a smooth depositional surface, from which pinnacles and ridges of live coral may protrude, but sand and gravel fans are formed when reef-derived sediment is washed in through entrances or over the bordering reef (Kench, 1998).

Atolls are well developed in the Indonesian region, especially in the Flores Sea, and are also found off the north coast of Australia, where barrier reefs have not formed. Seringapatam, 460 km north of Broome, is an example, rising abruptly from a depth of almost 550 m near the outer edge of a broad, sloping continental shelf. The enclosing reef is about 900 m wide, and the lagoon, 9.6 km long and 6.4 km broad, has an average depth of 35 m. Scott Reefs, not far away, consist of a similar enclosed atoll and a second atoll that is incomplete, with the superficial form of a horseshoe reef. Water more than 180 m deep surrounds and separates the two reefs, but the enclosed and partly enclosed lagoons have floors at a depth of 35–45 m.

Compound atolls are found in the Houtman Abrolhos, a group of ring-shaped reef platforms and patches with remnants of emerged Pleistocene reefs rising sharply from a depth of 55 m on the outer part of the continental shelf off the west coast of Australia. These southernmost (29°S) coral reefs in the Indian Ocean show

Figure 13.4 Kapingamarangi Atoll in the Caroline Islands, Pacific Ocean east of the Philippines

vigorous coral and algal growth in water warmed by the south-flowing Leeuwin current. Pelsart Island is 11 km long, a few hundred metres wide, and up to 2 m high, an emerged Pleistocene coralline reef on the eastern limb of the V-shaped Pelsart Group, essentially a tilted atoll. The reefs are penetrated by deep sinkholes (blue holes), produced by subaerial solution processes when they were dissected during Pleistocene low sea level phases.

In the Pacific Ocean there are a number of sea mounts, generally extinct volcanoes that rise steeply from the deep ocean floor, several of which have been planed off to form flat-topped guyots, platforms at depths of between 550 and 900 m. Some bear fossil corals of Cretaceous age, but had evidently subsided tectonically below the limit of coral reef growth during Tertiary and Quaternary times. Those that remained within the range of coral growth have been built up as atolls, but some have failed to maintain upward growth and form drowned atolls, notably in the Caroline Islands.

Small-scale atoll-like reefs up to a kilometre in diameter and enclosing lagoons less than 20 m deep are known as faros in equatorial regions (especially in the Maldive Islands, 2°S to 7°N), where they may have originated from coral reefs that lived through the Last Glacial low sea level phase in these low latitudes. During Pleistocene low sea level phases, ocean temperatures fell by up to 6°C, so that coral growth could then have continued only in the warmer parts of the oceans, probably within latitudes 10–15°N and S.

Still smaller are microatolls, intertidal ring-shaped organic reef structures typically 1–6 m in diameter, with slightly raised rims of living coral, built up to about low neap tide level, and thus standing above the general level of a reef platform (Stoddart and Scoffin, 1979).

13.8 Emerged coral reefs

Coral reef platforms are built up to a level just above low tide, and emerged reefs, consisting of dead coral and associated reef organisms standing above low tide level, can result from tectonic uplift or a lowering of sea level. Emerged reefs have been reported from several sites in the Pacific, notably in the Society, Tuamoto and Cook Islands, many with solution notches indicating phases of stability separated by tectonic uplift or sea level fall. An emerged reef forms part of the Pelsart Group in Houtman Abrolhos, off the coast of Western Australia, which was evidently built when the sea stood 3–8 m above its present level during a Late Pleistocene still-stand. It is bordered by modern reef platforms, which have grown up to just above low tide level.

Similar emerged reefs have been found farther south, on Rottnest Island, beyond the present range of reef-building. Emerged reefs extending up to 3 m above present low tide level have been reported at various other places around the Australian coast, including the shores of Melville Island, islands in Torres Strait, and Raine Island off the Queensland coast. They have generally been regarded as reef platforms built up to a slightly higher sea level, then laid bare as a consequence of a ensuing fall in sea level, and they may correspond to emerged shore platforms cut in dune calcarenite and other formations around the Australian coast.

Near Tokyo uplifted early Holocene coral formations are found well to the north of living coastal reefs, implying that the limits of Holocene coral growth in the north Pacific then extended further north, while the emerged Holocene reef that borders Peel Island, in Moreton Bay (27° 30′S), south of the present limit of coral growth in Queensland coastal waters, may indicate a similar southward extension of reef growth (Hopley, 1982).

Examples of emerged Quaternary fringing reefs that have been raised by tectonic movements are found on the mountainous slopes of the Huon peninsula, in NE New Guinea, where they form a stairway of reef terraces, each representing a phase of still-stand between episodes of uplift (Section 3.6). They are being rapidly consumed by subaerial processes, notably solution by rainwater. The reef terraces diverge on either side of an axis of upwarping, on which they attain a maximum elevation of about 750 m in the vicinity of the deeply incised Tewai gorge (Chappell, 1974). The structure of the fringing reef terraces can be seen on exposures along the sides of this gorge.

Islands formed by the emergence of coral reefs include Christmas Island, 320 km south of Java, which stands 390 m high and has been uplifted tectonically to such an extent that underlying Eocene marine limestones are exposed. The Loyalty Islands, NE of New Caledonia, include uplifted atolls such as Maré, where the former lagoon floor is an interior plain surrounded by an even crested rim of reef limestone 30 m high, the steep outer coastline showing 15 notched terraces indicative of stages in its uplift, and Uvéa, which has been intermittently tilted during upheaval so that the terraced eastern rim has emerged while the western rim, submerged, has a newer loop of developing reefs. Rennell Island is another emerged atoll, with high ridges

Figure 13.5 Notch and visor profile on the edge of an uplifted coral reef limestone on the Isle of Pines, New Caledonia

of coral limestone enclosing a lake that is linked to the sea by way of a cavern.

Emerged coral reefs are attacked by rainwater and sea spray solution, and develop a pitted and hollowed karstic topography, with pinnacles and crevices, known on Pacific islands as *makatea*. Their coasts show cliffs with basal notches, fronted by low tide shore platforms similar to those on other limestone and dune calcarenite coasts. In the Bismarck Archipelago cliffs at the edge of emerged reefs show two notches, one formed when the sea stood about 1.5 m above present mean tide level, where the modern notch has a pitted basal slope descending to the modern fringing reef.

The Isle of Pines, south of New Caledonia, is fringed by emerged coral reefs that are cliffed, with typical notch-and-visor profiles (Figure 13.5). The emerged reefs are bordered by shore platforms cut to present low tide level and ex- tending seaward by the growth of a modern fringing reef at the same level.

13.9 Islands on coral reef platforms

Fragments of coral broken off during occasional storms are cast up on to the reef platform as coralline sand and gravel, and occasionally larger boulders. Blocks measuring several cubic metres thrown up during a storm surge may persist on a reef platform for decades, even centuries. They develop features typical of limestone coasts, being gradually undermined by notches formed by solution and becoming rugged and pitted as they are dissolved by rainwater and sea spray.

Low islands (known as cays, and also as cayes or keys) have been built on reef platforms by the accumulation of sand, shingle and boulders

Figure 13.6   A sand cay on a reef platform in the Maldive Islands, Indian Ocean. Photograph by Dr K.G. Boston

formed from reef debris eroded by wave action and thrown up on the platform (Figure 13.6). These carbonate sediments include large lumps of broken reef limestone, broken cylindrical sticks of stag-horn coral, algae such as the sand-sized *Halimeda*, discoidal shingle and the gritty sands into which coralline reef material disintegrates. Pumice, a light and porous rock derived from volcanic eruptions, floats in the sea, and is washed up on many cay beaches.

Cays are found on the Belize barrier reef in the Caribbean, where term caye or key indicates an island of coralline sand and gravel deposited on a coral reef. Off the Queensland coast there are cays on coral platforms in the Bunker and Capricorn reefs, at Fife Island north of Port Stewart and at Green Island, off Cairns. Waves generated by the prevailing SE winds have washed sediment across the reef platform at high tide to build the cays generally near the NW (leeward) edge. Waves refracted round the reef plat-

form converge on the lee side in such a way as to prevent reef debris from being swept over the lee edge of the platform. Generally cays have steeper and narrower beaches on their windward than on their leeward shores.

Algal sand is plentiful on coral cays. A cay is initiated as a sandbank or heap of coral shingle awash at high tide, but as sediment accumulates, and once it is built above high tide level, it is colonised by grasses and shrubs, then coconut trees, palms (*Pisonia* and *Pandanus* spp.) and casuarinas.

Most cays consist mainly of wave-deposited sediment, but Heron Island, off SE Queensland, has low dunes of coralline sand built by the SE trade winds. On reefs more exposed to strong wave action the cays are gravelly, with a predominance of coarse shingle over sand, as on Lady Musgrave Island, near the southern end of the Great Barrier Reef, and on the Gili Islands NE of Lombok in Indonesia.

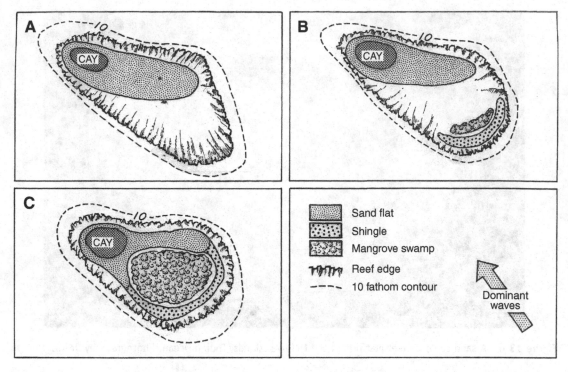

Figure 13.7   Stages in the formation of a cay (a) and low wooded island (b), (c) on a reef platform

Low wooded islands are more depositional islands found on reef platforms, as off NE Queensland (Figure 13.7). Low Isles, off Port Douglas, is an example. It consists of a sand cay that formed on the leeward (NW) side, then shingle ridges that were thrown up as ramparts on the windward side, built from debris eroded from the edge of the reef, with mangroves (mainly *Rhizophora*) in the intervening depression. Mangroves acidify seawater and hollows are formed where this dissolves underlying coral limestone. Changes occur as the result of sand accretion during relatively calm weather and erosion (sometimes with the addition of coral shingle) during storms. Some low wooded islands have several successively built parallel shingle ridges, the older ridges being dark in colour and largely cemented as conglomerate while the younger are white or cream, and unconsolidated. Addition of freshly formed

coralline debris occurs during tropical cyclones, widening the existing rampart or adding a new shingle ridge.

Cays and low wooded islands are related to wave energy. On the outer reefs high wave energy has prevented the formation of such islands, but the inner reefs, partly protected from ocean waves, have moderate wave energy that leads to sand cay formation by refracted waves generated by local winds. The reef platforms off NE Queensland are in deep water with a relatively broad fetch, over which the SE trade winds generate waves strong enough to build shingle ramparts on the windward side and cays on the leeward side. Cays and low wooded islands occur in similar conditions of wave energy variation in the reefs off British Honduras.

Cays change in configuration as erosion and deposition modify their shores. Between high and low tide levels cay beaches may be cemented by secondary deposition of calcium carbonate in the zone of repeated wetting and drying, sandy beaches forming the compact layered sandstone known as beach rock (Section 6.6), whereas shingle becomes a lithified beach conglomerate. The greater resistance of these formations impedes erosion of cay shores, but waves generated during tropical cyclones may expose beach rock on the cay shores as they modify the outline of the cay. It may be swept away altogether, leaving only patterns of eroded beach rock or shingle to conglomerate to commemorate the coastlines of vanished cays. Remnants of beach rock on a reef platform can thus be taken as an indication of the former presence of a cay when there is no longer any depositional island.

In 1961 Hurricane Hattie swept westward across the atolls and barrier reef off British Honduras towards the coast south of Belize, accompanied by gusts exceeding 320 km/hr. A 72 km wide storm surge raised sea level briefly by up to 4.5 m. Cays that had a dense vegetation cover were less modified than those on which the natural vegetation had been cleared and replaced by coconut plantations. The former were eroded on the windward side, with shingle and coral debris being piled up against the vegetation, but the latter were more severely damaged, and seven were swept away completely (Stoddart, 1965). The sandy cay of Makuluva, off Suva in Fiji, has migrated eastward across the coral reef platform, the windward (SW) coast having been cut back, so that lighthouse foundations and water tanks stand offshore in front of a sandy cliff capped by dune woodland, while the leeward coast has a broad prograding beach on to which grasses and shrubs are spreading. Aves Island, off Venezuela, is another example of a migrating cay.

Cays in the Great Barrier Reef region have shown a prevalence of erosion on their coastlines in recent decades, confirmed by extensive outlying exposures of beach rock, so that they have been diminishing in size, perhaps as the result of a rising sea level. In many respects the cays are similar to the spits and barriers built by coastal deposition as described in Chapter 8, their distinctive features largely a response to the special environment of the reef platforms on which they have been built.

13.10 Algal and other biogenic reefs

The commonest non-coralline reefs are those built by algae on rocky shores or in the intertidal zone at the base of a cliff. Calcareous algae withdraw carbonates from seawater and build solid reefs or encrustations. Some algae also have sediment-trapping structures. On the coasts of the Caribbean, the Mediterranean and southern Australia calcareous algae such as *Lithophyllum* and *Porolithon* form solid ripples and rims on shore platforms, particularly those cut in dune calcarenite. On some shore platforms algae produce rimmed terracettes that descend from a zone of higher water level, maintained by frequent large waves, seaward or sideways, and occasionally landward. Waterfalls form as wave swash floods over these platforms.

Reefs built largely by algae occur on subtropical coasts close to, and beyond, the limits of coral growth, as in Florida and Brazil, around the Mediterranean and in the hypersaline waters of Shark Bay, Australia. In the absence of corals, or where coral growth is poor, algae can construct reefs, or at least form reefs surmounting submerged ridges of sandstone or dune calcarenite. Algal reefs of this kind are extensive in the Cape Verde Islands, in NE Brazil.

The growth of calcareous algae and associated marine invertebrates can form a veneer on a rocky shelf, known as a trottoir, at about high tide level, notably on sectors of coast sheltered from strong wave action in the Mediterranean. Corniches are projections built by calcareous organisms such as *Lithophyllum* in similar situations. Occasionally waves break off the outer edge. Similar features have been seen in the Pacific, as in the Mariana Islands.

Reference has been made (Section 3.10.3) to calcareous structures built by tubeworms (*Galeolaria*) to form ledges on rocky shores and harbour walls, or cushions around pier supports in SE Australia (Figure 3.10). Their upper limits are just above mid-tide level.

Oyster reefs are found on many coasts. On the shores of the Gulf and Atlantic coasts of the United States oyster reefs built by *Crassostrea virginica* are extensive, and there are also small mounds built by bryozoans, serpulids and mussels on rocky or sandy foundations. Vermetid (long, coiled) gastropods have built reefs in the Mediterranean and Atlantic, and form biodepositional sandrocks along the seaward margins of mangroves on the low wave energy coast of western Florida. They form in turbid water as tubular concretionary structures that become aggregated as minor reefs. Sabellariid worms build shore ledges on beaches in eastern Florida.

Non-biogenic reef structures include tufa deposits built up around submarine springs, as in Ikka Fjord in Greenland, where numerous columnar towers have grown up from a sea floor seepage zone. They consist of the mineral ikaite, and their growth can be rapid: up to 0.5 m/yr.

The towers have summits that are occasionally exposed at sea level on calm days.

13.11 Summary

Coral reefs form in warm seas as the result of the growth of corals and associated organisms, notably algae, which require sunlight. Their growth is impeded by turbidity, as off river mouths. Their present form and distribution, attained during the Holocene marine transgression, was influenced by subsidence of the ocean floors. Fringing reefs have formed along present coastlines; barrier reefs originated on ancient coastlines that have subsided and atolls commemorate the outlines of subsided islands. Emerged coral reefs are subject to the weathering and erosion processes that affect other limestones. There are also islands of coralline sand and gravel, derived from reefs and deposited on reef platforms as cays or low wooded islands. Often the coralline sediment has been cemented as beach rock.

Reefs and rims are also built entirely by algae, particularly beyond the limits of coral growth. Other biogenic reefs have been built by oysters, bryozoans, serpulids, mussels, vermetid gastropods and tubeworms such as *Galeolaria*.

14

Future coasts

14.1 Introduction

The changes that will take place on the world's coastline over the coming century cannot be predicted simply by extrapolating the gains, losses and modifications that occurred during the 20th century. Sea level has been relatively stable on much of the world's coastline during the past 6000 years, apart from minor oscillations, but there are sectors where the coastal land has continued to rise or fall within this period: some coasts are responding to a slow sea level rise (as on the Gulf and Atlantic seaboards of the United States), others to a lowering of sea level, notably where the land is being uplifted as the result of deglaciation and isostatic recovery (as in Scandinavia and parts of northern Canada). Coastal landforms have been changing as the result of erosion and accretion in response to coastal processes with the sea at or near its present level, and there will be additional responses to relative sea level changes.

The present chapter considers possible responses of coastal landforms and associated features to forecast climatic changes and sea level rise. These should be taken into account in devising future policies for integrated coastal management.

14.2 Greenhouse effect and sea level rise

Global climatic changes are likely to result from the human-induced accumulation in the Earth's atmosphere of such gases as carbon dioxide, nitrous oxide and methane, produced mainly by industry and agriculture. In particular, the burning of fossil fuels (coal, oil and natural gas) is returning to the atmosphere carbon dioxide that was withdrawn from earlier atmospheres by plant photosynthesis and retained in swamp forests that became fossil fuel deposits in the geological past.

Atmospheric monitoring initiated during the International Geophysical Year in 1957 has shown that concentrations of carbon dioxide and other gases in the atmosphere have been increasing. The carbon dioxide concentration, for example, increased from 315 parts per million (ppm) in 1958 to almost 380 ppm in 2005. Such an increase will enhance the natural greenhouse effect, whereby the atmosphere intercepts some of the solar radiation reflected into space from the Earth's surface, and so maintains global temperatures at a higher level than would otherwise prevail. It is expected that the mean temperature of the lower atmosphere will increase

Coastal Geomorphology: An Introduction, Second Edition. Eric Bird.
© 2008 by John Wiley & Sons, Ltd

by between 1.5 and 4.5°C over the coming century. Such human-induced global warming will lead to expansion of the oceans (the steric effect) and some melting of the world's snowfields, ice sheets and glaciers, resulting in a world-wide rise of sea level. As has been noted, there is evidence from tide gauge records suggesting that the sea has been rising at the rate of between 1 and 2 mm/yr around much of the world's coastline (Pirazzoli, 1996), but as the majority of tide gauges are located in port areas and may show local anomalies the pattern and scale of a contemporary marine transgression require confirmation, notably from continuing global satellite altimetric surveys of sea level. Although there has been extensive melting of glaciers and ice sheets, many coasts still show little if any sea level rise so far. This response lag has led to speculation about compensating effects such as sea floor subsidence and the deepening of ocean basins.

Calculations by the Intergovernmental Panel on Climatic Change (IPCC, 2007) indicated that average global sea level will probably rise between 18 and 59 cm by the end of the 21st century. There has already been extensive reduction of glaciers and ice sheets, and it may now be too late to prevent large areas of polar ice from melting. Melting of the Greenland ice sheet could raise average global sea level by about 7 m, and large-scale breakaways of the West Antarctic ice shelf could also cause average global sea level to rise several metres (IPCC, 2007).

A sea level rise will initiate or accelerate coastal changes around the world (Bird, 1993c). An obvious outcome will be that submerging coastlines, currently confined mainly to sectors where the land has been subsiding (Figure 3.12), will become more extensive, and that emerging coastlines will become rarer. A rising sea level will reduce the effects of land emergence, slowing down the advance of coastlines such as those bordering the Gulf of Bothnia. An accelerating sea level rise will eventually equal, then exceed, the rate of land uplift, and in due course all of

the world's coastline will be submerging as the result of sea level rise.

Depletion of the world's upper atmospheric ozone layer, which intercepts much of the ultraviolet radiation arriving from the sun, is thought to be due to the effects of chlorine monoxide produced from chlorofluorocarbon (CFC) emissions, generated by aerosol propellants, refrigerators and various industrial processes. In addition to depleting ozone, CFCs also contribute to the enhancement of the greenhouse effect, global warming and sea level rise, but ozone depletion leads to increased ultraviolet radiation, which has adverse effects on the growth and health of plants and animals, and is harmful to humans.

14.3 General effects of a rising sea level

If global sea level rises in the manner predicted there will certainly be extensive marine submergence of low lying coastal areas (Figure 14.1). High and low tide lines will advance landward, and at least part of the present intertidal zone will become completely submerged. On steep hard rock coasts where there is little or no marine erosion in progress and on solid artificial structures such as vertical sea walls, a sea level rise will simply raise the high and low tide lines and the coastline will remain in its present position.

It is possible that there will be a slight increase in tide ranges around the world's coastline as the oceans deepen, the rise that actually occurs being modified as tidal amplitude is adapted to the changing coastal and nearshore configuration. Deepening of nearshore seas may increase tide range at the coast. On many coasts the extent to which the high tide line moves landward will be augmented by an increase in erosion as nearshore waters deepen and larger and more destructive waves break upon the shore. As sea

Figure 14.1 Changes in response to a sea level rise: (a) on a cliffed coast with a shore platform; (b) on beach ridges; (c) on a marsh terrace; (d) on a barrier-fringed lagoon; (e) on a coral reef

level rises, erosion will begin on coasts that are at present stable, and accelerate on coasts that were already receding. On coasts that had been prograding the seaward advance of the land will be curbed, and erosion may begin. This has been seen on the coasts of the Caspian Sea, where the proportion eroding was 10 per cent in 1977 after several years of falling sea level and widespread progradation, but has increased to over 40 per cent in the ensuing sea level rise (Figure 3.11).

Erosion will increase still further where the climatic changes that accompany the rising sea level lead to more frequent and severe storms, generating surges that penetrate further inland than they do now. Coasts already subject to recurrent storm surges (e.g. the hurricane-prone Gulf and Atlantic coasts of the United States) will have more frequent and extensive marine flooding as submergence proceeds, and more severe erosion and structural damage where larger waves reach the coast through deepening nearshore waters. These effects could be more damaging than the direct impact of a sea level rise.

The extent of coastal erosion will also depend on how the nearshore sea floor is modified by the rising sea as the coastline moves landward. Although there will generally be deepening of nearshore waters and a consequent increase in wave energy approaching the coastline, some coasts may receive an augmented sediment supply, derived from increasing fluvial or alongshore sources, and this could maintain, or even shallow, the nearshore profile as the sea rises. On such coasts wave energy will not intensify, and there may be little, if any, coastline erosion; there may even be some progradation.

Coastal land reclamation and coastal protection have rendered many coastlines artificial because of sea wall construction. As the sea rises alongside such structures there is an increased risk that they will be overwashed or breached by storm surges, and the associated rise in the water table will lead to waterlogging and flooding of reclaimed areas. In recent years a number of areas of reclaimed coast have been deliberately abandoned, a procedure called managed retreat or coastal realignment. Freiston in Lincolnshire and Porlock Bay in Somerset are examples of coastal lowlands abandoned to marine invasion. More broadly, some coastal structures have been moved inland in response to the threat of coastline recession (McGlashan, 2003). As sea level rises great modifications will have to be made

to cities, ports and low lying areas around the world's coastline. So far Galveston in Texas is one of the few cities that have raised land levels and buildings to prevent submergence by a rising sea.

Attempts have been made to predict the effects of a sea level rise on coastlines (Tooley and Jelgersma, 1992; Warrick, Barrow and Wigley, 1993; Bird, 1993c; Cracknell, 2003), but they remain speculative because of limited monitoring of recent and continuing changes and the imprecision of available models. A major difficulty is that the prediction is an accelerating sea level rise, with no early prospect of stabilisation. If the sea continues to rise, coastal erosion will accelerate and become more widespread as any compensating sedimentation declines. The rising sea will encounter, re-shape and in due course submerge the 'raised beaches' and other emerged shore features that formed during Pleistocene interglacial phases (Section 3.6). Revival at higher levels of features similar to those now seen on the world's coastline, with some parts eroding while others are stable or prograding, must await the establishment of a new sea level still-stand. This could be when the greenhouse effect has been brought under control, or when all the world's glaciers, ice sheets and snowfields have melted and the water contained in them has flowed back into the oceans, producing a global sea level rise of more than 60 m (Section 3.3.4). It would be possible to assess the nature of coastal landforms, sediments and ecological features on coastlines after the sea has risen to a specified level, but with present knowledge this is highly speculative. Pictures have been published of St Paul's Cathedral and the Eiffel Tower standing in the sea, and maps showing the extent of submergence expected in various regions with sea level raised by up to 60 m (flooding most of the world's capital cities) have drawn attention to scale of the problem, but a more realistic approach would concentrate on what is actually happening on coasts where subsidence has already led to a rising sea level.

14.4 Effects of a changing climate

A global sea level rise will be accompanied by other responses to the increase in atmospheric temperature. As well as raising temperatures in coastal regions, global warming will cause a migration of climatic zones, tropical sectors expanding as temperate sectors migrate poleward and the domain of arctic coasts shrinks. Tropical cyclones may become more frequent and severe, extending into higher latitudes and bringing storm surges and torrential downpours to coasts that now lie outside their range. Coastal climates will also be modified by changes in ocean currents, as already shown by the El Niño Southern Oscillation, which has led to heavy rain and river flooding in China, Ecuador and Peru and droughts in Australia. Changes in the Gulf Stream could have major impacts on the climate of NW Europe, leading to cooling that might at least partly offset the effects of global warming in that region.

As warming proceeds, some regions are expected to receive more rainfall while others become drier. Where rainfall increases there will be more frequent and persistent river flooding, and the water table rise will be added to that caused by the rising sea level. Some low lying parts of coastal plains will become permanent swamps or lagoons, the salinity of which will depend on interactions between increasing marine incursion (and perhaps an upwards movement of subterranean salt) and any offsetting effects of augmented rainfall and freshwater runoff. With increased rainfall, coastal vegetation may become more luxuriant; with drier conditions it is likely to be depleted. Coastal regions that become drier will also have more extensive marine salinity penetration into both surface and underground water. Coastal lagoons will become more brackish, and some will dry out as saline flats. Desiccation will also reduce the vegetation cover and allow wind action to be more erosive, especially in coastal dune areas, whereas more vigorous and extensive vegetation will tend to stabilise dunes that are now bare and drifting. Vegetation may also be impoverished by the effects of increasing ultraviolet radiation resulting from atmospheric ozone depletion, and thus less able to trap sediment on dunes or in salt marshes and mangrove swamps. On the other hand, increasing atmospheric carbon dioxide may enhance photosynthesis and plant growth, and produce vegetation that is more effective in trapping dune sand or intertidal sediment.

Most marine plants and animals have specific latitudinal ranges, usually depending on maximum or minimum sea temperatures, which will rise as global warming proceeds. Mangroves, now largely confined to tropical coasts, will extend their range poleward along coasts whereby there are suitable habitats, and corals will extend northward and southward beyond their present latitudinal limits, providing there are suitable substrates and available coral larvae. Many temperate salt marsh species will also migrate poleward, whereas the distribution of kelp, a plant restricted to cooler waters, will contract to higher latitudes. Global warming is also likely to affect coastal ecosystems by way of thermal stress, a factor already obvious where coral bleaching has occurred.

14.5 Effects on cliffs and shore platforms

Existing cliffs and rocky shores are largely a consequence of the relatively stable sea level around much of the world's coastline during the past 6000 years. If sea level rises nearshore waters will deepen, submerging shore platforms and rocky shores, and the deeper water will allow larger waves to reach the coast and attack the base of cliffs and bluffs, accelerating their erosion (Figure 14.1(a)). On some coasts the rock outcrops are so resistant that the high and low tide lines simply move up the existing cliff face, but elsewhere increasing erosion will prompt demands for coastal defence works, notably the extension and elaboration of sea walls.

Where the coast consists of vegetated bluffs a rising sea level will increase basal wave erosion and slumping, and the bluffs will develop into retreating cliffs. Existing cliffs will generally become more unstable, and recede more rapidly. Cliff erosion accelerated on cliffs cut in soft sediments on the coasts of the Great Lakes in North America during phases of rising water level, and a rising sea level resulting from land subsidence following oil extraction around Long Beach, California, has intensified wave attack, accelerating the retreat of soft clay cliffs at Huntington (Bird, 1993c). Clayton (1989) estimated that in Britain cliffs that are already retreating a metre per year will show an accelerated retreat of 0.35 m/yr for every millimetre rise in sea level. On the volcanic island of Nii-jima, off the Japanese coast south of Tokyo, Sunamura (1992) calculated that cliffs cut in poorly consolidated volcanic gravel and ash, now retreating at 1.2 m/yr, will recede at between 2.3 and 2.9 m/yr with a sea level rise of a metre by the year 2100, or more if the accompanying climatic change increases storminess in coastal waters. It will be difficult to recognise an acceleration of cliff retreat, at least in the early stages of rising sea level, unless existing rates of recession have been measured with sufficient accuracy to provide a basis for comparison. Such monitoring is currently only available for very limited parts of the world's cliffed coastline.

Shore platforms that have developed in front of cliffs or bluffs will be submerged for longer periods, and become permanently inundated when the sea level rise exceeds present tide range. Waves that now reach the base of the cliff only at high tide will attack more consistently, thereby accelerating cliff erosion. Again, the eroded material will usually be carried away alongshore, but if it remains as a beach in front of the cliff the shore platform may be protected from further erosion. This will also be the case where submerging shore platforms acquire a mantle of accretionary growths of nearshore plant and animal communities. Algal encrustations on formerly eroding shore platforms near Port Hedland in Western Australia and Nyali in Kenya (Bird and Guilcher, 1982) may be a consequence of a slight sea level rise. Cementation of sandy ripples on the surface of a shore platform cut in Pleistocene dune calcarenite on the coast near Sorrento, in SE Australia may also indicate a rising sea level. Such changes may at least partly offset the effects of nearshore deepening and intensification of wave attack on backing cliffs.

Cliff-base notches formed by abrasion or solution will be enlarged upward by a rising sea level. The formation of a new notch at a higher level would require a rapid sea level rise followed by a new still-stand.

On soft rock formations waves have shaped the nearshore sea floor into a concave profile that declines seaward from the cliff base, and as the cliffs recede this nearshore sea floor is lowered in such a way as to move the concave profile landward. This has occurred where the cliffs and sea floor are cut into Pleistocene glacial drift on the east coast of England, as at Holderness. A sea level rise is likely to accelerate landward migration of the concave sea floor profiles as the cliff base retreats. Bray and Hooke (1997) have considered adapting the Bruun rule (Section 7.1.1) to the prediction of soft cliff retreat on the south coast of England.

Cliffs cut in soft rock formations on coasts sheltered from strong wave action are often dominated by gulleying, slumping and other features resulting from subaerial weathering and the effects of runoff and seepage, rather than by marine processes. A vertical cliff profile may be formed briefly after storm waves have removed basal talus and downwashed deposits, but subaerial processes soon restore the degraded profile. On such cliffs a rising sea level is likely to increase marine erosion, reducing and eventually suppressing the features developed by subaerial processes, as undercutting of the cliff base maintains a steeper or vertical receding cliff, but an increase in rainfall could maintain subaerial erosion by runoff.

A sea level rise is likely to increase the frequency of coastal landslides and produce new and more extensive slumping along coasts, especially where the rock formations dip seaward. The frequency of slumping will accelerate where a cliff is cut back through horizontal or landward-dipping strata into an area where the rocks dip seaward. Coastal landslides will become more frequent if rainfall increases, and less if the coastal climate becomes drier. A byproduct of increased landsliding will be an augmented supply of sediment to adjacent and downdrift beaches, perhaps offsetting the effects of a sea level rise on these beaches.

Rates of cliff recession are influenced by the availability of rocky debris, including beach material, which can be mobilised by wave action and used as ammunition for cliff-base abrasion. Cliffs fringed by a narrow beach have retreated more rapidly than cliffs on the same formation where the beach is wide and high (so that waves do not reach the cliff base), or where there is no beach material (so that the cliff base is attacked only by the hydraulic action of breaking waves). Debris eroded from cliffs is generally carried away alongshore by wave action, but if during a sea level rise this sediment is retained as a persistent talus apron of sand and gravel, as a protective beach in front of the cliff or as bars and terraces in the nearshore zone, wave energy will diminish, and cliff retreat will slow down and perhaps be halted. On the other hand, the depletion and narrowing of a beach along the base of a cliff will result in more vigorous

wave abrasion and accelerated cliff retreat, but if the beach is completely removed wave abrasion will diminish, and with purely hydraulic action cliff retreat could decelerate.

Rates of cliff retreat as sea level rises may be estimated by comparing recession rates on similar rock formations with differing exposure. The retreat of a cliff fronted by a seaward-sloping shore platform and subject to only brief episodes of basal wave attack at high tide can be compared with a cliff cut in similar material where wave attack is stronger and more persistent because nearshore water is deeper. Such a contrast can be found where a resistant formation dips alongshore, so that it protects a cliff from sustained wave attack on one sector, but not on the next. On the coast of Kimmeridge Bay, in southern England, cliffs cut in Jurassic shale and limestone are being cut back more rapidly behind shores where the sea floor declines steeply than where the cliff base is protected by a flat structural shore platform. After a sea level rise has submerged the structural platform the rate of cliff recession will increase, while that of the unprotected sectors will also accelerate as the water deepens.

There have been few measurements of cliff recession rates on subsiding coasts, but in southern England, which is believed to have been subsiding at about 2 mm/yr, surveys by Brunsden and Jones (1980) indicated that the slumping cliffs of Lyme Bay have been retreating at an average rate of 40 m per century. If sea level rise accelerates these cliffs could steepen, with recession proceeding too quickly for mass movement to maintain the existing topography, or the present jumbled coastal morphology could persist as the coastline retreats. The second possibility is more likely if the sea level rise is accompanied by a climatic change that increases rainfall, seepage and runoff on this coast.

The response to cliff retreat where it threatens or destroys structures such as roads or buildings is generally to attempt to halt recession by building sea walls or other protective structures

Figure 14.2 Desperate remedies. A boulder bank has been built on the shingle beach at Fairlight Cove, Sussex, in an attempt to exclude wave action and so halt the cliff recession which has destroyed several houses. As sea level rises, this kind of problem will become more widespread

(Figure 14.2). Some enlightened coastal municipalities have used beach nourishment (Section 7.4) to protect eroding cliffs.

14.6 Effects on beaches, spits and barriers

There is evidence (Chapter 3) from tide gauge records that a world-wide sea level rise of about 1–2 mm/yr has been taking place during the past few decades, offset on some coasts by equal or greater land uplift, and complicated by geophysical factors that have raised or lowered the surface of the oceans (Pirazzoli, 1996). The beach erosion problem (Chapter 7) will intensify if, as is predicted, a global sea level rise increases in the next few decades. A sea level rise will generally result in a deepening of nearshore water, so that larger waves break upon the shore,

initiating erosion on beaches or accelerating it where it is already taking place. Beach erosion will become even more extensive and severe than it is now.

On beaches that had been prograding the seaward advance will be curbed, and erosion may begin (Figure 14.1(b)). Where beach-ridge plains and coastal barriers have shown Holocene progradation a sea level rise is likely to initiate or accelerate erosion along their seaward margins (Figure 8.12). Coastal barriers that are already transgressive (migrating landward) will continue to migrate as sea level rises, and some barriers that have remained stationary, or have prograded during the Holocene still-stand, may become transgressive as the result of erosion along their seaward margins, with washovers and landward drifting of dunes. This has happened on the coasts of the Caspian Sea during the marine transgression that began in 1977.

In the SW, where the nearshore gradient is very gentle (about 1:1000) the rising sea level has led to beach erosion accompanied by the formation of transgressive barriers in front of shallow lagoons that are moving diachronously landward (Leontiev and Veliev, 1990).

On coasts where the beach fringe is narrow, backed by high ground, beaches will soon disappear unless the sea level rise increases nearby cliff erosion and generates additional sediment to maintain them. There will still be beaches that persist as they retreat through beach ridge plains or coastal barriers of sand and shingle. Beaches that front salt marshes or mangrove swamps are likely to be eroded and overwashed, with the sea invading the land that lies behind them. Where sea walls have been built to halt coastline recession, beaches fronting them will be depleted or removed altogether by scour due to reflection of incident waves.

Beaches will be maintained where there is a continuing supply of sand and shingle, or where the supply is increased as the result of accelerated nearby cliff erosion, greater sediment yield from rivers because of heavier or more effective rainfall, catchment devegetation or disturbance by tectonic uplift or volcanic activity. If the nearshore profile is maintained or made shallower by accretion or tectonic uplift, or if the longshore sediment supply continues at a sufficient level (with sediment supplied by rivers or erosion of cliffs and shore outcrops, especially where these supplies increase as the result of a wetter or stormier climate), beaches may persist, or even be prograded, as sea level continues to rise. In most cases the high tide line will move landward as the result of submergence, and accompanying erosion as the nearshore profile also migrates landward.

The effects of a sea level rise can already be seen on beaches where there has been submergence as a result of land subsidence in recent decades, as on the Atlantic seaboard of the United States (Leatherman, 1990), on the Frisian Islands in the southern North Sea (Eitner, 1996) and in the NW Adriatic. Submergence has already contributed to beach profile erosion on these coasts, much as predicted by the Bruun rule (Section 7.1.1). According to this rule the extent of recession of the transverse beach profile will be 50–100 times the dimensions of the rise in sea level, so that a one metre rise would cause the beach to retreat by 50–100 m. Since many seaside resort beaches are no more than 30 m wide, the implication is that these will have disappeared by the time the sea has risen 15–30 cm (IPCC predictions indicate that this will be between the years 2040 and 2085), unless they are artificially replaced.

The Bruun rule states that on beaches that have attained a profile of equilibrium (Section 6.10.11) the response to a sea level rise will be erosion of the upper beach and withdrawal of sediment from the beach to the adjacent sea floor in such a way as to restore the previous transverse profile in relation to the higher sea level (Figure 7.1). There has been much discussion of its applicability to precise forecasting of the extent of coastline retreat after a sea level rise (SCOR Working Group, 1991). A pre-condition of the Bruun rule is that the beaches are initially in equilibrium (Pilkey et al., 1993), which has been interpreted in various ways. Beach erosion is already widespread and few of the world's sandy beaches are currently in equilibrium.

In order to apply the Bruun rule it is necessary to determine a seaward boundary of the profile that will be restored at a higher level. Bruun (1988) suggested that the boundary should be at the line where predominantly coarser nearshore sediment gives place to generally finer offshore material, where the water has become too deep for waves to move material of beach calibre, but this requires detailed sedimentological surveys of the sea floor, which were not necessarily available before a sea level rise began. On many coasts there are gradual transitions, and commonly the nearshore topography is variable, with sand bars, rocky outcrops, biogenic structures such as coral reefs or a muddy substrate offshore.

Restoration of the transverse profile can only be completed after the sea has become stable at a higher level, coastline recession coming to an end as a new equilibrium is attained. The Bruun rule is likely to underestimate the extent of beach recession while sea level is actually rising, and is difficult to apply to a prediction of an accelerating sea level rise without any indication of the level at which it will eventually stabilise. The Bruun rule takes no account of changes in processes or rates of sediment supply and removal resulting from the climatic variations that may accompany a sea level rise, such as increased rainfall and runoff, stronger winds and stormier seas.

The Bruun rule deals only with sediment interchange between the beach face and the nearshore sea floor, and omits gains or losses as the result of longshore drifting, and their effects on changing beach profiles. It is thus more likely to apply to swash-dominated rather than drift-dominated beaches. Many beaches also lose sediment blown to backshore dunes, washed into tidal inlets, or swept over barriers or spits into lagoons and swamps. It is difficult to predict the proportions of sediment that will be lost seaward, alongshore and landward from an eroding beach: these proportions may or may not maintain budgets of the kind shown in Figure 7.7 on an already-submerging coast.

If climatic changes accompanying a sea level rise lead to increased rainfall, beaches may be receive more sand or gravel washed down coastal slopes, and changes in coastal wind regimes may increase sand blown from hinterland dunes to beaches, or modify incident wave regimes to increase longshore drifting. An abundant sediment supply from alongshore could maintain or prograde a beach during a phase of rising sea level, either directly, or by shallowing the nearshore zone so that shoreward drifting ensues.

Where the transverse beach and nearshore gradient is low, the effects of a rising sea level are complicated because storm waves will steepen and re-shape the upper beach profile by throwing sand or gravel up above high tide level to form a barrier in front of a low lying area that becomes submerged as a lagoon. The barrier and lagoon are then driven landward as sea level continues to rise. As it does so, the steepened beach on the seaward side may maintain its profile as sediment is lost, partly by landward overwash and partly by losses seaward (Ignatov *et al.*, 1993). Analysis of transgressive Caspian coast barriers as possible models for responses to accelerating sea level rise elsewhere showed that the Bruun rule should be extended to a third dimension, and should include longshore drifting, washovers and the effects of wind action on beaches (Kaplin and Selivanov, 1995).

Where beaches are bordered by nearshore sand bars a sea level rise may be accompanied by upward growth and landward movement of these bars as the beach is cut back, providing there is a sufficient supply of suitable sediment available to maintain the nearshore profile. The Bruun rule could be extended to predict that the landward retreat of the outer limit of the sand bars will be equal to the extent of beach recession, so that the overall profile is preserved (Dubois, 1992). A complication is that sand bars usually consist of finer material than is present on the beach face, so that erosion and seaward movement of beach sand may not provide sediment of suitable calibre for their maintenance, at least in their existing form. On the shores of Lake Michigan, Dubois (1977) found that beach face erosion as lake level rose was matched by accretion on the landward side of the nearshore bar, which widened as its outer slope remained unchanged.

There are variations in the applicability of the Bruun rule in differing wave energy environments. On the generally low to moderate wave energy coasts of New Jersey and Maryland,

Everts (1985) found that the Bruun rule over-estimated measured beach recession, but on the Pacific coast, where wave energy is greater, erosion was between two and four times as much as predicted. Prediction of the extent of beach recession as sea level rises thus faces a number of difficulties. As long as the sea is rising, beach-fringed coastlines will continue to recede, as erosion accompanies submergence, and if the sea rises at an increasing rate the beach erosion will accelerate. This is already a major problem on subsiding coasts, as in the Venice region in Italy, where sea walls have replaced some eroded beaches (Figure 14.3).

A century hence, most beach-fringed coastlines will have retreated substantially, and (in the absence of protective structures or artificial nourishment) they will be eroding more rapidly. Prediction of the position of the coastline would be possible if the sea level rise were followed by a still-stand, which could occur after the human modifications of the atmosphere that enhance the greenhouse effect were brought under control.

14.6.1 Evidence from past marine transgressions

The prediction that beach erosion will be initiated or accelerated by a rising sea level has to be reconciled with evidence that in the geological past some marine transgressions were accompanied by shoreward drifting of sea floor sediment, and that beaches formed and prograded on coastlines as sea level rise slackened and came to an end. This was the case on many coasts during the Holocene marine transgression, which brought the sea up to, or close to, its present

Figure 14.3 An artificial coastline formed after a sea wall and groynes failed to retain the beach at Litorale di Pellestrina, near Venice (location shown in Figure 11.10) on the coast of Italy, and limestone blocks were dumped on the shore

level about 6000 years ago. Some of the beaches that were then formed have since prograded as Holocene beach ridge plains and barriers, although where sea level is still rising (as on the Gulf and Atlantic coasts of the United States) many beaches are now the seaward fringes of barriers migrating intermittently landward.

The Holocene marine transgression advanced across a land surface that had previously emerged from beneath the sea during a marine regression that occurred about 80 000 years ago. This surface had been strewn with Pleistocene beach and dune deposits left stranded as the emergence took place and to these were added fluvial and aeolian deposits, as well as glacial and periglacial deposits in high latitudes. There were also unconsolidated materials where shelf rock outcrops had been subaerially weathered during the prolonged low sea level phase. This shoaly topography was the source from which sand and gravel deposits were derived and swept landward to form beaches and barriers, many of which subsequently prograded as the marine transgression came to an end.

14.6.2 The future of beaches

Some existing coastal lowlands have features similar to those of continental shelves in the Late Pleistocene, notably where there are extensive coastal dunes, but many do not have sediment mantles that would provide beach sediment for a rising sea, and of course many have been developed and urbanised. It is unlikely that many existing beaches will be maintained or prograded during a phase of rising sea level, but as sea level rises across existing coastal plains, especially those with dunes, there may eventually be formation of beaches by shoreward drifting of sediment up to the new coastline, along the contour on which submergence comes to an end. By then present day beaches will have been submerged, buried or destroyed, so that the short term response to a rising sea level will be the onset or acceleration of erosion on existing beaches.

14.7 Effects on coastal dunes

Most coastal dunes lie behind beaches, and as sea level rises beach erosion will lead to increased backshore cliffing of dunes by storm waves, which will occur more frequently where the climate becomes more boisterous. As coastal dune fringes are cut back, more blowouts will be initiated, and some of these may grow into large transgressive dunes as sand is excavated and blown landward. A rising sea level will thus accentuate the development of transgressive dune formations.

If climatic changes accompanying a sea level rise result in drier and windier conditions, coastal dunes that are at present stable and vegetated may become unstable as the vegetation cover is weakened and sand mobilised, and dunes that are already active will become more mobile. On the other hand a wetter or calmer climate could facilitate vegetation growth and dune stabilisation.

14.8 Effects on intertidal wetlands

A rising sea level rise will submerge existing intertidal areas, including sandflats, mudflats, salt marshes and mangroves. As the nearshore water deepens, stronger wave action will initiate or accelerate erosion on microcliffs along the seaward margins of salt marsh and mangrove terraces (Phillips, 1986). It is doubtful whether the Bruun rule, devised to explain changes on beach profiles as sea level rises, can be applied to submerging salt marshes, because the response of coherent peat and clay formations are unlikely to be the same as that of unconsolidated beach sand. Tidal creeks that intersect salt marshes and mangrove areas will widen and deepen, and extend headward as they are submerged. As submergence proceeds, the retreat of the seaward margin will be matched by a diachronous transgression of the landward margins of salt marsh or mangroves on to the hinterland at a rate related to the transverse gradient.

Where the hinterland is low lying, new intertidal areas will form to landward: salt marshes or mangroves will migrate to displace freshwater or terrigenous vegetation communities, or invade backing salt flats. The vegetation zones will thus move landward to maintain their position in relation to the shifting intertidal zone (Figure 14.1(c)). Some may widen, others become narrower or coalescent, in relation to variations in the transverse profile (Titus, 1988). Landward migration of salt marshes and mangroves will be impeded where the hinterland rises steeply, the vegetation zones being compressed (coastal squeeze) as sea level rises on coasts where the transverse profile is concave. The salt marsh or mangrove fringe will disappear completely on the many sectors of coast where these communities are backed by a sea wall. Similar changes will occur on coasts bordered by freshwater swamps, such as the reedswamp bordering parts of the Baltic coast. These changes will be countered in areas where sedimentation continues (or is augmented by climatic changes leading to increased fluvial sediment yields) at a sufficient rate for the depositional terrace to be maintained by vertical accretion as sea level rises. Much will depend on whether mud from the bordering mudflats is swept into the mangroves, retained on the mudflats or carried away seaward (Figure 14.4).

There is the possibility of artificial nourishment of salt marsh and mangrove fringes by dumping mud in the nearshore area, where it can be washed onshore and into the vegetated areas. Where the shore profile is maintained the seaward margin will remain in place, and existing vegetation patterns can persist (Pethick, 1981), with the variations in species zonation and dis-

tribution that are related to the transverse profile (Reed, 2002). There will be landward migration of inner salt marsh and mangrove communities, providing there are suitable low lying habitats. The outcome will be a widening of the aggrading salt marsh or mangrove terrace.

There is also a possibility that as sea level rises some of the sediment stirred from bordering mudflats by strengthening wave action will be washed up into the salt marsh or mangrove fringe. This has happened in the Lagoon of Venice, where the salt marshes have continued to aggrade while their area was reduced by erosion. Coasts with wide bordering mudflats, such as the Bay of Saint Michel in France and Bridgwater Bay in western England, may well maintain their salt marshes as the result of shoreward drifting of muddy sediment as sea level rises, and a similar response could occur where wide mudflats front mangroves, as in the gulfs of northern Australia, but the extent to which this will happen is still a matter for debate. In the Gulf of Papua, Pernetta and Osborne (1988) decided that fluvial deposition on the Purari delta would provide sufficient sediment to maintain the mangrove ecosystem as sea level rose, whereas in the tidally dominated estuaries of the adjacent Kikori coast, where sedimentation is slower, mangroves would retreat landward and become reduced in area. Similar conclusions have been reached in northern Australia (Woodroffe, 1995).

Variations in the growth rates of salt marsh plants, in sedimentation and peat accretion, in tidal regimes and in species response all complicate attempts to predict exactly how salt marshes and mangroves will respond to a rising sea level. Where intertidal areas are enriched with nutrients derived from eroding sediments or enhanced runoff from the hinterland, it is possible that invigorated growth of salt marshes or mangroves could match a sea level rise by building up the depositional terrace with accumulating peat. In the absence of sustaining sediment accretion, it is thought that mangroves could maintain

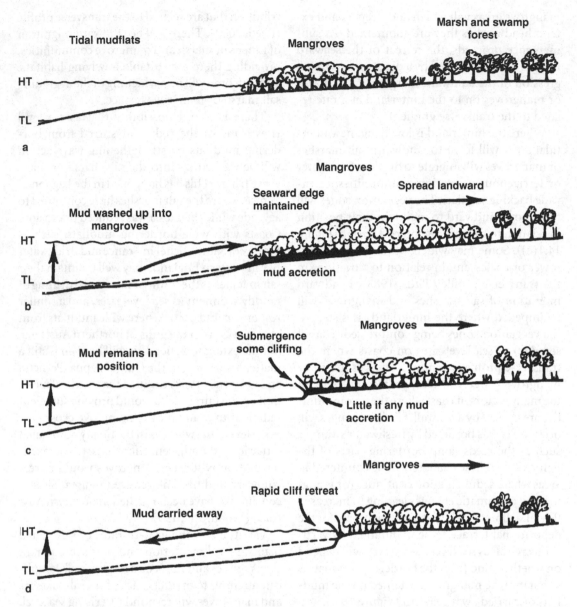

Figure 14.4 Changes on a mangrove-fringed coast (a) as sea level rises will be influenced by the movement of sediment on bordering mudflats. If mud is washed up into the mangroves their seaward edge is likely to be maintained by vertical accretion keeping pace with the sea level rise (b). If the mud remains on the mudflats the nearshore water will deepen, and submergence of the mangrove fringe will be accompanied by erosion, forming a microcliff (c). If the mud is carried away seaward the nearshore deepening will increase, and stronger wave action will cause rapid erosion of the mangrove fringe (d)

themselves on their accumulating peat with sea level rising at up to 9 cm per century, but that they would be impeded by a faster submergence, and suffer widespread destruction and erosion of their substrates when the rise exceeded 12 cm per century (Ellison and Stoddart, 1991).

Evidence of the response of salt marshes and mangroves to a rising sea level can be found on coasts that are already submerging, as on the Atlantic seaboard of the United States. Chesapeake Bay has salt marsh islands that have diminished in area while others have already disappeared. The rising sea level has also resulted in salt marsh plants invading backing meadowland (Kearney and Stevenson, 1991).

As global warming proceeds, mangroves are likely to spread to suitable habitats northward and southward of their present latitudinal limits, but this is likely to be a slow response because of impeding factors resulting from human impacts.

Marine organisms that are vertically zoned in relation to specific intertidal levels will migrate upward on rocky shores and artificial structures such as sea walls. Reference has been made (Sections 3.10.3 and 13.10) to the tubeworm *Galeolaria caespitosa* in SE Australia, which in recent years has moved upward on pier supports and the edges of shore platforms (Bird, 1988).

Intertidal areas, comprising sandflats, mudflats and rocky shores, will be modified as sea level rises. The outer part of the existing intertidal zone will become permanently submerged, and, as backing salt marshes and mangroves are eroded and coastal lowland fringes cut back, areas previously occupied by this vegetation will become mudflats or sandflats, and underlying rocky areas may be exposed. Again, landward migration of this kind will not be possible where sea walls have been built along the coast, so that the area of sandflats, mudflats and rocky shores will be reduced in width (squeezed), and eventually submerged by the rising sea. Sea grasses and marine algae will tend to migrate shoreward, their inner limits spreading on to the sandy

and muddy substrates that form as beaches are submerged and salt marshes and mangroves are cut back, their outer margins dying away as the water becomes too deep. As in salt marsh and mangrove terraces, changes in the area of sandflats, mudflats and rocky shores as sea level rises will depend on the extent to which submergence is offset by continuing sediment accretion. The sediment supply is more likely to be maintained in the vicinity of river mouths and may increase where fluvial sediment yields are augmented by larger or more intensive rainfall and runoff; it could also be maintained where there are shallow sea areas from which sediment can drift shoreward as submergence proceeds.

14.9 Effects on estuaries and lagoons

As sea level rises estuaries and lagoons will generally widen and deepen, and may transgress inland. Tides will penetrate farther upstream, tide ranges may increase, and there will be changes in patterns of shoal deposition (Shennan and Sproxton, 1990). The discharge of river floods will be impeded by the rising sea, so that flooding becomes more extensive and persistent, and a higher proportion of fluvial sediment will be retained within the submerging estuaries, instead of being delivered to the sea floor or adjacent coasts. There may be contrasts within an estuary, as between the windward and leeward shores of Chesapeake Bay (Stevenson and Kearney, 1996). However, if accompanying climatic changes increase rainfall in coastal regions or in the hinterland, some of these changes may be at least partly offset by greater fluvial discharge and sediment supply to the coast.

An analysis of the effects of a sea level rise averaging 6 mm/yr in the Blackwater estuary in Essex, using a formula that relates channel width (w) to discharge (Q) ($w = aQ^{0.21}$, where a is a constant) indicated that (in the absence of preventive structures such as shore walls) the augmented discharge would increase channel width between high and low water mean spring tide by between 400 and 500 m at the mouth, diminishing to zero 10 km upstream (Pethick, 1998). It is more difficult to apply this approach in estuaries where channels split around islands.

Coastal lagoons will generally be enlarged and deepened as sea level rises (Figure 14.1(d)), with submergence and erosion of their shores and fringing swamp areas, and widening and deepening of tidal entrances, increasing the inflow of sea water during rising tides and drought periods. Erosion of enclosing barriers may lead to breaching of new lagoon entrances, and continuing submergence may eventually remove the coastal barriers and reopen the lagoons as marine inlets and embayments. On the other hand, new lagoons may be formed by sea water incursion into low lying areas behind dune fringes on coastal plains, or where depressions are flooded as the water table rises to form seasonal or permanent lakes and swamps.

As sea level rises the currents that flow through existing tidal entrances and gaps between barrier islands, as on the Dutch and German North Sea coasts, may be strengthened, augmenting the inflow of water and sediment. In parts of the Wadden Sea increased sediment inflow has been building up intertidal and nearshore sandflats and mudflats and maintaining vertical accretion on salt marshes, even though coastal subsidence is in progress.

There will be much variation in the nature and extent of changes in coastal lagoons as sea level rises, depending on their existing configuration and dynamics, and the extent to which they have been modified by human activities. The Lagoon of Venice (Figure 11.10), for example, has been maintained partly by continuing subsidence in the NW Adriatic region. and partly by the diversion of rivers such as the Po and Brenta, which had been carrying sediment into it. Recent changes in the Lagoon of Venice

have been documented from successive air photographs, beginning with those taken from Airship Parseval in 1913 (Cavazzoni, 1983). A sea level rise of about a metre during the past century has led to more frequent storm surges (*acqua alta*, Section 3.10).

The deepening and enlargement of estuaries and coastal lagoons may be countered where there is a supply of sediment, arriving at a sufficient rate to offset the effects of submergence. The Holocene stratigraphy of the New Jersey coast shows that sea level rise (due to land subsidence) during the past 2500 years was accompanied by the upward growth and expansion of salt marshes so that coastal lagoons backing sandy barrier islands were reduced in area (Psuty, 1986). Tropical estuaries may be modified as sediment yields change in response to greater or lesser rainfall and runoff, and variations in catchment vegetation related to climate (Wolanksi and Chappell, 1996). Rapid sedimentation could prevent the enlargement of coastal lagoons as sea level rises, but submergence will postpone infilling of the residual lagoon basin.

14.10 Effects on deltaic coasts

A rising sea level will cause submergence and erosion of low lying deltas and coastal plains, especially where there is little or no compensating sediment accretion. As sea level rises, progradation of most deltas will be curbed, and erosion will become more extensive and more rapid. Deltaic coastlines that are already receding because of submergence and erosion will show accelerated retreat, but progradation will continue around the mouths of rivers that continue to supply sufficient sediment to the coast, including those where the sediment yield has been increased as the result of human activities such as deforestation, farming and mining.

Most large deltas already show isostatic subsidence under the weight of accumulating

Figure 14.5 Subsidence on part of the Mamberamo Delta on the north coast of Irian Jaya, Indonesia, has been accompanied by the spread of mangroves back from a shore fringe into the subsiding area

sediment, leading to submergence of areas no longer being maintained by sedimentation. This is obvious on the Mississippi delta, where former subdelta lobes no longer maintained by sedimentation have subsided beneath the sea (Figure 12.2). Coastal subsidence has curtailed the seaward growth of large deltas such as those of the Rhine, Guadalquivir, Colorado and Amazon, and a rising sea level will increase the effects of subsidence, extending submergence and intensifying erosion. Fringing vegetation is likely to spread inland as subsidence proceeds (Figure 14.5).

The extent to which these effects will be countered if sea level rise is accompanied by increasing water and sediment yields from rivers depends on changes within the river catchments. Deposition of delta sediments will accelerate if rainfall and runoff from the river catchments increases as a consequence of climatic changes associated with global warming. It has been estimated that the sediment yield from Javanese rivers will increase by up to 43 per cent as the

result of greater runoff from their catchments during the coming century (Parry, Magalhaes and Nguyen, 1991). Under these circumstances sedimentation may build some deltas upward to maintain their area as sea level rises, and sedimentation may continue to aggrade natural levees alongside river channels and extend them seaward. Submerging deltas that are now lobate or arcuate in form may become digitate, like the modern Mississippi delta, with intervening areas of submerging swamp and deepening sea. It should be noted that an increase in water and sediment yields from rivers because of greater runoff could be partly offset if warmer and wetter conditions increase the extent and luxuriance of catchment vegetation.

14.11 Effects on coral and algal reefs

Corals and algae on the surface of intertidal reef platforms will be stimulated by a rising sea level, leading to a revival of upward reef growth (Figure 14.1(e)), initiated by the expansion and dispersal of currently sparse and scattered living corals on reef platforms. If global sea level has indeed been rising during the past century at 1.0–1.2 mm per year, and present sea level is therefore 10–12 cm higher than it was in the 1900s, it could be expected that coral reefs would be among the first features to show indications of such a sea level rise, but there have not yet been reports of the general spread of living corals or accelerating coral growth. Few reefs have been mapped and monitored with sufficient accuracy for such a change to be measured, but there is likely to be a widespread revival of coral growth on reef platforms as sea level continues to rise. That reef platforms are quickly responsive to sea level changes is shown by the renewed upward growth of coral on reefs in Houtman Abrolhos, Western Australia, as a result of a relative rise of sea level.

Revival of coral growth will be strongly influenced by ecological factors that influence the ability of coral species to recolonise submerging reef platforms. Global warming will modify the distribution of corals by increasing sea temperature and salinity in areas that become more arid. Recent reports of widespread coral bleaching (Brown, 1990) may be a consequence of higher temperatures in tropical seas during the El Niño events in 1982–83, 1987 and 1996–98, when coral mortality was extensive on Pacific Ocean reefs. This suggests that increasing sea temperatures are likely to impede coral growth and reef aggradation.

Climatic changes that result in higher rainfall and greater discharges of fresh water will also impede coral growth, especially where runoff from the hinterland produces more extensive turbidity in coastal waters around coral reefs.

The response of coral reefs to a rising sea level will depend on the rate at which the sea rises. A slowly rising sea should stimulate the revival of coral growth on reef platforms. Some reefs are able to maintain upward growth with a rising sea level, while others will survive to grow up to the surface when the sea level rise slackens or comes to an end. A rapid sea level rise may lead to the drowning of corals and the submergence of inert reef formations. Neumann and Macintyre (1985) pictured the relationship between rates of sea level rise and rates of upward reef growth in terms of a colloquial classification: the first category as keep-up reefs, the second as catch-up reefs and the third as give-up reefs. They noted that there will be variations in response related to ecological composition.

As measurements of mean upward growth rates of existing coral reefs are up to 1 cm/yr, it is likely that reef formations will grow upward to match sea level rise as long as it is within this range. Theoretically, coral reefs could also expand northward and southward beyond their present latitudinal limits as sea temperatures rise, but this is likely to be a slow response because of impeding factors resulting from human impacts.

Carbon dioxide, a major greenhouse gas, is absorbed by algae in the course of photosynthesis and corals fix it as carbonates in reef structures. Coral and algal reefs have imbibed large quantities of carbon dioxide, and it is possible that artificial reefs will be nurtured to enhance their role in absorbing carbon dioxide to offset the greenhouse effect.

A slackening sea level rise could permit some surviving corals to grow up, and others to recolonise as the reef shallows. Reefs submerged during a period of rising sea level could catch up if the rate of submergence then diminished, and especially if a new still-stand were to ensue.

The impacts of a sea level rise on existing coral reefs can also be considered in terms of stratigraphic evidence of what happened to coral reefs during the Holocene marine transgression. It should be borne in mind, however, that the ecology and geomorphology of existing reef formations differ in various ways from those when

Late Pleistocene reef revival commenced on submerging coastlines and pre-existing dissected reef limestones, and a much cooler climate had started to ameliorate. The most notable differences are the impacts of human activities on existing coral reefs during the past few centuries, which may have made the world's coral reefs much less capable than they were under the natural conditions of the Pleistocene of responding to a sea level rise (Yap, 1989).

Spencer (1995) reviewed the literature on coral reef growth and found that reefs were likely to keep up with a sea level rise of less than about 1 cm/yr, to be growing upward with 1–2 cm/yr and to be drowned when sea level rise exceeded 2 cm/yr. However, the actual response will vary with accompanying changes in sea temperature and salinity, the incidence and effects of tropical cyclones and biophysical constraints (including human impacts). There will also be variations depending on the nature of existing reefs, coral gardens of Caribbean type being more easily maintained as sea level rises than solid Indo-Pacific reef platforms, which may grow upward as coral gardens rather than solid structures.

Measurements can be made on reefs that are subsiding at known rates. On tilting coral reefs and atolls, such as Uvéa in the Loyalty Islands the submerged portion shows rapid, if patchy, upward growth of corals, with an inner zone of slow submergence where upward growth is being maintained, passing laterally to a central zone of moderate submergence where the corals are growing, but are failing to keep up with the rising sea, and an outer zone of more rapid submergence where the corals have died and the reefs are drowned (Figure 14.6). The inner submerging portion of the Uvéa atoll shows keep-up growth of coral gardens, rich in corals and accumulating sediment, but forming fragile structures rather than a solid reef platform. Keep-up reefs are thus likely to be coral garden structures that cannot be walked over. As existing Pacific and Indian Ocean reef platforms are submerged by the rising sea, reviving coral growth is likely to form coral gardens similar to those found in the Caribbean, or in shallow water to the lee of existing reefs, rather than consolidated reef platforms, which require a sea level still-stand for their completion, or even a phase of growth

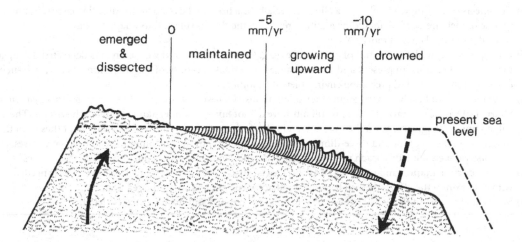

Figure 14.6   Tilting reefs (such as Uvéa in the Loyalty Islands, New Caledonia) show the response of corals to a sea level rise. Where sea level rises up to 5 mm/yr corals are likely to maintain the reef, where the rise is between 5 and 10 mm/yr they will continue to grow upward but will fail to maintain the reef and where the rise exceeds 10 mm/yr they will be drowned

Panel 14.1   Problems in coastal geomorphology

There are many unsolved problems in coastal geomorphology, and in concluding this book it may be useful to draw attention to some of them, in the hope that coastal researchers will deal with them, and so advance the subject.

(a) Coastal landforms have been strongly influenced by the history of Quaternary land and sea level changes, which has varied from one sector of coast to another, and requires more precise mapping. On some coasts sectors only a few kilometres apart may have had very different land and sea level histories. The Holocene marine transgression, complicated by upward or downward movements of the land and variations in sea surface geometry, provides the background for studies of coastal evolution.

(b) The patterns and rates of change on coastlines require further mapping and documentation over specific periods, particularly the past century, and explanation in terms of processes: much more information is needed on coastlines that have been little studied (i.e. outside North America, Britain and Europe, and Australasia).

(c) It is necessary to explain why particular landforms (e.g. cliff and shore morphology, notably seaward-sloping and subhorizontal shore platforms) occur on some sectors of the coast and not on other, essentially similar sectors (e.g. slope-over-wall profiles, extensive on the Atlantic coast of Europe, are poorly developed in equivalent latitudes on the Pacific coast of North America)

(d) More attention should be given to the documentation and analysis of variations in transverse shore and nearshore profiles, particularly on beaches and coasts that have shore platforms, with a critical review of the idea that these tend towards some kind of equilibrium.

(e) There is much variation in the nature and composition of beach materials, and more detailed research is needed into their origins and mode of delivery to shore sectors, with particular attention to sediment supply (past and present) from the sea floor.

(f) Artificially nourished beaches should be regarded as field experiments and monitored to elucidate process-response relationships (a procedure that can improve subsequent beach nourishment projects).

(g) Analysis of the variability of coastal environments is necessary before research findings on one sector of coast can be applied to other sectors. Thus the work of Californian engineers on beaches as rivers of sand is not universally applicable on the world's coastline: indeed, more than half the sand on Californian beaches has come from cliff erosion.

(h) The effects of a rising sea level on coastlines, especially beaches, can be studied on subsiding coasts, but more precise models are needed, including a critical review of the Bruun rule and its adaptation to three-dimensional coastal morphology.

(i) Attempts to extrapolate evidence of past changes (e.g. from the stratigraphy of sediments deposited during a marine transgression) to present and future changes should take account of the environments in which such changes occurred, and of preceding environmental conditions.

(j) Several aspects of beach systems await elucidation. In his *Handbook of Beach and Shoreface Morphodynamics* Short (1999, p. 20) remarked 'There is still much we do not know or understand about beach systems. There is still no universal agreement on the existence and role of edge waves, on the breaking wave hypothesis, on the formation of beach cusps and three-dimensional beach topography, on the outer limit of beach processes, on the concept of equilibrium beach profiles, and the presence and role of such phenomena as shear waves.'

(k) Further documentation and mapping of artificial coast structures and coastal reclamation areas is needed, with attention to the effects of such features on adjacent coastlines and on the sea floor.

(l) Substantial sectors of the world's coastline remain to be mapped and documented in detail.

to a higher Holocene sea level followed by down-cutting and planation of the reef limestone.

Islands on coral reefs will be modified by a rising sea level. Many cays and low wooded is-lands will be eroded by larger waves approach-ing through deepening waters, and may disap-pear, overwashed by storm surges, especially if there is an increasing frequency and severity of tropical cyclones as sea and atmospheric temperatures rise. Inhabitants of low islands such as Kiribati in the Pacific are already concerned about erosion and submergence by a rising sea level, and are facing the prospect of eventual evacuation.

On the other hand, there is a possibility that where submergence is slow enough, reviving coral growth on the surrounding reef platforms will at least partly offset erosion of cay shores by impeding wave attack. Low islands may even be enlarged by accretion of coralline material derived by stronger wave action from the growing reef gardens.

14.12 Conclusion

Discussion of the possible effects of global warming and a rising sea level on the world's coasts is speculative in the absence of reliable meteorological and sea level change data and the inadequacy of existing predictive models. Although there is increasing acceptance of the idea of global warming and acknowledgement that coastal erosion is already a widespread problem, there is still uncertainty about contemporary changes in global sea levels. It will certainly be difficult to achieve global support for the control of pollutants that are causing global warming, but it may be still more difficult to adapt and retreat as sea level rises to submerge coastal lowlands.

14.13 Summary

Predicted global warming implies a world-wide sea level rise, which will cause coastal submergence, inundating lowlands and enlarging estuaries and lagoons. It will also initiate or accelerate erosion of beaches, cliffs, coastal dunes, marshes and swamps and deltas. The response of existing coral reefs to warmer climates and rising sea levels is explored: some may be maintained, expanding poleward, others submerged. Coastal changes will be much affected by human impacts, past, present and future.

References

Aagard, T. (1991) Multiple-bar morphodynamics and its relations to low frequency edge waves. *Journal of Coastal Research*, **7**: 801–813.

Adam, P. (1990) *Salt Marsh Ecology*. Cambridge University Press, Cambridge.

Ager, D.V. (1980) *The Geology of Europe*. Wiley, New York.

Allen, J.R.L. (1965) Coastal geomorphology of eastern Nigeria: beach ridge, barrier islands and vegetated tidal flats. *Geologie en Mijnbouw*, **44**: 1–21.

Allen, J.R.L. (1968) *Current Ripples*. North-Holland, Amsterdam.

Allen, J.R.L. (1989) Evolution of salt marsh cliffs in muddy and sandy systems: a qualitative comparison of British west coast estuaries. *Earth Surface Processes and Landforms*, **14**: 85–92.

Allen, J.R.L. (1996) Shoreline movement and vertical textural patterns in salt marsh deposits. *Proceedings of the Geologists' Association*, **107**: 15–23.

Allen, J.R., Psuty, N.P., Bauer, B.O. and Carter, R.W.G. (1996) A field data assessment of contemporary models of beach cusp formation. *Journal of Coastal Research*, **12**: 622–629.

Allen, J.R.L. and Pye, K. (eds) (1991) *Saltmarshes: Morphodynamics, Conservation and Engineering Problems*. Cambridge University Press, Cambridge.

Allison, M.A. (1998) Historical changes on the Ganges–Brahmaputra delta front. *Journal of Coastal Research*, **14**: 1269–1275.

Allison, R.J. and Brunsden, D. (1990) Some mudslide movement patterns. *Earth Surface Processes and Landforms*, **15**: 297–312.

Atlas Okeanov (1974) *Pacific (Tikhii) Ocean*. Moscow.

Atlas Okeanov (1977) *Atlantic and Indian Oceans*. Moscow.

Atwater, B.F. (1987) Evidence for Great Holocene earthquakes along the outer coast of Washington State. *Science*, **236**: 942–944.

Atwater, B.F. and Hemphill-Haley, E. (1997) *Recurrence Intervals for Great Earthquakes of the Past 3,500 Years at Northeastern Willapa Bay, Washington*, United States Geological Survey Professional Paper 1576.

Augustinus, P.G.E.F. (ed.) (1989) Cheniers and chenier plains. *Marine Geology*, **198**: 219–351.

Augustinus, P.G.E.F. (1995) Geomorphology and sedimentology of mangroves. In: G.M.E. Perillo (ed.) *Geomorphology and Sedimentology of Estuaries*. Elsevier, Amsterdam, pp. 333–357.

Badr, A.A. and Lotfy, M.E. (1999) Tracing beach sand movement using fluorescent quartz along the Nile delta promontories, Egypt. *Journal of Coastal Research*, **15**: 261–265.

Baker, G. (1943) Features of a Victorian limestone coastline. *Journal of Geology*, **51**: 359–386.

Baker, G. (1956) Sand drift at Portland Harbour, Victoria. *Proceedings, Royal Society of Victoria*, **68**: 151–198.

Baker, G. (1958) Stripped zones at cliff edges along a high wave current energy coast, Port Campbell, Victoria. *Proceedings, Royal Society of Victoria*, **71**: 175–179.

Bakker, T.W.M., Jungerius, P.D. and Klijn, J.A. (eds) (1990) European coastal dunes. *Catena Supplement* **18**.

Barnes, R.S.K. (1980) *Coastal Lagoons, the Natural History of a Neglected Habitat*. Cambridge University Press, Cambridge.

Barrow, G. (1906) *The Geology of the Isles of Scilly*. Memoirs of the Geological Survey, London.

Bascom, W.N. (1951) Relationship between sand size and beach face slope. *Transactions, American Geophysical Union*, **32**: 866–874.

Bascom, W.N. (1954) The control of stream outlets by wave refraction. *Journal of Geology*, **62**: 600–605.

Bauer, B.O., Sherman, D.J., Nordstrom, K.F. and Gares, P.A. (1990) Aeolian transport measurement and prediction across a beach and dune at Castroville, California. In: K.F. Nordstrom, N. Psuty and R.W.G. Carter (eds) *Coastal Dunes: Form and Process*. Wiley, Chichester, pp. 39-55.

Belperio, A.P. (1993) Land subsidence and sea level rise in the Port Adelaide estuary: Implications for monitoring the greenhouse effect. *Australian Journal of Earth Sciences*, **40**: 359–368.

Berendsen, H.J.A. (1998) Birds-eye view of the Rhine–Meuse delta. *Journal of Coastal Research*, **14**: 740–753.

Bird, E.C.F. (1967a) Depositional features in estuaries on the south coast of New South Wales. *Australian Geographical Studies*, **5**: 113–124.

Bird, E.C.F. (1967b) Coastal lagoons of south-eastern Australia. In: J.N. Jennings and J.A. Mabbutt (eds) *Landform Studies from Australia and New Guinea*. Australian National University Press, Canberra, pp. 365–385.

Bird, E.C.F. (1978a) *The Geomorphology of the Gippsland Lakes Region*, Ministry for Conservation, Victoria, Environmental Studies Series 186.

Bird, E.C.F. (1978b) The nature and source of beach materials on the Australian coast. In: J.L. Davies and M.A.J. Williams (eds) *Landform Evolution in Australasia*. Australian National University Press, Canberra, pp. 144–157.

Bird, E.C.F. (1985a) *Coastline Changes*. Wiley, Chichester.

Bird, E.C.F. (1985b) Recent changes on the Somers–Sandy Point coastline. *Proceedings, Royal Society of Victoria*, **97**: 115–128.

Bird, E.C.F. (1986) Mangroves and intertidal morphology in Westernport Bay. *Marine Geology*, **77**: 327–331.

Bird, E.C.F. (1988) The tubeworm *Galeolaria caespitosa* as an indicator of sea level rise. *Victorian Naturalist*, **105**: 98–104.

Bird, E.C.F. (1989) The beaches of Lyme Bay. *Proceedings, Dorset Natural History and Archaeological Society*, **111**: 91–97.

Bird, E.C.F. (1993a) *The Coast of Victoria*. Melbourne University Press, Melbourne.

Bird, E.C.F. (1993b) Physical setting and geomorphology of coastal lagoons. In: B. Kjerfve (ed.) *Coastal Lagoon Processes*. Elsevier, Amsterdam, pp. 9–39.

Bird, E.C.F. (1993c) *Submerging Coasts: the Effects of a Rising Sea Level on Coastal Environments*. Wiley, Chichester.

Bird, E.C.F. (1995) *Geology and Scenery of Dorset*. Ex Libris, Bradford on Avon.

Bird, E.C.F. (1996a) *Beach Management*. Wiley, Chichester.

Bird, E.C.F. (1996b) Lateral grading of beach sediments: a commentary. *Journal of Coastal Research*, **12**: 774–785.

Bird, E.C.F. (1997) *The Shaping of the Isle of Wight*. Ex Libris, Bradford on Avon.

Bird, E.C.F. (1998) *The Coasts of Cornwall*. Alexander, Fowey.

Bird, E.C.F. (2003) *The World's Coasts: Online. An Electronic Encyclopedia*. Springer, Dordrecht.

Bird, E.C.F. (2005) An evolving electronic encyclopedia—The World's Coasts: Online. *Journal of Coastal Research*, **21**: 618–623.

Bird, E.C.F. (2006) The effects of a higher sea level on the coasts of Port Phillip Bay. *Victorian Naturalist*, **123**: 49–54.

Bird, E.C.F. and Christiansen, C. (1982) Coastal progradation as a by-product of human activity: an example from Hoed, Denmark. *Geografisk Tidsskrift*, **82**: 1–4.

Bird, E.C.F. and Dent, O.F. (1966) Shore platforms on the south coast of New South Wales. *Australian Geographer*, **10**: 71–80.

Bird, E.C.F., Dubois, J.P. and Iltis, J.A. (1984) *The Impacts of Opencast Mining on the Rivers and Coasts of New Caledonia*. United Nations University, Tokyo.

Bird, E.C.F. and Fabbri, P. (1987) Archaeological evidence of coastline changes illustrated with reference to Latium, Italy. *Déplacement des Lignes de Rivage en Méditeranée*, Colloques Internationaux CNRS, pp. 107–113.

Bird, E.C.F. and Guilcher, A. (1982) Observations préliminaires sur les récifs frangeants actuels du Kenya et sur les formes littorales associées. *Revue de Géomorphologie Dynamique*, **31**: 113–135.

Bird, E.C.F. and Hopley, D. (1969) Geomorphological features on a humid tropical sector of the Australian coast. *Australian Geographical Studies*, **7**: 89–108.

Bird, E.C.F. and Jones, D.J.B. (1988) The origin of foredunes of the coast of Victoria, Australia. *Journal of Coastal Research*, **4**: 181–192.

Bird, E.C.F. and Ongkosongo, O.S.R. (1980) *Environmental Changes on the Coasts of Indonesia*. United Nations University, Tokyo.

Bird, E.C.F. and Ranwell, D.S. (1964) The physiography of Poole Harbour, Dorset. *Journal of Ecology*, **52**: 355–366.

Bird, E.C.F. and Rosengren, N.J. (1984) The changing coastline of the Krakatau Islands, Indonesia. *Zeitschrift für Geomorphologie*, **28**: 346–366.

Bird, E.C.F. and Rosengren, N.J. (1987) Coastal cliff management: an example from Black Rock Point, Melbourne, Australia. *Journal of Shoreline Management*, **3**: 39–51.

Bird, E.C.F. and Schwartz, M.L. (eds) (1985) *The World's Coastline*. Van Nostrand Reinhold, New York.

Bird, E.C.F. and Schwartz, M.L. (2000) Shore platforms at Cape Flattery, Washington. *Washington Geology*, **28**: 21–26.

Bird, J.F. (2000) The impact of mining waste on the rivers draining into Georges Bay, Northeast Tasmania. In: S. Brizga and B.L. Finlayson (eds) *River Management: the Australian Experience*. Wiley, Chichester, pp. 151–172.

Black, K.A. and Rosenberg, M.A. (1992) Natural stability of beaches around a large bay. *Journal of Coastal Research*, **8**: 385–397.

Bodge, K.R. (1992) Representing equilibrium beach profiles with an exponential expression. *Journal of Coastal Research*, **8**: 47–55.

Borowca, M. and Rotnicki, K. (1994) Intensity, directions and balance of aeolian transport on the beach barrier and the problem of sand nourishment. In: K. Rotnicki (ed.) *Changes of the Polish Coastal Zone*. Quaternary Research Institute, Adam Mickiewicz University, Poznan, pp. 108–115.

Borowca, R.K. (1990) The Holocene development and present morphology of the Leba dunes, Baltic coast of Poland. In: K.F. Nordstrom, N. Psuty and R.W.G. Carter (eds) *Coastal Dunes: Form and Process*. Wiley, Chichester, pp. 289–313.

Bourman, R.P. (1990) Artificial beach progradation by quarry waste disposal at Rapid Bay South Australia. *Journal of Coastal Research*, Special Issue 6: 69–76.

Bourman, R.P. and Harvey, N. (1983) The Murray mouth flood tidal delta. *Australian Geographer*, **15**: 403–406.

Bray, M.J. (1997) Episodic shingle supply and the modified development of Chesil Beach, England. *Journal of Coastal Research*, **13**: 1035–1049.

Bray, M.J., Carter, D.J. and Hooke, J.M. (1995) Littoral cell definition and budgets for Central South England. *Journal of Coastal Research*, **11**: 381–400.

Bray, M.J. and Hooke, J.M. (1997) Prediction of soft-cliff retreat with accelerating sea-level rise. *Journal of Coastal Research*, **13**: 453–467.

Brown, B.E. (1990) Coral bleaching. *Coral Reefs*, **8**: 153–232.

Brown, E.H. (1960) *The Relief and Drainage of Wales*. University of Wales Press, Cardiff.

Brown, M.J.F. (1974) A development sequence: disposal of mining waste on Bougainville, Papua New Guinea. *Geoforum*, **1**: 19–27.

Brunsden, D. and Jones, D.K.C. (1980) Relative time scales in formative events in coastal landslide systems. *Zeitschrift für Geomorphologie, Supplementband* **34**: 1–19.

Brunsden, D. and Prior, D.B. (1984) *Slope Instability*. Wiley, Chichester.

Bruun, P. (1962) Sea level rise as a cause of shore erosion. *Proceedings, American Society of Civil Engineers, Waterways and Harbour Division*, **88**: 117–130.

Bruun, P. (1988) The Bruun Rule of erosion by sea level rise: a discussion of large-scale two- and three-dimensional usages. *Journal of Coastal Research*, **4**: 627–248.

Bryant, E.A. (1985) Rainfall and beach erosion relationships, Stanwell Park, Australia, 1895–1980: worldwide implications for coastal erosion. *Zeitschrift für Geomorphologie, Supplementband* 57: 51–65.

Bryant E.A., Young, R.S. and Price, D.M. (1996) Tsunami as a major control on coastal evolution, southeastern Australia. *Journal of Coastal Research*, **12**: 831–840.

Buddemeier, R.W. and Smith, S.V. (1988) Coral reef growth in an era of rapidly rising sea level. *Coral Reefs*, **7**: 51–56.

Burges, A. and Drover, D.P. (1957) The rate of podsol development in the sands of the Woy Woy district, New South Wales. *Australian Journal of Botany*, **1**: 83–94.

Cailleux, A. (1948) Lithologie des dépôts émergés actuels de l'embouchure du Var au Cap d'Antibes. *Bulletin de l'Institut Océanographique de Monaco*, **94**: 1–11.

Cambers, G. (1976) Temporal scales in coastal erosion systems. *Transactions, Institute of British Geographers*, **1**: 246–256.

Carey, A.E. and Oliver, F.W. (1918) *Tidal Lands*. Blackie, Edinburgh.

Carlton, J.M. (1974) Land-building and stabilization by mangroves. *Environmental Conservation*, 1: 285–294.

Carr, A.P. (1965) Shingle spit and river mouth: short-term dynamics. *Transactions, Institute of British Geographers*, 36: 117–130.

Carter, C.H. and Guy, D.E. (1988) Coastal erosion: processes, timing and magnitude at the bluff toe. *Marine Geology*, 84: 1–17.

Carter, R.W.G., Orford, J.D., Forbes, D.L. and Taylor, R.B. (1990) Morphosedimentary development of drumlin-flank barriers with rapidly rising sea level, Story Head, Nova Scotia. *Sedimentary Geology*, 69: 117–138.

Carter, R.W.G. and Wilson, P. (1990) The geomorphological, ecological and pedological development of coastal foredunes at Magilligan Point, Northern Ireland. In: K.F. Nordstrom, N. Psuty and R.W.G. Carter (eds) *Coastal Dunes: Form and Process*. Wiley, Chichester, pp. 129–157.

Carter, R.W.G. and Woodroffe, C.D. (eds) (1994) *Coastal Evolution. Late Quaternary Morphodynamics*. Cambridge University Press, Cambridge.

Cartwright, P.E. (1974) Years of peak astronomical tides. *Nature*, 248: 656–657.

Castaing, P. and Guilcher, A. (1995) Geomorphology and sedimentology of rias. In: G.M.E. Perillo (ed.) *Geomorphology and Sedimentology of Estuaries*. Elsevier, Amsterdam, pp. 69–111.

Cavazzoni, S. (1983) Recent erosive processes in the Venetian Lagoon. In: E.C.F. Bird and P. Fabbri (eds) *Coastal Problems in the Mediterranean Sea*. University of Bologna, Bologna, pp. 19–22.

Cencini, C. (1998) Physical processes and human activities in the Po delta, Italy. *Journal of Coastal Research*, 14: 774–793.

Chapman, D.M., Geary, M., Roy, P.S. and Thom, B.G. (1982) *Coastal Evolution and Coastal Erosion in New South Wales*. Coastal Council of New South Wales, Sydney.

Chapman, V.J. (1976) *Mangrove Vegetation*. Cramer, Vaduz.

Chappell, J. (1974) Geology of coral terraces, Huon Peninsula, New Guinea. *Bulletin, Geological Society of America*, 85: 553–570.

Chappell, J. (1983) A revised sea-level record for the last 300,000 years. *Search*, 14: 99–101.

Christiansen, C., Christoffersen, H., Dalsgard, H. and Nornberg, P. (1981) Coastal and nearshore changes associated with die-back in eel grass (*Zostera marina*). *Sedimentary Geology*, 28: 163–173.

Chun-Hsin, Chung (1985) The effects of introduced *Spartina* grass on coastal morphology in China. *Zeitschrift für Geomorphologie, Supplementband* 57: 169–174.

Clapp, C.H. (1913) Contraposed shoreline. *Journal of Geology*, 21: 537–542.

Clayton, K.M. (1989) Sediment input from the Norfolk cliffs, eastern England—a century of coast protection and its effect. *Journal of Coastal Research*, 5: 433–442.

Clayton, K.M. and Shamoon, N. (1998) A new approach to the relief of Great Britain, II. A classification of rocks based on relative resistance to denudation. *Geomorphology*, 25: 155–171.

Codignotto, J.O., Kokot, R.R. and Marcomini, S.C. (1992) Neotectonism and sea level changes in the coastal zone of Argentina. *Journal of Coastal Research*, 8: 125–133.

Coleman, J.M. (1981) *Deltas*. Burgess, Minneapolis, MN.

Coleman, J.M., Roberts, H.H. and Stone, G.W. (1998) Mississippi River delta: an overview. *Journal of Coastal Research*, 14: 698–717.

Colombo, G. (1977) Lagoons. In: R.S.K. Barnes (ed.) *The Coastline*. Wiley, New York, pp. 63–81.

Cooper, J.A.G. (1994) Lagoons and microtidal coasts. In: R.W.G. Carter and C.D. Woodroffe (eds) *Coastal Evolution. Late Quaternary Shoreline Morphodynamics*. Cambridge University Press, Cambridge, pp. 219–265.

Cooper, W.S. (1958) *Coastal Sand Dunes of Oregon and Washington*, Memoirs, Geological Society of America 72.

Cooper, W.S. (1967) *Coastal Dunes of California*. Memoirs, Geological Society of America 104.

Costa, S., Hénaff, A. and Lageat Y. (2006) The gravel beaches of north-west France and their contribution to the dynamics of the coastal cliff–shore platform system. *Zeitschrift für Geomorphologie, Supplementband* 144: 199–214.

Cotton, C.A. (1956) Rias sensu stricto and sensu lato. *Geographical Journal*, 122: 360–364.

Cotton, C.A. (1974) *Bold Coasts*. Wellington, New Zealand.

Cowell, P.J. and Thom, B.G. (1994) Morphodynamics of coastal evolution. In: R.W.G. Carter and C.D.

Woodroffe (eds) *Coastal Evolution. Late Quaternary Shoreline Morphodynamics.* Cambridge University Press, Cambridge, pp. 33–86.

Cracknell, B.E. (2003) *Outrageous Waves: Global Warming and Coastal Change in Britain.* Phillimore, Chichester.

Cressard, A.P. and Augris, C. (1982) Etudes des phénomènes d'érosion cotière liès a l'extraction de materiaux sur le plateau continentale. *Proceedings, 4th Conference, International Association of Engineering Geology,* 7: 203–211.

Cruz, O., Coutinho, P.N., Duarte, G.M., Gomes, A. and Muehe, D. (1985) Brazil. In: E.C.F. Bird and M.L. Schwartz (eds) *The World's Coastline.* Van Nostrand Reinhold, New York, pp. 85–91.

Dalrymple, R.W., Zaitlin, B.A. and Boyd, R. (1992) Estuarine facies models: conceptual basis and stratigraphic implications. *Journal of Sedimentary Petrology,* 62: 1130–1146.

Davidson-Arnott, R. (2005) Beach and nearshore instrumentation. In M.L. Schwartz (ed.) *Encyclopedia of Coastal Science.* Springer, Dordrecht, pp. 130–138.

Davidson-Arnott, R.G.D. and Law, M.N. (1990) Seasonal patterns and controls on sediment supply to coastal foredunes, Long Point, Lake Erie. In: K.F. Nordstrom, N. Psuty and R.W.G. Carter (eds) *Coastal Dunes: Form and Process.* Wiley, Chichester, pp. 177–200.

Davidson-Arnott, R.G.D., Van Proosdij, D., Ollerhead, J. and Schostak, L. (2002) Hydrodynamics and sedimentation in salt marshes: examples from a macrotidal marsh, Bay of Fundy. *Geomorphology,* 48: 209–231.

Davies, J.L. (1957) The importance of cut and fill in the development of sand beach ridges. *Australian Journal of Science,* 20: 105–111.

Davies, J.L. (1974) The coastal sediment compartment. *Australian Geographical Studies,* 12: 139–151.

Davies, J.L. (1980) *Geographical Variation in Coastal Development* (2nd edition). Longman, London.

Davies, P.J. (1983) Reef growth. In: D.J. Barnes (ed.) *Perspectives on Coral Reefs.* Manuka, Clouston, pp. 69–106.

Davis, J.H. (1940) The ecology and geological rôle of mangroves in Florida. *Publications of the Carnegie Institution,* 524: 303–412.

Davis, R.A. (1989) Texture, composition and provenance of beach sands, Victoria, Australia. *Journal of Coastal Research,* 5: 37–47.

Davis, W.M. (1896) The outline of Cape Cod. *Proceedings, American Academy Arts and Sciences,* 31: 303–332.

Davis, W.M. (1928) *The Coral Reef Problem,* American Geographical Society Special Publication 9.

Dawson, A.G. (1994) Geomorphological effects of tsunami run-up and backwash. *Geomorphology,* 10: 83–94.

Dawson, A.G. and Shi, S. (2000) Tsunami deposits. *Pure and Applied Geophysics,* 157: 857–897.

Day, J.W., Scarton, F., Rismondo, A. and Are, D. (1998) Rapid deterioration of a salt marsh in Venice Lagoon, Italy. *Journal of Coastal Research,* 14: 583–590.

Dean, R.G. (1991) Equilibrium beach profiles: characteristics and applications *Journal of Coastal Research,* 7: 53–84.

Dean, R.G. (2005) Dynamic equilibrium of beaches. In M.L. Schwartz (ed.) *Encyclopedia of Coastal Science.* Springer, Dordrecht, pp. 399–400.

Delft Hydraulics Laboratory (1987) *Manual on Artificial Beach Nourishment.* Centre for Civil Engineering Research, Codes and Specifications, Rijkwaterstaat, the Netherlands.

Dietz, R.S. (1963) Wave-base, marine profile of equilibrium, and wave-built terraces: a critical appraisal. *Bulletin, Geological Society of America,* 74: 971–990.

Dionne, J.C. (1972) Caractéristiques des schorres des régions froides. *Zeitschrift für Geomorphologie,* 13: 131–162.

Dionne, J.C. (2003) The boulder barricade at Cap à la Baleine, north shore of Gaspé peninsula (Québec): nature of boulders, origin and significance. *Journal of Coastal Research,* 18: 652–661.

Dolan, R. and Davis, R.E. (1992) An intensity scale for Atlantic coast northeast storms. *Journal of Coastal Research,* 8: 840–853.

Donn, W.L., Farrand, W.R. and Ewing, J.R. (1962) Pleistocene ice volumes and sea level lowering. *Journal of Geology,* 70: 206–214.

Donoghue, J.P. and White, N.M. (1995) Late Holocene sea level change and delta migration, Apalachicola River region, Northwest Florida. *Journal of Coastal Research,* 11: 651–663.

Dornbusch, U., Moses, C.A., Robinson, D.A. and Williams, R.B.G. (2004) Laboratory simulation tests on beach flint shingle. In R. N. Mortimore and A Duperret (eds) *Coastal Chalk Cliff Instability,* Geological Society Engineering Geology Special Publication 20, pp. 131–138.

Dornbusch, U., Robinson, D.A., Moses, C.A., Williams, R.B.G. and Costa, S. (2006) Retreat of chalk cliffs in the eastern English Channel during the last century. *Journal of Maps*, 2006: 1–8.

Dornbusch, U., Williams, R.B.G., Moses, C.A. and Robinson, D.A. (2002) Life expectancy of shingle beaches: measuring in situ abrasion. *Journal of Coastal Research*, Special Issue 36: 249–255.

Douglas, B.C., Kearney, M.S. and Leatherman, S.P. (2000) *Sea Level Rise: History and Consequences*. Academic, New York.

Dubois, G. (1924) Recherches sur les terrains quaternaires du Nord de la France. *Memoirs du Société Géologique du Nord*, **8**: 1–356.

Dubois, R.N. (1977) Predicting beach erosion as a function of rising water level. *Journal of Geology*, **85**: 470–476.

Dubois, R.N. (1992) A re-evaluation of Bruun's Rule and supporting evidence. *Journal of Coastal Research*, **8**: 618–628.

Duller, G.A.T. (2003) Dating methods. In A. Goudie (ed.) *Encyclopedia of Geomorphology*. Routledge, London, pp. 220–224.

Eitner, V. (1996) Geomorphological response of the Frisian barrier islands to sea level rise. *Geomorphology*, **15**: 57–65.

Ellison, J.C. and Stoddart, D.R. (1991) Mangrove ecosystem collapse during predicted sea-level rise: Holocene analogues and implications. *Journal of Coastal Research*, **7**: 151–165.

Emery, K.O. (1960) *The Sea off Southern California*. Wiley, New York.

Emery, K.O. (1961) Submerged marine terraces and their sediments. *Zeitschrift für Geomorphologie, Supplementband* **3**: 17–29.

Emery, K.O. (1969) *A Coastal Pond Studied by Oceanographic Methods*. Elsevier, New York.

Emery, K.O. and Aubrey, D.G. (1991) *Sea Levels, Land Levels and Tide Gauges*. Springer, New York.

Emery, K.O. and Kuhn, G.G. (1982) Sea cliffs: their processes, profiles and classification. *Bulletin, Geological Society of America*, **93**: 644–654.

Emery, K.O. and Stevenson, R.E. (1957) Estuaries and lagoons. *Memoirs, Geological Survey, America*, **67**: 673–750.

Empsall, B. (1989) Workington ironworks reclamation. *Proceedings, Conference Institute of Engineers, Bournemouth*, pp. 279–292.

Evans, G. and Bush, P.R. (1969) Some oceanographical observations on a Persian Gulf Lagoon. In: A.

Castañares and F.B. Phleger (eds) *Lagunas Costeras, un Simposio*, National University of Mexico, pp. 155–170.

Everard, C.E. (1962) Mining and shoreline evolution near St Austell, Cornwall. *Transactions of the Royal Geological Society of Cornwall*, **19**: 199–219.

Everts, C.H. (1985) Sea-level rise effects on shoreline position. *Journal of Waterway, Port, Ocean and Coastal Engineering*, **111**: 985–999.

Fairbridge, R.W. (1967) Coral reefs of the Australian region. In: J.N. Jennings and J.A. Mabbutt (eds) *Landform Studies of Australia and New Guinea*. Australian National University Press, Canberra, pp. 386–417.

Fairbridge, R.W. (1980) The estuary: its definition and geodynamic cycle. In: E. Olausson and I. Cato (eds) *Chemistry and Biogeochemistry of Estuaries*. Wiley, New York, pp. 1–36.

Finkl, C. (2005) Nearshore geomorphological mapping. In M.L. Schwartz (ed.) *Encyclopedia of Coastal Science*. Springer, Dordrecht, pp. 685–698.

Fisher, J.J. (1980) Shoreline erosion, Rhode Island and North Carolina coasts. In: M.L. Schwartz and J.J. Fisher (eds) *Proceedings of the Per Bruun Symposium*, University of Rhode Island, pp. 32–54.

Fisher, R.L. (1955) Cuspate spits of St. Lawrence Island, Alaska. *Journal of Geology*, **63**: 133–142.

Fonseca, M.S. (1996) The role of seagrasses in nearshore sedimentary processes: a review. In: Nordstrom K.F. and Roman C.T. (eds) *Estuarine Shores*, Wiley, Chichester, pp. 261–286.

Fotheringam, D.G. and Goodwins, D.R. (1990) Monitoring the Adelaide beach system. *Proceedings 1990 Workshop on Coastal Zone Management*, Yeppoon, Queensland, pp. 118–132.

Fox W.T., Haney, R.L. and Curran, H.A. (1995) Penouille Spit, evolution of a complex spit, Gaspé, Quebec, Canada. *Journal of Coastal Research*, **11**: 478–493.

French, J.R. and Spencer, T. (1993) Dynamics of sedimentation in a tide-dominated backbarrier salt marsh, Norfolk, U.K. *Marine Geology*, **110**: 315–331.

French J.R. and Stoddart, D.R. (1992) Hydrodynamics of salt marsh creek systems: implications for marsh morphological development and material exchange. *Earth Surface Processes and Landforms*, **17**: 235–252.

French, P.W. (1996) Implications of a salt marsh chronology for the Severn estuary based on independent lines of dating evidence. *Marine Geology*, **135**: 115–125.

French, P.W. (1997) *Coastal and Estuarine Management.* Routledge, London.

Frey. R.W. and Basan, P.B. (1985) Coastal salt marshes. In: R.A. Davis (ed.) *Coastal Sedimentary Environments.* Springer, New York, pp. 225–301.

Fuenzalida, H., Cooke, R., Paskoff, R., Segerstrom, K. and Weischet, W. (1965) High stands of Quaternary sea level along the Chilean Coast. *Geological Society of America,* Special Paper 84: 473–496.

Galvin, C.J. (1972) Waves breaking in shallow water. In: R. Meyer (ed.) *Waves on Beaches.* Academic, London, pp. 413–455.

Gares, P.A. (1990) Eolian processes and dune changes at developed and undeveloped sites, Island Beach, New Jersey. In: K.F. Nordstrom, N. Psuty and R.W.G. Carter (eds) *Coastal Dunes: Form and Process.* Wiley, Chichester, pp. 361–380.

Gell, R.A. (1978) Shelly beaches on the Victorian coast. *Proceedings, Royal Society of Victoria,* **90**: 257–269.

Gibbs, R.J. (1970) Circulation in the Amazon estuary and adjacent Atlantic Ocean. *Journal of Marine Research,* **28**: 113–123.

Gilbert, G.K. (1890) *Lake Bonneville,* US Geological Survey Monograph 1.

Gill, E.D. (1973) Rate and mode of retrogradation on rocky coasts in Victoria, Australia. *Boreas,* **2**: 143–171.

Gill, E.D. and Lang J.G. (1983) Micro-erosion meter measurements of rock wear on the Otway coast of south-eastern Australia. *Marine Geology,* **42**: 141–156.

Gimingham, C.H., Ritchie, W., Willetts, B.B. and Willis, A.J. (eds) (1989) Coastal sand dunes. *Proceedings, Royal Society of Edinburgh,* **B96**: 1–313.

Goldsmith, P. and Hieber, S. (1991) Space techniques support the monitoring of sea level. In: R. Frassetto (ed.) *Impacts of Sea Level Rise on Cities and Regions.* Marsilio, Venice, pp. 220–226.

Goldsmith V. (1985) Coastal dunes. In: R.A. Davis (ed.) *Coastal Sedimentary Environments.* Springer, New York, pp. 303–378.

Goodfellow, B.W. and Stephenson, W.J. (2005) Beach morphodynamics in a strong-wind bay: a low-energy environment? *Marine Geology,* **214**: 101–116.

Goudie, A. (ed.) (2004) *Encyclopedia of Geomorphology.* Oxford University Press, Oxford.

Gresswell, R.K. (1953) *Sandy Shores in South Lancashire.* University of Liverpool Press, Liverpool.

Griggs, G.B. and Trenhaile, A.S. (1994) Coastal cliffs and platforms. In: R.W.G. Carter and C.D. Woodroffe (eds) *Coastal Evolution. Late Quaternary Shoreline Morphodynamics.* Cambridge University Press, Cambridge, pp. 425–450.

Guilcher, A. (1958) *Coastal and Submarine Morphology.* Methuen, London.

Guilcher, A. (1966) Les grandes falaises et megafalaises des cotes sud-ouest et ouest d'Irlande. *Annals de Géographie,* **75**: 26–38.

Guilcher, A. (1967) Origin of sediments in estuaries. In: G.H. Lauff (ed.) *Estuaries,* American Association for the Advancement of Science, Scientific Publication 83, pp. 149–157.

Guilcher, A. (1979) Marshes and estuaries in different latitudes. *Interdisciplinary Science Reviews* **4**: 158–168.

Guilcher, A. (1981) Shoreline changes in salt marshes and mangrove swamps (mangals) within the past century. In: E.C.F. Bird and K. Koike (eds) *Coastal Dynamics and Scientific Sites,* Komazawa University, Tokyo, pp. 31–53.

Guilcher, A. (1985) Retreating cliffs in the humid tropics: an example from Paraiba, north-eastern Brazil. *Zeitschrift für Geomorphologie, Supplementband* **57**: 95–103.

Guilcher, A. (1988) *Coral Reef Geomorphology.* Wiley, Chichester.

Guilcher, A. and Nicolas, J.P. (1954) Observations sur la langue de Barbarie et les bras du Sénégal aux environs de St. Louis. *Bulletin d'Information, Comité d'Océanographie et d'Etude des Côtes,* **6**: 227–242.

Guilcher, A., Vallantin, P., Angrand, J.P. and Galloy, P. (1957) Les cordons littoraux de la Rade de Brest. *Bulletin d'Information, Comité d'Océanographie et d'Etude des Côtes,* **9**: 21–54.

Guza, R.T. and Inman, D.L. (1975) Edge waves and beach cusps. *Journal of Geophysical Research,* **80**: 2997–3012.

Hansom, J. (1983) Shore platform development in the South Shetland Islands, Antarctica. *Marine Geology,* **53**: 211–229.

Hansom, J.D. (2005) Boulder pavements. In M. L. Schwartz (ed.) *Encyclopedia of Coastal Science.* Springer, Dordrecht, pp. 208–210.

Hardisty, J. (1990) *Beaches: Form and Process.* Unwin Hyman: London.

Hardisty, J. (1994) Beach and nearshore sediment transport. In: K. Pye (ed) *Sediment Transport and Depositional Processes.* Blackwell, Oxford, pp. 219–255.

Harris, R.L. (1955) *Restudy of Test Shore Nourishment by Offshore Deposition of Sand, Long Branch, New Jersey,* Beach Erosion Board, Technical Memorandum 62.

Healy, T.R. (1968) Bioerosion on shore platforms in the Waitemata formation, Auckland. *Earth Sciences Journal,* **2**: 26–37.

Healy, T.R. (1977) Progradation of the entrance to Tauranga Harbour, Bay of Plenty. *New Zealand Geographer,* **30**: 90–91.

Hegge, B., Eliot, I. and Hsu, J. (1996) Sheltered sandy beaches of southwestern Australia. *Journal of Coastal Research,* **12**: 748–760.

Hénaff, A., Lageat, Y. and Costa, S. (2006) Geomorphology and shaping processes of chalk shore platforms of the Channel coasts. *Zeitschrift für Geomorphologie, Supplementband* **144**: 61–91.

Herd, D.G., Yourd, T.L., Hansjurgen, C., Person, W.J. and Mendoza, C.. (1981) The great Tumaco, Colombia, earthquake of 12 December 1979. *Science,* **211**: 441–445.

Hesp, P.A. (1984) Foredune formation in southeast Australia. In: B.G. Thom (ed.) *Coastal Geomorphology in Australia.* Academic, Sydney, pp. 69–97.

Hesp, P.A. (1988) Surfzone, beach and foredune interactions on the Australian south east coast. *Journal of Coastal Research,* Special Issue 3: 15–25.

Hesp, P.A. and Hilton, M.J. (1996) Nearshore-surfzone system limits and the impacts of sand extraction. *Journal of Coastal Research,* **12**: 726–747.

Hesp, P.A. and Thom, B.G. (1990) Geomorphology and evolution of active transgressive dunefields. In: K.F. Nordstrom, N. Psuty and R.W.G. Carter (eds) *Coastal Dunes: Form and Process.* Wiley, Chichester, pp. 253–288.

Hicks, D.M. and Hume, T.M. (1996) Morphology and size of ebb-tidal deltas at natural inlets on open sea and pocket-bay coasts, North Island, New Zealand. *Journal of Coastal Research,* **12**: 47–63.

Hill, P.R., Barnes, P.W., Héquette, A. and Ruz, M.H. (1994) Arctic coastal plains shorelines. In: R.W.G. Carter and C.D. Woodroffe (eds) *Coastal Evolution. Late Quaternary Shoreline Morphodynamics.* Cambridge University Press, Cambridge, pp. 341–372.

Hills, E.S. (1970) Fitting, fretting and imprisoned boulders. *Nature,* **226**: 345–347.

Hills, E.S. (1971) A study of cliffy coastal profiles based on examples in Victoria, Australia. *Zeitschrift für Geomorphologie,* **15**: 137–180.

Hinton, A.C. (1998) Tidal changes. *Progress in Physical Geography,* **22**: 282–294.

Hodgkin, E.P. (1964) Rate of erosion of intertidal limestone. *Zeitschrift für Geomorphologie,* **8**: 385–392.

Holman, R.A. and Bowen, A.J. (1982) Bars, bumps and holes: models for the generation of complex beach topography. *Journal of Geophysical Research,* **87**: 457–468.

Hopley, D. (1971) The origin and significance of North Queensland island-spits. *Zeitschrift für Geomorphologie,* **15**: 371–389.

Hopley, D. (1982) *Geomorphology of the Great Barrier Reef.* Wiley Interscience, New York.

Hopley, D. (1994) Continental shelf reef systems. In: R.W.G. Carter and C.D. Woodroffe (eds) *Coastal Evolution. Late Quaternary Shoreline Morphodynamics.* Cambridge University Press, Cambridge, pp. 303–340.

Hopley, D. and Kinsey, D. W. (1988) The effects of a rapid short-term sea-level rise on the Great Barrier Reef. In: G.I. Pearman (ed.) *Greenhouse: Planning for Climatic Changes.* CSIRO Division of Atmospheric Research, Melbourne, pp. 189–201.

Horikawa, K. (ed) (1988) *Nearshore Dynamics and Coastal Processes: Theory, Measurement and Predictive Models.* University of Tokyo Press, Tokyo.

Hutchinson, J.N., Brunsden, D. and Lee, E.M. (1991) The geomorphology of the landslide complex at Ventnor, Isle of Wight. In: R.J. Chandler (ed.) *Slope Stability Engineering.* Telford, London, pp. 213–218.

Ignatov, Y., Kaplin, P.A., Lukyanova, S.A. and Soloveiva, G.D. (1993) Evolution of Caspian Sea coasts under conditions of sea level rise. *Journal of Coastal Research,* **9**: 104–111.

Ingle, J.C. (1966) *The Movement of Beach Sand.* Elsevier, Amsterdam.

Inman, D. L. and Jenkins, S. A. (2005) Accretion and erosion waves on beaches. In M. L. Schwartz (ed.) *Encyclopedia of Coastal Science.* Springer, Dordrecht, pp. 1–4.

Inman, D.L. and Nordstrom, C.E. (1971) On the tectonic and morphologic classification of coasts. *Journal of Geology,* **79**: 1–21.

Intergovernmental Panel on Climate Change (IPCC) (2007) *Climate Change.* World Meteorological Organisation, Geneva.

Isla, F.I. (1995) Coastal lagoons. In: G.M.E. Perillo (ed.) *Geomorphology and Sedimentology of Estuaries.* Elsevier, Amsterdam, pp. 241–272.

Jacobson, H.A. (1988) Historical development of the salt marsh at Wells, Maine. *Earth Surface Processes and Landforms,* **13**: 475–486.

James, P. (1995) *The Sunken Kingdom*. Cape, London.

Jenkin, J.J. (1966) *The Geomorphology and Upper Cainozoic Geology of South-East Gippsland*, Geological Survey of Victoria, Memoir 27.

Jennings, J.N. (1957) On the orientation of parabolic or U-dunes. *Geographical Journal*, **123**: 474–480.

Jennings, J.N. (1959) The coastal geomorphology of King Island, Bass Strait. *Records, Queen Victoria Museum, Launceston*, **11**: 1–39.

Jennings, J.N. (1963) Some geomorphological problems of the Nullarbor Plain. *Transactions, Royal Society of South Australia*, **87**: 41–62.

Jennings, J.N. (1964) The question of coastal dunes in humid tropical climates. *Zeitschrift für Geomorphologie*, **8**: 150–154.

Jennings, J.N. (1965) Further discussion on factors affecting coastal dune formation in the tropics. *Australian Journal of Science*, **28**: 166–167.

Jennings, J.N. (1967) Cliff-top dunes. *Australian Geographical Studies*, **5**: 40–49.

Jennings, J.N. and Bird, E.C.F. (1967) Regional geomorphological characteristics of some Australian estuaries. In: G.H. Lauff (ed.) *Estuaries*. American Association for the Advancement of Science, Washington, DC, pp. 121–128.

Johnson, D.W. (1919) *Shore Processes and Shoreline Development*. Wiley, New York.

Johnson, J.W. (1956) Dynamics of nearshore sediment movement. *Bulletin, American Association of Petroleum Geologists*, **40**: 2211–2232.

Jolliffe, I.P. (1961) The use of tracers to study beach movements and the measurement of longshore drift by a fluorescent technique. *Revue de Géomorphologie Dynamique*, **12**: 81–95.

Jolliffe, I.P. (1964) An experiment designed to compare the relative rates of movement of beach pebbles. *Proceedings of the Geologists' Association*, **75**: 67–86.

Jones, B. and Hunter, I.G. (1992) Very large boulders on the coast of Grand Cayman: the effects of giant waves on rocky coastlines. *Journal of Coastal Research*, **8**: 763–774.

Jones, J.R., Cameron, B. and Fisher, J.J. (1993) Analysis of cliff retreat and shoreline erosion, Thompson Island, Massachussetts, U.S.A. *Journal of Coastal Research*, **9**: 87–96.

Jutson, J.T. (1940) The shore platforms of Mount Martha, Victoria. *Proceedings, Royal Society of Victoria*, **52**: 164–174.

Kaplin, P.A. and Selivanov, A.O. (1995) Recent coastal evolution of the Caspian Sea as a model for coastal responses to the possible acceleration of global sea level rise. *Marine Geology*, **124**: 161–175.

Kaye, C.A. (1959) *Shoreline Features and Quaternary Shoreline Changes. Puerto Rico*, US Geological Survey Professional Paper 317-B, pp. 49–140.

Kaye, C.A. (1973) *Map Showing Changes on Shoreline of Martha's Vineyard, Massachusetts, During the Past 200 Years*, US Geological Survey Miscellaneous Field Studies, Map MF-534.

Kearney, M.S. and Stevenson, J.C. (1991) Island land loss and marsh vertical accretion rate evidence for historical changes in Chesapeake Bay. *Journal of Coastal Research*, **7**: 403–416.

Kelletat, D. and Zimmerman, L. (1991) *Verbreitung und Formtypen rezenter und subrezenter organischer Gesteinsbildungen an der Küsten Kretas*, Essener Geographische Arbeiten 23.

Kench, P.S. (1998) Physical controls on development of lagoon sand deposits and lagoon infilling in an Indian Ocean atoll. *Journal of Coastal Research*, **14**: 1014–1024.

Kidson, C. (1960) The shingle complexes of Bridgwater Bay. *Transactions of the Institute of British Geographers*, **28**: 75–87.

Kidson, C. (1963) The growth of sand and shingle spits across estuaries. *Zeitschrift für Geomorphologie*, **7**: 1–22.

Kidson, C. (1971) The Quaternary history of the coasts of south-west England, with special reference to the Bristol Channel coast. In: K.J. Gregory and W. Ravenhill (eds) *Exeter Essays in Geography*, University of Exeter, pp. 1–22.

Kidson, C. (1986) Sea level changes in the Holocene. In: O. Van de Plassche (ed.) *Sea Level Research: A Manual for the Collection and Evaluation of Data*. Geo Books, Norwich, pp. 27–64.

King, C.A.M. (1972) *Beaches and Coasts* (2nd edition). Arnold, London.

King, C.A.M. and McCullach, M.J. (1971) A simulation model of a complex recurved spit. *Journal of Geology*, **79**: 22–36.

King, L.C. (1962) *The Morphology of the Earth*. Oliver and Boyd, Edinburgh.

Kjerfve, B. (ed.) (1993) *Coastal Lagoon Processes*. Elsevier, Amsterdam.

Klein, G. de V. (1985) Intertidal flats and intertidal sand bodies. In: R.A. Davis (ed.) *Coastal Sedimentary Environments*, Springer, New York, pp. 187–224.

Kolb, C.R. and Van Lopik, J.R. (1966) Depositional environments of the Mississippi River deltaic plain. In:

M.L. Shirley (ed.) *Deltas in their Geologic Framework*. Houston Geological Society, Houston, TX, pp. 19–65.

Komar, P.D. (1976) *Beach Processes and Sedimentation*. Prentice-Hall, Englewood Cliffs, NJ.

Komar, P.D. (1983) *Handbook of Coastal Processes and Erosion*. CRC Press, Boca Raton, FL.

Komar, P.D. (1986) The 1982–83 El Niño and erosion on the coast of Oregon. *Shore and Beach*, **64**: 3–12.

Komar, P.D. and Shih, S. M. (1993) Cliff erosion along the Oregon Coast: a tectonic–sea level imprint plus local controls by beach processes. *Journal of Coastal Research*, **9**: 747–765.

Kraft, J.C., Aschenbrenner, S.E. and Rapp, G. (1988) Paleo-geographic reconstruction of coastal Aegean archaeological sites. *Science*, **195**: 941–947.

Kraus, N.C. and Pilkey, O.H. (eds) (1988) The effects of seawalls on the beach. *Journal of Coastal Research*, Special Issue 4.

Lakhan, V.C. and Pepper, D.A. (1997) Relationship between concavity and convexity of a coast and erosion and accretion patterns. *Journal of Coastal Research*, **13**: 226–232.

Landon, R.E. (1930) An analysis of pebble abrasion and transportation. *Journal of Geology*, **38**: 437–446.

Landsberg, S.Y. (1956) The orientation of dunes in Britain and Denmark in relation to the wind. *Geographical Journal*, **122**: 176–189.

Lauff, G.H. (ed.) (1967) *Estuaries*, American Association for the Advancement of Science, Publication 83.

Lawton, G. (2001) Monsters of the deep. *New Scientist*, **2297**: 28–32.

Le Bourdiec, P. (1958) Aspects de la morphogenèse plio-quaternaire en basse Côte d'Ivoire. *Revue de Géomorphologie Dynamique*, **9**: 33–42.

Leatherman, S.P. (1990) Modelling shore response to sea-level rise on sedimentary coasts. *Progress in Physical Geography*, **1**: 447–464.

Leatherman, S.P., Whitman, D. and Zhang, K. (2005) Airborne laser terrain mapping and light detection and ranging. In M.L. Schwartz (ed.) *Encyclopedia of Coastal Science*. Springer, Dordrecht, pp. 21–24.

Leontiev, O.K. and Nikiforov, L.G.. (1965) Reasons for the world-wide occurrence of barrier beaches. *Oceanology*, **5**: 61–67.

Leontiev, O.K. and Veliev, K.A. (1990) *Transgressive Coasts*, Proceedings, International Symposium, Baku, USSR, 1990, pp. 1–71.

Lewis, W.V. (1932) The formation of Dungeness foreland. *Geographical Journal*, **80**: 309–324.

Lewis, W.V. (1938) The evolution of shoreline curves. *Proceedings of the Geologists' Association*, **49**: 107–127.

Lewis, W.V. and Balchin, W.G.V. (1940) Past sea levels at Dungeness. *Geographical Journal*, **96**: 258–285.

Liu Cangzi and Walker, H.J. (1989) Sedimentary characteristics of cheniers and the formation of the chenier plain of East China. *Journal of Coastal Research*, **5**: 353–368.

Lotfy, M.F. and Frihy, O.E. (1993) Sediment balance in the nearshore zone of the Nile delta coast, Egypt. *Journal of Coastal Research*, **9**: 654–662.

Ly, C.K. (1980) The role of the Akosombo Dam on the Volta River in causing coastal erosion in central and eastern Ghana. *Marine Geology*, **37**: 323–332.

Lyell, C. (1833) *Principles of Geology*, Vol. 2. Murray, London.

Mahinda Silva, A.T. (1986) Ecological and socio-economic aspects of environmental changes in two mangrove-fringed lagoon systems in southern Sri Lanka. In: P. Kunstadter, E.C.F. Bird and S. Sabhasri (eds) *Man in the Mangroves*. United Nations University, Tokyo, pp. 79–86.

Mandelbrot, B. (1967) How long is the coast of Britain? *Science*, **155**: 636–638.

Marker, M.E. (1967) The Dee estuary: its progressive silting and salt marsh development. *Transactions, Institute of British Geographers*, **41**: 65–71.

Masselink, G. and Pattiaratchi, C. (1998) Morphodynamic impact of sea breeze activity on a beach with beach cusp morphology. *Journal of Coastal Research*, **14**: 393–406.

Masselink, G. and Short, A.D. (1993) The effect of tide range on beach morphodynamics and morphology: a conceptual beach model. *Journal of Coastal Research*, **9**: 785–800.

Mathers, S. and Zalasiewicz, J. (1999) Holocene sedimentary architecture of the Red River delta, Vietnam. *Journal of Coastal Research*, **15**: 314–325.

May, V.J. (1972) Earth cliffs. In: R.S.K. Barnes (ed.) *The Coastline*. Wiley, New York, pp. 215–235.

May, V.J. and Heeps, C. (1985) The nature and rates of change on chalk coastlines. *Zeitschrift für Geomorphologie, Supplementband* 57: 81–94.

McCann, S.B. (1964) The raised beaches of north-east Islay and western Jura, Argyll. *Transactions, Institute of British Geographers*, **35**: 1–10.

McCave, I.N. (1978) Grain size trends and transport along beaches: example from East Anglia. *Marine Geology*, **28**: M43–M57.

McCave, I.N. and Geiser, A.C. (1979) Megaripples, ridges and runnels on intertidal flats of The Wash, England. *Sedimentology*, **26**: 353–369.

McGlashan, D.J. (2003) Managed relocation: an assessment of its feasibility. *Geographical Journal*, **169**: 6–20.

McKenzie, P. (1958) The development of beach sand ridges. *Australian Journal of Science*, **20**: 213–214.

McLean, R.F. (1978) Recent coastal progradation in New Zealand. In: J.L. Davies and M.A.J. Williams (eds) *Landscape Evolution in Australasia*. Australian National University, Canberra, pp. 168–196.

McLean, R.F. and Woodroffe, C.D. (1994) Coral atolls. In: R.W.G. Carter and C.D. Woodroffe (eds) *Coastal Evolution. Late Quaternary Shoreline Morphodynamics*. Cambridge University Press, Cambridge, pp. 267–302.

Michel, D. and Howa, H.L. (1999) Short-term morphodynamic response of a ridge and runnel system on a mesotidal sandy beach. *Journal of Coastal Research*, **15**: 428–237.

Miller, D.J. (1960) *Giant Waves at Lituya Bay, Alaska*, United States Geological Service, Professional Paper 354-C, pp. 51–66.

Milliman, J.D. and Haq, B.U. (1996) *Sea-Level Rise and Coastal Subsidence: Causes, Consequences and Strategies*. Kluwer, Dordrecht.

Møller, J.T. (1963) Accumulation and abrasion in a tidal area. *Geografisk Tidsskrift*, **62**: 56–79.

Møller, J.T. (1985) Denmark. In: E.C.F. Bird and M.L. Schwartz (eds) *The World's Coastline*, Van Nostrand Reinhold, New York, pp. 325–383.

Molnia, B.F. (1985) Processes on a glacier-dominated coast, Alaska. *Zeitschrift für Geomorphologie, Supplementband* **57**: 141–153.

Morang A., Larson, R. and Gorman, L. (1997) Monitoring the coastal environment, part I: waves and currents. *Journal of Coastal Research*, **13**: 111–133.

Morton, J.K. (1957) Sand dune formation on a tropical shore. *Journal of Ecology*, **45**: 495–497.

Morton, R.A., Leach, M.P., Paine, J.G. and Cardoza, M.A. (1993) Monitoring beach changes using GPS techniques. *Journal of Coastal Research*, **9**: 702–720.

Moses, C.A., Robinson, D.A., Williams, R.B.G. and Marques, F. (2006) Predicting rates of shore platform downwearing from rock geotechnical properties and laboratory simulation of weathering and erosion processes. In D.A. Robinson and Y. Lagent (eds) *European Shore Platform Dynamics. Zeitschrift für Geomorphologie, Supplementband*, **144**: 19–37.

Mottershead, D.N. (1989) Rates and patterns of bedrock denudation by coastal salt spray weathering: a seven year record. *Earth Surface Processes and Landforms*, **14**: 383–398.

Mottershead, D.N. (1998) Coastal weathering of greenschist in dated structures, south Devon. *Quarterly Journal of Engineering Geology*, **31**: 343–346.

Neumann, A.C. and Macintyre, I. (1985) Reef response to a sea level rise: keep-up, catch-up or give-up. *Proceedings, 5th International Coral Reef Congress*, **3**: 105–110.

Nichols, M.M. and Biggs, R.B. (1985) Estuaries. In: R.A. Davis (ed.) *Coastal Sedimentary Environments*, Springer, New York, pp. 77–186.

Nordstrom, K.F. (1992) *Estuarine Beaches*. Elsevier, London.

Nordstrom, K.F. and Roman, C.T. (1996) *Estuarine Shores*. Wiley, Chichester.

Nordstrom, K.F., Psuty, N and Carter, R.W.G. (1990) *Coastal Dunes: Form and Process*. Wiley, Chichester.

Norrman, J.O. (1980) Coastal erosion and slope development in Surtsey Island, Iceland. *Zeitschrift für Geomorphologie, Supplementband* **34**: 20–38.

Nossin, J.J. (1965) The geomorphic history of the Padang delta. *Journal of Tropical Geography*, **20**: 54–64.

Oaks, R.A. and Coch, N.K. (1963) Pleistocene levels, southeastern Virginia. *Science*, **140**: 979–984.

O'Brien, M.P. and Dean, R.G. (1972) Hydraulic and sedimentary stability of tidal inlets. *Proceedings 13th Coastal Engineering Conference*. American Society of Civil Engineers, New York, pp. 761–780.

Olson, J.S. (1958) Lake Michigan dune development. *Journal of Geology*, **66**: 254–263, 345–351, 473–483.

Orford, J.D., Carter, R.W.G. and Jennings, S.C. (1996) Control domains and morphological phases in gravel-dominated coastal barriers of Nova Scotia. *Journal of Coastal Research*, **12**: 589–604.

Orford, J.D., Forbes, D.L. and Jennings, S.C (2002) Organisational controls, typologies and time scales of paraglacial gravel-dominated coastal systems. *Geomorphology*, **48**: 51–85.

Orme, A.R. (1985) California. In: E.C.F. Bird and M.L. Schwartz (eds) *The World's Coastline*, Van Nostrand Reinhold, New York, pp. 27–36.

Orme, A.R. (1990) The instability of Holocene coastal dunes: the case of the Morro dunes, California. In: K.F. Nordstrom, N. Psuty and R.W.G. Carter (eds)

Coastal Dunes: Form and Process. Wiley, Chichester, pp. 315–336.

Orviku, K., Bird, E.C.F. and Schwartz, M.L. (1995) The provenance of beaches on the Estonian islands of Hiiumaa, Saaremaa and Muhu, *Journal of Coastal Research,* **11**: 96–106.

Orviku, K., Jaagus, J., Kont, A., Ratas, U. and Rivis. R. (2003) Increasing activity of coastal processes associated with climate change in Estonia. *Journal of Coastal Research,* **19**: 364–375.

Otvos, E.G. (2000) Beach ridges—definition and significance. *Geomorphology,* **32**: 83–108.

Parker, B. (2001) Where is the shoreline? *Hydro International,* **5**(5): 6–9.

Parry, M., Magalhaes, A.R. and Nguyen Huu Ninh (eds) (1991) *The Potential Socio-Economic Effects of Climatic Change.* United Nations Environment Programme: Nairobi.

Paskoff, R. (1994) *Les Littoraux. Impact des Aménagements sur leur Évolution* (2nd edition). Masson, Paris.

Paskoff, R. and Petiot, R. (1990) Coastal progradation as a by-product of human activity: an example from Chañaral Bay, Atacama Desert, Chile. *Journal of Coastal Research,* Special Issue 6: 91–102.

Perillo, G.M.E. (ed) (1995) *Geomorphology and Sedimentology of Estuaries.* Elsevier, Amsterdam.

Pernetta, J.C. and Osborne, P.L. (1988) Deltaic floodplains: the Fly River and the mangroves of the Gulf of Papua. In: *Potential Impacts of Greenhouse Gas Generation on Climatic Change and Projected Sea Level Rise on Pacific Islands.* United Nations Environment Programme, Split, pp. 94–111.

Pethick, J.S. (1980) Salt marsh initiation during the Holocene transgression: the example of the North Norfolk marshes, England. *Journal of Biogeography,* **7**: 1–9.

Pethick, J. (1981) Long-term accretion rates on tidal salt marshes. *Journal of Sedimentary Petrology,* **51**: 571–577.

Pethick, J. (1984) *An Introduction to Coastal Geomorphology.* Arnold, London.

Pethick, J.S. (1992) Salt marsh geomorphology In: J.R.L. Allen and K. Pye (eds) *Salt Marshes.* Cambridge University Press, Cambridge, pp. 41–62.

Pethick, J.S. (1994) Estuaries and wetlands: function and form. In: R.A. Falconer and P. Goodwin (eds) *Wetland Management.* Institute of Civil Engineers, Telford, pp. 75–87.

Pethick, J.S. (1996) The geomorphology of mudflats. In: K.F. Nordstrom and C.T. Roman (eds), *Estuarine Shores.* Wiley, Chichester, pp. 185–211.

Pethick, J.S. (1998) Coastal management and sea level rise. In: S. Lane, K. Richards and J,. Chandler (eds) *Landform Monitoring, Modelling and Analysis,* Wiley, Chichester, pp. 403–419.

Philip, A.L. (1990) Ice-pushed boulders on the shores of Gotland, Sweden. *Journal of Coastal Research,* **6**: 661–676.

Phillips, J.D. (1986) Coastal submergence and marsh fringe erosion. *Journal of Coastal Research,* **2**: 427–436.

Phleger, F.B. (1969) Some general features of coastal lagoons. In: A Castañares and F.B. Phleger (eds) *Lagunas Costeras, un Simposio,* National University of Mexico, pp. 5–26.

Phleger, F.B. (1981) A review of some general features of coastal lagoons. *Coastal Lagoon Research, Past, Present and Future,* UNESCO Technical Papers in Marine Science 33, pp. 7–14.

Piccazzo, M., Firpo, M., Corradi, N. and Campi, F. (1992) Coastline changes on beaches affected by defence works: a case study in Albissola (Western Liguria, Italy). *Bolletino di Oceanologica Teorica ed Applicata,* **10**: 197–210.

Pierce, J.W. (1969) Sediment budget along a barrier island chain. *Sedimentary Geology,* **3**: 5–16.

Pilkey, O.H. and Clayton, T.D. (1989) Summary of beach replenishment experience on US East Coast barrier islands. *Journal of Coastal Research,* **5**: 147–159.

Pilkey, O.H., Young, R.S., Riggs, S.R., Sam Smith, A.W., Wu, H. and Pilkey, W.D. (1993) The concept of shoreface profile of equilibrium: a critical review. *Journal of Coastal Research,* **9**: 255–278.

Pirazzoli, P.A. (1986) Secular trends of relative sea-level changes indicated by tide-gauge records. *Journal of Coastal Research,* Special Issue 1: 1–26.

Pirazzoli, P.A. (1989) Recent sea-level changes in the North Atlantic. In: D.B. Scott, P.A. Pirazzoli and C.A. Honig (eds) *Late Quaternary Sea-level Correlations and Applications.* Kluwer, Dordrecht, pp. 153–167.

Pirazzoli, P.A. (1996) *Sea-Level Changes: the Last 20,000 Years.* Wiley, Chichester.

Prager, E.J. and Halley, R.B. (1999) The influence of seagrass on shell layers and Florida Bay mudbanks. *Journal of Coastal Research,* **15**: 1151–1162.

Prêcheur, C. (1960) *Le Littoral de la Manche: de Ste Andresse à Ault. Etude Morphologique*, Norois special volume.

Price, W.A. (1947) Equilibrium of form and forces in tidal basins of the coast of Texas and Louisiana. *Bulletin of the American Association of Petroleum Geologists*, **31**: 1619–1663.

Pringle, A. (1981) Beach development and coastal erosion in Holderness, North Humberside. In: J. Neale and J Flenley (eds) *The Quaternary in Britain*. Pergamon, Oxford, pp. 194–205.

Pritchard, D.W. (1967) Observations of circulation in coastal plain estuaries. In: G.H. Lauff (ed.) *Estuaries*, American Association for the Advancement of Science, Scientific Publication 83, pp. 37–44.

Psuty, N.P. (1965) Beach-ridge development in Tabasco, Mexico. *Annals of the Association of American Geographers*, **55**: 112–124.

Psuty, N.P. (1986) Impacts of impending sea level rise scenarios: the New Jersey barrier island responses. *Bulletin of the New Jersey Academy of Science*, **31**: 29–36.

Psuty, N.P. (ed.) (1988) *Dune–Beach Interaction, Journal of Coastal Research*, Special Issue 3.

Psuty, N.P. (1990) Foredune mobility and stability, Fire Island, New York. In: K.F. Nordstrom, N. Psuty and R.W.G. Carter (eds) *Coastal Dunes: Form and Process*. Wiley, Chichester, pp. 159–176.

Psuty, N.P. and Moreira, M.E. (1990) Nourishment of a cliffed coastline, Praia da Rocha, the Algarve, Portugal. *Journal of Coastal Research*, Special Issue 6: 21–32.

Pugh, D.T. (1987) *Tides, Surges and Sea Level. A Handbook for Engineers and Scientists*. Wiley, Chichester.

Pugh, D.T. (1991) Cities on water: sea level measurements and planning. In: R. Frassetto (ed.) *Impacts of Sea Level Rise on Cities and Regions*. Marsilio, Venice, pp. 217–219.

Pye, K. (1983) Formation and history of Queensland coastal dunes. *Zeitschrift für Geomorphologie, Supplementband* **45**: 175–204.

Pye, K. (1996) Evolution of the shoreline of the Dee estuary, United Kingdom. In: K.F. Nordstrom and C.T. Roman (eds) *Estuarine Shores*. Wiley, Chichester, pp. 15–37.

Quigley, R. M. and De Nardo, L. R. (1980) Cyclic instability modes of eroding clay bluffs, Lake Erie North Shore Bluffs at Port Bruce, Ontario, Canada. *Zeitschrift für Geomorphologie, Supplementband* **34**: 39–47.

Ranwell, D.S. (1972) *Ecology of Salt Marshes and Sand Dunes*. Chapman and Hall, London.

Reed, D.J. (2002) Sea level rise and coastal marsh sustainability: geological and ecological factors in the Mississippi delta plain. *Geomorphology*, **48**: 233–243.

Revelle, R. and Emery, K.O. (1957) *Chemical Erosion of Beach Rock and Exposed Reef Rock*, US Geological Survey Professional Paper 260-T, pp. 699–709.

Rey, V., Davies, A.G. and Belzons, M (1995) On the formation of bars by the action of waves on an erodible bed: a laboratory study. *Journal of Coastal Research*, **11**: 1195–1203.

Rink, W.J. (1999) Quartz luminescence as a light-sensitive indicator of sediment transport in coastal processes. *Journal of Coastal Research*, **15**: 148–154.

Ristaniemi, O., Eronen, M., Glückert, G. and Rantala, P. (1997) Holocene and recent shoreline changes on the rapidly uplifting coast of Western Finland. *Journal of Coastal Research*, **13**: 397–406.

Ritchie, W. (1977) Machair development and chronology in the Uists and adjacent islands. *Proceeedings of the Royal Society of Edinburgh*, **77B**: 107–122.

Ritchie, W. (1983) The Sands of Forvie. *North-East Scotland Field Guide and Geographical Essays*. University of Aberdeen, pp. 12–19.

Ritchie, W. (2006) Scotland—Angus and Aberdeenshire. In Bird, E.C.F. (ed.) *The World's Coasts: Online*. Springer, Dordrecht, 7.24.3.

Ritchie, W. and Mather, A.S. (1984) *The Beaches of Scotland*. Scottish Countryside Commission, Perth.

Robertson-Rintoul, M.J. (1990) A quantitative analysis of the near-surface wind flow pattern over coastal parabolic dunes. In: K.F. Nordstrom, N. Psuty and R.W.G. Carter (eds) *Coastal Dunes: Form and Process*, Wiley, Chichester, pp. 57–78.

Robinson, A.H.W. (1955) The harbour entrances of Poole, Christchurch, and Pagham. *Geographical Journal*, **121**: 33–50.

Robinson, A.H.W. (1966) Residual currents in relation to shoreline evolution of the East Anglian coast. *Marine Geology*, **4**: 57–84.

Robinson, D.A. and Jerwood, L.C. (1987) Sub-aerial weathering of chalk shore platforms during harsh winters in southeast England. *Marine Geology*, **77**: 1–14.

Robinson, L.A. (1976) The micro-erosion meter technique in a littoral environment. *Marine Geology*, **22**: M51–M58.

Robinson, L.A. (1977) Marine erosive processes at the cliff foot. *Marine Geology*, **23**: 257–271.

Rooney, J.J. and Smith, S.V. (1999) Watershed land use and bay sedimentation. *Journal of Coastal Research*, **15**: 478–485.

Rosen, P.S. (2005) Boulder barricades. In M. L. Schwartz (ed.) *Encyclopedia of Coastal Science*. Springer, Dordrecht, pp. 204–206.

Roy, P.S., Cowell, P.J., Ferland, M.A. and Thom, B.G. (1994) Wave-dominated coasts. In: R.W.G. Carter and C.D. Woodroffe (eds) *Coastal Evolution. Late Quaternary Shoreline Morphodynamics*. Cambridge University Press, Cambridge, pp. 121–186.

Russell, R.J. (1967). *River plains and sea coasts*. University of California Press, Berkeley, CA.

Russell, R.J. and Howe, H.V. (1935) Cheniers of southwestern Louisiana. *Geographical Review*, **25**: 449–461.

Rutin, J. (1992) Geomorphic activity of rabbits on a coastal sand dune, De Blink dunes, the Netherlands. *Earth Surface Processes and Landforms*, **17**: 85–94.

Ruz, M.H. (1987) Recent evolution of the southeast barrier coast of Ireland. *Journal of Coastal Research*, **5**: 523–540.

Ruz, M.H., Allard, M., Michaud, Y. and Héquette. A. (1998) Sedimentology and evolution of subarctic tidal flats along a rapidly emerging coast, Eastern Hudson Bay, Canada. *Journal of Coastal Research*, **14**: 1242–1254.

Saenger, P., Hegerl, E.J. and Davie, J.D.S. (eds) (1983) Global status of mangrove ecosystems. *The Environmentalist*, **3**: 1–88.

Sanderson, P.G. and Eliot, I. (1996) Shoreline salients, cuspate forelands and tombolos on the coast of Western Australia. *Journal of Coastal Research*, **12**: 761–773.

Sanderson, P.G., Eliot, I. and Fuller, M. (1998) Historical development of a foredune plain at Desperate Bay, Western Australia. *Journal of Coastal Research*, **14**: 1187–1201.

Savigear, R.A.G. (1952) Some observations on slope development in South Wales. *Transactions, Institute of British Geographers*, **18**: 31–52.

Savigear, R.A.G. (1962) Some observations on slope development in North Devon and North Cornwall. *Transactions, Institute of British Geographers*, **31**: 23–42.

Sayles, R.W. (1931) Bermuda during the ice age. *Proceedings, American Academy, Arts and Science*, **66**: 381–467.

Scheidegger, A.E. (1991) *Theoretical Geomorphology* (3rd edition). Springer, Berlin.

Schofield, J.C. (1975) Sea level fluctuations cause periodic postglacial progradation, South Kaipara barrier. *New Zealand Journal of Geology and Geophysics*, **18**: 295–316.

Scholl, D.W. (1968) Mangrove swamps: geology and sedimentology. In: R.W. Fairbridge (ed.) *Encyclopaedia of Geomorphology*. Reinhold, New York, pp. 683–688.

Schou, A. (1945) *Det Marine Forland*. Copenhagen.

Schou, A. (1952) Direction determining influence of the wind on shoreline simplification and coastal dunes. *Proceedings, 17th Conference, International Geographical Union*, Washington, pp. 370–373.

Schwartz, M.L. (1965) Laboratory study of sea level rise as a cause of erosion. *Journal of Geology*, **73**: 568–534.

Schwartz, M.L. (1967) The Bruun theory of sea level rise as a cause of shore erosion. *Journal of Geology*, **75**: 76–92.

Schwartz, M.L. (1971) The multiple causality of barrier islands. *Journal of Geology*, **79**: 91–93.

Schwartz, M.L. (ed.) (1972) *Spits and Bars*. Dowden, Hutchinson and Ross, Stroudsburg, PA.

Schwartz, M.L. (ed.) (1973) *Barrier Islands*. Dowden, Hutchinson and Ross, Stroudsburg, PA.

Schwartz, M.L. (ed.) (2005) *Encyclopedia of Coastal Science*. Springer, Dordrecht.

Schwartz, M.L. and Bird, E.C.F. (eds) (1990) *Artificial Beaches, Journal of Coastal Research*, Special Issue 6.

Schwartz, M.L., Fabbri, P. and Scott Wallace, R. (1987) Geomorphology of Dungeness Spit, Washington, USA. *Journal of Coastal Research*, **3**: 451–455.

Schwartz, M.L. and Terich, T.A. (1985) Washington. In: E.C.F. Bird and M.L. Schwartz (eds) *The World's Coastline*. Van Nostrand Reinhold, New York, pp. 17–22.

SCOR Working Group 89 (1991) The response of beaches to sea-level changes: a review of predictive models. *Journal of Coastal Research*, **7**: 895–921.

Sestini, G. (1992) Implications of climatic change for the Po delta and Venice lagoon. In: L. Jeftic, J.F. Milliman and G. Sestini (eds) *Climatic Changes and the Mediterrean*. Arnold, London, pp. 428–494.

Shackleton, N.J. and Opdyke, N.D. (1973) Oxygen isotope and palaeomagnetic stratigraphy of Equatorial Pacific core V-28-238. *Quaternary Research*, **3**: 39–55.

Shennan, I. and Sproxton, I.W. (1990) Possible impacts of sea level rise: a case study from the Tees estuary. In: J.C. Doornkamp (ed.) *The Greenhouse Effect and Rising Sea Levels in the U.K.* M1, Nottingham, pp. 109–133.

Shepard, F.P. (1952) Revised nomenclature for coastal depositional features. *Bulletin of the American Association of Petroleum Geologists*, **36**: 1802–1812.

Shepard, F.P. (1973) *Submarine Geology* (3rd edition). Harper and Row, New York.

Shepard, F.P. and Wanless, H.R. (1971) *Our Changing Coastlines.* McGraw-Hill, New York.

Sherman, D.J. and Hotta, S. (1990) Aeolian sediment transport: theory and measurement. In: K.F. Nordstrom, N. Psuty and R.W.G. Carter (eds) *Coastal Dunes: Form and Process.* Wiley, Chichester, pp. 17–37.

Short, A.D. (1992) Beach systems of the central Netherlands coast: processes, morphology and structural impacts in a storm driven multi-bar system. *Marine Geology*, **107**: 103–137.

Short, A.D. (ed.) (1999) *Handbook of Beach and Shoreface Morphodynamics.* Wiley, Chichester.

Shuisky, Y.D. (1994) An experience of studying artificial ground terraces as a means of coastal protection. *Ocean and Coastal Management*, **22**: 127–139.

Silvester, R. (1974) *Coastal Engineering.* Elsevier, Amsterdam.

Snedaker, S.C. (1982) Mangrove species zonation: why? In: D.N. Sen and K.S. Rajpurohit (eds) *Tasks for Vegetation Science.* Junk, The Hague, pp. 111–125.

So, C.L. (1965) Coastal platforms of the Isle of Thanet. *Transactions, Institute of British Geographers*, **37**: 147–152.

So, C.L. (1974) Some coast changes around Aberystwyth and Tanybwlch, Wales. *Transactions of the Institute of British Geographers*, **62**: 143–153.

So, C.L. (1982) Wind-induced movements of beach sand at Portsea, Victoria. *Proceedings, Royal Society of Victoria*, **94**: 61–68.

Soons, J.M. (1982) Westland: the west coast of the South Island. In J.M. Soons and M.J. Selby (eds) *Landforms of New Zealand.* Longman Paul, Auckland, pp. 299–315.

Spate, A.P., Jennings, J.N., Smith, D.I. and Greenaway, M.A. (1985) The micro-erosion meter: use and limitations. *Earth Surface Processes and Landforms*, **10**: 427–440.

Spencer, T. (1988) Coastal biogeomorphology In: H.A. Viles (ed.) *Biogeomorphology.* Blackwell, Oxford, pp. 255–318.

Spencer, T. (1995) Potentialities, uncertainties and complexities in the response of coral reefs to future sea-level rise. *Earth Surface Processes and Landforms*, **20**: 49–64.

Sprigg, R.C. (1959) Stranded sea beaches and associated sand accumulations of the upper south-east of South Australia. *Transactions, Royal Society of South Australia*, **82**: 183–193.

Stanley, D.J. and Warne, A.G. (1998) Nile delta in its destruction phase. *Journal of Coastal Research*, **14**: 794–823.

Stapor, F.W., Mathews, T.D. and Lindfors-Kearns, F.E. (1991) Barrier island progradation and Holocene sea-level history in southwest Florida. *Journal of Coastal Research*, **7**: 815–838.

Stearns, H.T. (1974) Submerged shorelines and shelves in the Hawaiian islands. *Bulletin, Geological Society of America*, **85**: 795–804.

Steers, J.A. (ed.) (1960) *Scolt Head Island.* Heffer, Cambridge.

Steers, J.A. (1964) *The Coastline of England and Wales* (2nd edition). Cambridge University Press, Cambridge.

Steers, J.A. (1973) *The Coastline of Scotland.* Cambridge University Press, Cambridge.

Steers, J.A., Stoddart, D.R., Bayliss-Smith, T.P., Spencer, T. and Durbridge, P.M. (1979) The storm surge of January 1978 on the east coast of England. *Geographical Journal*, **145**: 192–205.

Stephenson, W.J. (1997) Improving the traversing micro-erosion meter. *Journal of Coastal Research*, **13**: 236–241.

Stephenson, W. (2001) Shore platform width: a fundamental problem. *Zeitschrift für Geomorphologie*, **45**: 511–527..

Stephenson, W.J. and Kirk, R.M. (2000) Development of shore platforms on Kaikoura Perninsula, South Island, New Zealand. *Geomorphology*, **32**: 21–41, 43–56.

Stephenson, W.J, Taylor, A.J., Hemmingsen, M.A., Tsujimoto, H. and Kirk, R.M. (2004) Short-term microscale topographic changes of coastal bedrock on shore platforms. *Earth Surface Processes and Landforms*, **29**: 1663–1673.

Sterr, H., Schwarzer, K. and Kliewe, H. (1998) Holocene evolution and morphodynamics of the German Baltic Sea coast—recent research advances. In: D. Kelletat (ed.) *German Geographical Coastal Research: the Last Decade.* Institute for Scientific Co-Operation, Tübingen, Germany, pp. 107–134.

Stevenson, J.C. and Kearney, M.S. (1996) Shoreline dynamics on the windward and leeward shores of a large temperate estuary. In: K.F. Nordstrom and C.T. Roman (eds) *Estuarine Shores*. Wiley, Chichester, pp. 233–259.

Stoddart, D.R. (1965) Re-survey of hurricane effects on the British Honduras reefs and cays. *Nature*, **207**: 589–592.

Stoddart, D.R. and Cann, J.R. (1965) Nature and origin of beach rock. *Journal of Sedimentary Petrology*, **35**: 243–247.

Stoddart, D.R. and Scoffin, T.P. (1979) Micro-atolls: review of form, origin and terminology. *Atoll Research Bulletin* **224**.

Stumpf, R.P. (1983) The process of sedimentation on the surface of a salt marsh. *Estuarine, Coastal and Shelf Science*, **17**: 495–508.

Suanez, S. and Provansal, M. (1998) Large scale evolution of the littoral of the Rhône delta (southeast France). *Journal of Coastal Research*, **14**: 493–501.

Suess, E. (1906) *The Face of the Earth*. Clarendon, Oxford.

Sunamura, T. (1992) *Geomorphology of Rocky Coasts*. Wiley, Chichester.

Syvitski, J.P.M. and Shaw, J. (1995) Sedimentology and geomorphology of fiords. In: G.M.E. Perillo (ed.) *Geomorphology and Sedimentology of Estuaries*. Elsevier, Amsterdam, pp. 113–178.

Tait, J.F. and Griggs, G.B. (1990) Beach response to the presence of a sea wall: a comparison of field observations. *Shore and Beach*, **58**: 11–28.

Tanner, W.F. (1960) Florida coastal classification. *Transactions of the Gulf Coast Association of Geological Societies*, **10**: 259–266.

Tanner, W.F. (1982) Ripple marks In: Schwartz M.L. (ed.) *The Encylopaedia of Beaches and Coastal Environments*. Hutchinson Ross, Stroudsburg, PA, pp. 695–698.

Tanner, W.F. (1995) Origin of beach ridges and swales. *Marine Geology*, **129**: 141–161.

Tanner, W.F. and Stapor, F. (1972) Accelerating crisis in beach erosion. *International Geography*, **2**: 1020–1021.

Taylor, M. and Stone, G.W. (1996) Beach ridges: a review. *Journal of Coastal Research*, **12**: 612–621.

Teh Tiong Sa (1985) Pensinsular Malaysia. In: E.C.F. Bird and M.L. Schwartz (eds) *The World's Coastline*. Van Nostrand Reinhold, New York, pp. 789–795.

Teh Tiong Sa (1992) The permatang series of Peninsular Malaysia: an overview. In: H.D. Tjia and Sharifah Mastura Syed Abdullah (eds), *The Coastal Zone of Peninsular Malaysia*. Penerbit Universiti Kebangsaan Malaysia, Bangi, pp. 42–62.

Thom, B.G. (1967) Mangrove ecology and deltaic geomorphology: Tabasco, Mexico. *Journal of Ecology*, **55**: 301–343.

Thom, B.G. (1984) Sand barriers of eastern Australia: Gippsland—a case study. In: B.G. Thom (ed.) *Coastal Geomorphology in Australia*. Academic, Sydney, pp. 233–261.

Thom, B.G. and Wright, L.D. (1983) The geomorphology of the Purari delta. In: T. Petr (ed.) *The Purari: Tropical Environment of a High Rainfall River Basin*. Junk, Amsterdam, pp. 47–65.

Thompson, C.H. (1983) Development and weathering of large parabolic dune systems along the subtropical coast of eastern Australia. *Zeitschrift für Geomorphologie*, **45**: 205–225.

Thornton, L. E. and Stephenson, W.J. (2006) Rock strength: a control of shore platform elevation. *Journal of Coastal Research*, **22**: 224–231.

Tietze, W. (1962) Ein beitrag zum geomorphologischen problem der strandflate. *Petermanns Geographichen Mitteilungen*, **106**: 1–20.

Titus, J.G. (ed.) (1988) *Greenhouse Effect, Sea Level Rise and Coastal Wetlands*. US Environment Protection Agency, Washington, DC.

Tooley, M.J. and Jelgersma, S. (1992) *Impacts of Sea-Level Rise on European Coastal Lowlands*. Institute of British Geographers, Special Publication 27. Blackwell, Oxford.

Trenhaile, A.S. (1987) *The Geomorphology of Rock Coasts*. Clarendon, Oxford.

Trenhaile, A.S., Pérez Alberti, A., Martinez Cortizas, A, Costa Casais, M. and Blanco Chao, R. (1999) Rock coast inheritance: an example from Galicia, northwest Spain. *Earth Surface Processes and Landforms*, **24**: 605–621.

Tricart, J. (1959) Problèmes géomorphologiques du littoral oriental du Brésil. *Cahiers Océanographiques*, **11**: 276–308.

Turner, I.L. and Leatherman, S.P. (1997) Beach dewatering as a soft engineering solution to coastal erosion—a history and critical review. *Journal of Coastal Research*, **13**: 1050–1063.

US Army Corps of Engineers (1984) *Shore Protection Manual* (two volumes, 4th edition).

Valentin, H. (1952) *Die Küsten der Erde*. Petermanns Geographische Mitteilungen 246.

Van Straaten, L.M.J.U. (1953) Megaripples in the Dutch Wadden Sea and in the Basin of Arcachon. *Geologie en Mijnbouw*, 15: 1–15.

Van Straaten, L.M.J.U. (1959) Littoral and submarine morphology of the Rhône delta. *Proceedings, Second Coastal Geography Conference*, pp. 233–264.

Van Straaten, L.M.J.U. (1965) Coastal barrier deposits in south and north Holland. *Mededelingen van de Geologische Stichting*, 17: 41–75.

Vanney, J.R., Menanteau, L. and Zazo, C. (1979) Physiographie et évolution des dunes de Basse-Andalousie. In: A. Guicher (ed.) *Les Côtes Atlantiques de l'Europe*. University of Western Brittany, Brest, pp. 277–285.

Vaughan, T. (1910) The geologic work of mangroves in southern Florida. *Smithsonian Miscellaneous Collection*, 52: 461–464.

Verger, F. (1968) *Marais et Wadden du Littoral Francais*. Biscaye, Bordeaux.

Vermeer, D.E. (1985) Mauritania. In: E.C.F. Bird and M.L. Schwartz (eds) *The World's Coastline*. Van Nostrand Reinhold, New York, pp. 549–553.

Viles, H. and Spencer, T. (1995) *Coastal Problems*. Arnold, London.

Walker, D.J. and Jessup, A. (1992) Analysis of the dynamic aspects of the River Murray mouth, South Australia. *Journal of Coastal Research*, 8: 71–76.

Walker, H.J. (1988) *Artificial structures and shorelines*. Kluwer, Dordrecht.

Walton, K. (1956) Rattray: a study on coastal evolution. *Scottish Geographical Magazine*, 72: 85–96.

Wanless, H.R. (1982) Sea level is rising—so what? *Journal of Sedimentary Petrology*, 52: 1051–1054.

Warrick R.A., Barrow E.M. and Wigley T.M.L. (1993) *Climate and Sea Level Change*. Cambridge University Press, Cambridge.

Watson, J.D. (1928) Mangrove forests of the Malay Peninsula. *Malay Forest Records*, 6: 1–275.

Wells, J.T. (1995) Tide-dominated estuaries and tidal rivers. In: G.M.E. Perillo (ed.) *Geomorphology and Sedimentology of Estuaries*. Elsevier, Amsterdam, pp. 179–205.

Wells, J.T., Adams, C.E., Park, Y.A. and Frankenberg, E.W. (1990) Morphology, sedimentology and tidal channel processes on a high-tide-range mudflat, west coast of South Korea. *Marine Geology*, 95: 111–130.

Wilcock, P.R., Miller, D.S. and Shea, R.H. (1998) Frequency of effective wave activity and the recession of coastal bluffs: Calvert Cliffs, Maryland. *Journal of Coastal Research*, 14: 256–268.

Williams, M.A.J., Dunkerley, D.L., De Deckker, P., Kershaw, A.P. and Stokes, T. (1991) *Quaternary Environments*. Arnold, London.

Williams, R.B.G., Robinson, D.A., Dornbusch, U., Foote, Y.L.M., Moses, C.A. and Saddleton, P.R. (2004) A sturzstrom-like cliff fall on the chalk coast of Sussex. In R. N. Mortimore and A Duperret (eds), *Coastal Chalk Cliff Instability*, Geological Society Engineering Geology Special Publication 20, pp. 89–97.

Williams, W.W. (1956) An east coast survey: some recent changes in the coast of East Anglia. *Geographical Journal*, 122: 317–334.

Wilson, G. (1952) The influence of rock structures on coastline and cliff development around Tintagel, North Cornwall. *Proceedings of the Geologists' Association*, 63: 20–48.

Wolanski, E. and Chappell, J. (1996) The response of tropical Australian estuaries to a sea level rise. *Journal of Marine Systems*, 7: 267–279.

Wolanski, E. and Gibbs, R.J. (1995) Flocculation of suspended sediment in the Fly River estuary, Papua New Guinea. *Journal of Coastal Research*, 11: 754–762.

Woodroffe, C.D. (1993) Late Quaternary evolution of coastal and riverine plains of south-east Asia and northern Australia – an overview. *Sedimentary Geology*, 83: 163–175.

Woodroffe, C.D. (1995) Response of tide-dominated mangrove shorelines in northern Australia to anticipated sea level rise. *Earth Surface Processes and Landforms*, 20: 65–85.

Woodroffe, C.D. (1996) Late Quaternary infill of macrotidal estuaries in northern Australia. In: K.F. Nordstrom and C.T. Roman (eds) *Estuarine Shores*. Wiley, Chichester, pp. 89–114.

Woodworth, P.L. (1987) Trends in UK mean sea level. *Marine Geodesy*, 11: 57–87.

Woodworth, P.L. (1991) The Permanent Service for mean sea level and the Global Sea Level Observing System. *Journal of Coastal Research*, 7: 699–719.

Wright, L.D., Coleman, J.M. and Thom, B.G. (1973) Processes of channel development in a high-tide-range environment: Cambridge Gulf–Ord River delta. *Journal of Geology*, 81: 15–41.

Wright, L.D. and Short, A.D. (1984) Morphodynamic variability of surf zones and beaches: a synthesis. *Marine Geology*, 56: 93–118.

Wright, L.D., Thom, B.G. and Higgins, R. (1980) Sediment transport and deposition at wave-dominated river mouths. Examples from Australia and Papua New Guinea. *Estuarine and Coastal Marine Science*, **11**: 263–277.

Wright, L.W. (1970) Variation in the level of the cliff/shore platform junction along the south coast of Great Britain. *Marine Geology*, **9**: 347–353.

Yang, S. (1999) Tidal wetland sedimentation in the Yangtze delta. *Journal of Coastal Research*, **15**: 1091–1099.

Yap, H.T. (1989) Implications of expected climatic changes on natural coastal ecosystems. In: Chou Loke Ming (ed.) *Implications of Climatic Change in the East Asian Seas Region*. University of Singapore, Singapore, pp. 128–148.

Yasso, W.E. (1965) Fluorescent tracer particle determination of the size–velocity relation for foreshore sediment transport, Sandy Hook, New Jersey. *Journal of Sedimentary Petrology*, **35**: 989–993.

Zenkovich, V.P. (1959) On the genesis of cuspate spits along lagoon shores. *Journal of Geology*, **76**: 169–177.

Zenkovich, V.P. (1967) *Processes of Coastal Development* (trans. O.G. Fry; ed. J.A. Steers). Oliver and Boyd, Edinburgh.

Zhuang, W.Y. and Chappell, J. (1991) Effects of seagrass beds on tidal flat sedimentation. In: D.G. Smith *et al.* (eds) *Clastic Tidal Sedimentology*, Canadian Society of Petroleum Geologists Memoir 16, pp. 291–300.

Index

Coastal Geomorphology: An Introduction, Second Edition. Eric Bird.
© 2008 by John Wiley & Sons, Ltd